IFIP Advances in Information and Communication Technology 377

IFIP – The International Federation for Information Processing

IFIP was founded in 1960 under the auspices of UNESCO, following the First World Computer Congress held in Paris the previous year. An umbrella organization for societies working in information processing, IFIP's aim is two-fold: to support information processing within ist member countries and to encourage technology transfer to developing nations. As ist mission statement clearly states,

> IFIP's mission is to be the leading, truly international, apolitical organization which encourages and assists in the development, exploitation and application of information technology for the bene t of all people.

IFIP is a non-profitmaking organization, run almost solely by 2500 volunteers. It operates through a number of technical committees, which organize events and publications. IFIP's events range from an international congress to local seminars, but the most important are:

- The IFIP World Computer Congress, held every second year;
- Open conferences;
- Working conferences.

The flagship event is the IFIP World Computer Congress, at which both invited and contributed papers are presented. Contributed papers are rigorously refereed and the rejection rate is high.

As with the Congress, participation in the open conferences is open to all and papers may be invited or submitted. Again, submitted papers are stringently refereed.

The working conferences are structured differently. They are usually run by a working group and attendance is small and by invitation only. Their purpose is to create an atmosphere conducive to innovation and development. Refereeing is less rigorous and papers are subjected to extensive group discussion.

Publications arising from IFIP events vary. The papers presented at the IFIP World Computer Congress and at open conferences are published as conference proceedings, while the results of the working conferences are often published as collections of selected and edited papers.

Any national society whose primary activity is in information may apply to become a full member of IFIP, although full membership is restricted to one society per country. Full members are entitled to vote at the annual General Assembly, National societies preferring a less committed involvement may apply for associate or corresponding membership. Associate members enjoy the same benefits as full members, but without voting rights. Corresponding members are not represented in IFIP bodies. Affiliated membership is open to non-national societies, and individual and honorary membership schemes are also offered.

Andrew M. Dienstfrey   Ronald F. Boisvert (Eds.)

Uncertainty Quantification in Scientific Computing

10th IFIP WG 2.5 Working Conference, WoCoUQ 2011
Boulder, CO, USA, August 1-4, 2011
Revised Selected Papers

 Springer

Volume Editors

Andrew M. Dienstfrey
National Institute of Standards and Technology
Information Technology Laboratory
325 Broadway, Mail Stop 891, Boulder, CO 80305, USA
E-mail: andrew.dienstfrey@nist.gov

Ronald F. Boisvert
National Institute of Standards and Technology
Applied and Computational Mathematics Division
100 Bureau Drive, Mail Stop 8910, Gaithersburg, MD 20899, USA
E-mail: boisvert@nist.gov

ISSN 1868-4238   e-ISSN 1868-422X
ISBN 978-3-642-32676-9   e-ISBN 978-3-642-32677-6
DOI 10.1007/978-3-642-32677-6
Springer Heidelberg Dordrecht London New York

Library of Congress Control Number: 2012944033

CR Subject Classification (1998): F.1.2, I.6, D.2, H.4.2, J.1, G.1, I.2.3

Typesetting: Camera-ready by author, data conversion by Scientific Publishing Services, Chennai, India

Printed on acid-free paper

Springer is part of Springer Science+Business Media (www.springer.com)

Preface

Computing has become an indispensable component of modern science and engineering research. As has been repeatedly observed and documented, processing speed measured in floating point operations per second has experienced exponential growth in the last few decades. These hardware efficiencies have been accompanied by innovations in mathematical algorithms, numerical software, and programming tools. The result is that, by any measure, the modern computer is many orders of magnitude more powerful than its early predecessors, capable of simulating physical problems of unprecedented complexity.

Given the success of scientific computation as a research tool, it is natural that scientists, engineers, and policy makers strive to harness this immense potential by using computational models for critical decision making. Increasingly, computers are being used to supplement experiments, to prototype engineering systems, or to predict the safety and reliability of high-consequence systems. Such use inevitably leads one to question: "How good are these simulations? Would you bet your life on them?" Unfortunately, most computational scientists today are ill equipped to address such important questions with the same scientific rigor that is routine in experimental science.

The International Federation of Information Processing (IFIP) Working Conference on Uncertainty Quantification in Scientific Computing was convened as a means to address these questions. Participants in the working conference consisted of experts in mathematical modeling, numerical analysis, numerical software engineering, and statistics, as well as policy analysts from a range of application domains to assess our current ability to quantify uncertainty in modeling and simulation (UQ), to raise awareness of this issue within the numerical software community, and to help envision a research agenda to address this critical need. The conference was held in serial plenary sessions organized around four thematic areas: needs, theory, tools, and practice. Keynote speakers introduced each thematic area in broad strokes, followed by invited speakers presenting targeted studies. An additional "Hot Topics" session was organized in real time to provide participants with a venue to expand upon discussions generated by conference presentations, and to present late-breaking material. Finally, adding another dimension, a panel consisting of high-level representatives from government agencies and academia were invited to discuss present practice and future opportunities for uncertainty quantification in scientific computing in the context of the missions of their respective organizations.

The conference Program Committee hoped to generate active engagement on a range of topics both broad and deep. From questions about floating-point compliance and exception handling to computations of 10,000-year risk assessments of nuclear waste repositories, no scale of time, space, and numerical accuracy was beyond scope. Theoretical treatments reflected the full range of uncertainty and risk analysis from approaches recommended by international guidance documents of measurement institutes worldwide, to Bayesian analyses, to a presentation and energetic discussion of the axiomatic foundations of a relatively new theory of uncertainty quantification referred to as probabilistic bounds analysis. Finally, the applications and needs were equally diverse, covering topics of judiciary and regulatory constraints on the use of predictive computation for environmental and reactor safety models, to simulation-based engineering of the electrodeposition paint application process as used by prominent automotive manufacturers. Underlying this diversity, however, the common thread binding all participants was the shared commitment to better achieve the promise offered by numerical computation as a means not only to scientific discovery, but to reliable decision making in matters of importance for society at large.

Of the 24 talks given at the conference, 20 authors contributed papers for these proceedings. Keeping with the tradition of past IFIP Working Conferences, each conference talk was followed by a lively discussion session. During these sessions, assigned discussants presented forms to participants on which they recorded their questions for the speakers. These forms were collected and distributed to speakers with the request that they respond in writing. The resulting record of the discussions appears after each chapter.

As with any activity of this scope many acknowledgments are due. First the success of the conference can largely be attributed to an unmatched Program Committee drawn from a global network of leaders. Many thanks to them for actively participating in conference calls spanning multiple time zones. It was by these efforts that conference topics were defined and associated speakers identified. The fruits of this labor are represented in the pages that follow. Behind the scenes there were too many moving parts to thank all parties. We draw attention to the incredible logistical and planning support provided by Wendy McBride and other members of the Public and Business Affairs Office at the National Institute of Standards and Technology (NIST) in Boulder. Furthermore, numerous speakers and participants would have never made it to Boulder had it not been for the tireless efforts of Lorna Buhse and Robin Bickel of NIST in navigating the cross-cutting constraints mandated for international travel under sponsorship of the United States government. The financial support of the NIST Applied and Computational Mathematics Division is gratefully acknowledged, as is the in-kind support provided by the International Federation of Information Processing's Working Group 2.5 on Numerical Software, the Society of Industrial and Applied Mathematics, and the United States Department of Energy. Finally, we thank our wives and families for stepping in to fill the gaps created by our limitations, and remaining steadfast as we traveled the ups and downs associated with planning and executing such an event. We are happy to report that the ride is over (for now).

April 2012   Andrew Dienstfrey
                                                              Ronald F. Boisvert

Organization

The IFIP Working Conference on Uncertainty Quantification in Scientific Computing was organized by the US National Institute of Standards and Technology (NIST) on behalf of the International Federation for Information Processing (IFIP) Working Group 2.5 on Numerical Software, in cooperation with the Society for Industrial and Applied Mathematics (SIAM).

Organizing Committee

Ronald F. Boisvert	NIST, USA, *Chair*
Andrew Dienstfrey	NIST, USA
James C.T. Pool	CalTech, USA (retired)

Program Committee

Andrew Dienstfrey	NIST, USA, *Chair*
Ronald F. Boisvert	NIST, USA
Maurice Cox	National Physical Laboratory, UK
Bo Einarsson	Linköping University, Sweden (retired)
Brian Ford	Numerical Algorithms Group Ltd., UK (retired)
James (Mac) Hyman	Tulane University, US, SIAM Representative
William L. Oberkampf	W.L. Oberkampf Consulting, USA
Tony O'Hagan	University of Sheffield, UK
Michael Oberguggenberger	University of Innsbruck, Austria

Program

The IFIP Working Conference on Uncertainty Quantification in Scientific Computing was held in Boulder, Colorado, USA on August 1–4, 2011.

Monday August 1

Opening Session

08:15   Welcoming Remarks
        Ronald Boisvert, National Institute of Standards and Technology, US
        Andrew Dienstfrey, National Institute of Standards and Technology, US

Session I: UQ Need: Risk, Policy, and Decision Making (Part 1)
Chair: Andrew Dienstfrey, NIST, US
Discussant: Bo Einarsson, Linköping University, SE

08:30   **Keynote Address**
        Uncertainties in Using Genomic Information to Make Regulatory Decisions
        Pasky Pascual, Environmental Protection Agency, US
09:30   Considerations of Uncertainties in Regulatory Decision Making
        Mark Cunningham, Nuclear Regulatory Commission, US
10:15   Break
10:45   An Industrial Viewpoint on Uncertainty Quantification in Simulation: Stakes, Methods, Tools, Examples
        Alberto Pasanisi, Electricité de France, France
11:30   Living with Uncertainty
        Patrick Gaffney, Bergen Software Services International, Norway
12:15   Lunch

Session II: UQ Need: Risk, Policy, and Decision Making (Part 2)
Chair: Tony O'Hagan, University of Sheffield, UK
Discussant: Tim Hopkins, University of Kent, UK

13:15   Uncertainty and Sensitivity Analysis: From Regulatory Requirements to Conceptual Structure and Computational Implementation
        Jon Helton, Sandia National Laboratories, US
14:00   Interpreting Regional Climate Predictions
        Doug Nychka, National Center for Atmospheric Research, US
14:45   Weaknesses and Failures of Risk Assessment
        William Oberkampf, W.L. Oberkampf Consulting, US
15:30   Break

16:00   Panel Discussion: UQ and Decision Making
        Mac Hyman, Tulane University, US (Moderator)
        Sandy Landsberg, Department of Energy, US
        Larry Winter, University of Arizona, US
        Charles Romine, NIST, US
17:30   Adjourn
18:00   Reception

Tuesday August 2

Session III: UQ Theory (Part 1)
Chair: Richard Hanson, Rogue Wave Software, US
Discussant: Peter Tang, The D.E. Shaw Group, US

08:30   **Keynote Address**
        Bayesian Analysis for Complex Physical Systems Modeled by Computer
        Simulators: Current Status and Future Challenges
        Michael Goldstein, Durham University, UK
09:30   Scientific Computation and the Scientific Method: A Tentative Road
        Map for Convergence
        Les Hatton, Kingston University, UK
10:15   Break
10:45   Overview of Uncertainty Quantification Algorithm R&D in the
        DAKOTA Project
        Michael Eldred, Sandia National Laboratories, US
11:30   A Compressive Sampling Approach to Uncertainty Propagation
        Alireza Doostan, University of Colorado
12:15   Lunch

Session IV: UQ Theory (Part 2)
Chair: Ronald Cools, Katholieke Universiteit Leuven, Belgium
Discussant: Shigeo Kawata, Utsunomiya University, Japan

13:15   **Keynote Address**
        Verified Computation with Probability Distributions and Uncertain
        Numbers
        Scott Ferson, Applied Biomathematics, US
14:15   Parametric Uncertainty Computations with Tensor Product
        Representations
        Hermann Matthies, Technische Universitt Braunschweig, Germany
15:00   Break

Hot Topics Session
Moderator: Brian Ford, NAG Ltd., UK

1.   UQ for Life Cycle Assessment Indicators
     Mark Campanelli, NIST, US

2. Assessing Uncertainties in Prediction of Models Validated using
    Experimental Data Distantly Related to Systems of Interest
    William Oberkampf, W.L. Oberkampf Consulting,
    and Wayne King, Lawrence Livermore National Laboratory
3. For UQ: What Software Tools, What Computing Languages, etc. are
    Required?
    Richard Hanson, Rogue Wave, US
4. Expert Elicitation
    Anthony O'Hagan, Sheffield University, UK
5. Software Security Attacks and Security Failures
    Mladen Vouk, North Carolina State University, US
6. What Hinders the Enjoyment of Interval Arithmetic by its Potential
    Beneficiaries
    William Kahan (University of California at Berkeley)
7. Choosing Between Suspects — Availability of the Human Mind to Handle
    Information
    Mladen Vouk, North Carolina State University, US

Wednesday August 3

Session V: UQ Tools
Chair: Bo Einarsson, Linköping University, Sweden
Discussant: Van Snyder, NASA Jet Propulsion Laboratory, US

08:30   **Keynote Address**
        Desperately Needed Remedies for the Undebugability of Large-scale
        Floating-point Computations in Science and Engineering
        William Kahan, University of California at Berkeley, US
09:30   Accurate Prediction of Complex Computer Codes via Adaptive Designs
        William Welch, University of British Columbia, Canada
10:15   Break
10:45   Using Emulators to Estimate Uncertainty in Complex Models
        Peter Challenor, National Oceanography Centre, UK
11:30   Measuring Uncertainty in Scientific Computations Using the Test
        Harness
        Brian Smith, Numerica 21 Inc., US
12:15   Lunch
13:00   Tour of NIST and NOAA Laboratories
17:15   Reception and Banquet (Chautauqua Dining Hall)

Thursday August 4

Session VI: UQ Practice (Part 1)
Chair: Michael Thuné, University of Uppsala, Sweden
Discussant: Wilfried Gansterer, University of Vienna, Austria

08:30 **Keynote Address**
      Numerical Aspects of Evaluating Uncertainty in Measurement
      Maurice Cox, National Physical Laboratory, UK
09:30 Model-based Interpolation, Approximation, and Prediction
      Antonio Possolo, NIST, US
10:15 Break
10:45 Uncertainty Quantification for Turbulent Reacting Flows
      James Glimm, State University of New York at Stony Brook, US
11:30 Visualization of Error and Uncertainty
      Chris Johnson, University of Utah, US
12:15 Lunch

Session VII: UQ Theory (Part 2)
Chair: Ronald Boisvert, NIST
Discussant: Jennifer Scott, Rutherford Appleton Laboratory, UK

13:15 Uncertainty Reduction in Atmospheric Composition Models
      by Chemical Data Assimilation
      Adrian Sandu, Virginia Tech, US
14:00 Emerging Architectures and UQ: Implications and Opportunities
      Michael Heroux, Sandia National Laboratory, US
14:45 Interval Based Finite Elements for Uncertainty Quantification
      in Engineering Mechanics
      Rafi Muhanna, Georgia Tech Savannah, US
15:30 Break
16:00 Reducing the Uncertainty When Approximating the Solution of ODEs
      Wayne Enright, University of Toronto, Canada
16:45 Closing Remarks
      Andrew Dienstfrey, NIST, US
17:00 Conference Adjourns

Participants

Ron Bates	Rolls-Royce PLC	UK
Ronald Boisvert	NIST	USA
Mark Campanelli	NIST	USA
Peter Challenor	National Oceanography Centre	UK
Kevin Coakley	NIST	USA
Ronald Cools	Katholieke Universiteit Leuven	BE
Maurice Cox	National Physical Laboratory	UK
Mark Cunningham	Nuclear Regulatory Commission	USA
Andrew Dienstfrey	NIST	USA
Alireza Doostan	University of Colorado Boulder	USA
Bo Einarsson	Linköping University	SE
Michael Eldred	Sandia National Laboratories	USA
Ryan Elmore	National Renewable Energy Lab	USA
Wayne Enright	University of Toronto	CA
Jonathan Feinberg	Simula Research Laboratory	NO
Scott Ferson	Applied Biomathematics	USA
Jeffrey Fong	NIST	USA
Brian Ford	NAG/University of Oxford	UK
Patrick Gaffney	Bergen Software Services International AS	NO
Wilfried Gansterer	University of Vienna	AT
James Gattiker	Los Alamos National Laboratory	USA
James Glimm	Stony Brook University	USA
Michael Goldstein	Durham University	UK
John Hagedorn	NIST	USA
Richard Hanson	Rogue Wave	USA
Leslie Hatton	Kingston University, London	UK
Jon Helton	Arizona State University	USA
Michael Heroux	Sandia National Laboratories	USA
Kyle Hickmann	Tulane University	USA
Charles Hogg	NIST	USA
Tim Hopkins	University of Kent	UK
Shang-Rou (Henry) Hsieh	Lawrence Livermore National Laboratory	USA
Mac Hyman	Tulane University	USA
Ruth Jacobsen	NIST /University of Maryland	USA
Chris Johnson	University of Utah	USA
Russell Johnson	NIST	USA
Raghu Kacker	NIST	USA
William Kahan	University of California at Berkeley	USA
Shigeo Kawata	Utsunomiya University	JP
Peter Ketcham	NIST	USA

Wayne King	Lawrence Livermore National Laboratory	USA
Emanuel Knill	NIST	USA
Ulrich Kulisch	Universität Karlsruhe	DE
Sandy Landsberg	Department of Energy	USA
Eric Machorro	Nevada National Security Site	USA
Hermann G. Matthies	Technische Universität Braunschweig	DE
Ian Mitchell	University of British Columbia	CA
Felipe Montes	USDA Agricultural Research Service	USA
Rafi Muhanna	Georgia Institute of Technology	USA
Katherine Mullen	NIST	USA
Douglas Nychka	National Center for Atmospheric Research	USA
Abbie O'Gallagher	NIST	USA
Anthony O'Hagan	University of Sheffield	UK
William Oberkampf	W.L. Oberkampf Consulting	USA
Michal Odyniec	National Security Technologies	USA
Alberto Pasanisi	Électricité de France R&D	FR
Pasky Pascual	U.S. Environmental Protection Agency	USA
Adele Peskin	NIST	USA
Antonio Possolo	NIST	USA
John Reid	JKR Associates	UK
John Rice	Purdue University	USA
Adrian Sandu	Virginia Tech	USA
Jennifer Scott	Rutherford Appleton Laboratory	UK
David Sheen	NIST	USA
Brian Smith	Numerica 21 Inc.	USA
William Van Snyder	NASA Jet Propulsion Laboratory	USA
Jolene Splett	NIST	USA
Philip Starhill	IBM	USA
Ping Tak Peter Tang	Intel Corporation	USA
Judith Terrill	NIST	USA
Michael Thuné	Uppsala University	SE
Mladen Vouk	North Carolina State University	USA
William Wallace	NIST	USA
Jack C.M. Wang	NIST	USA
William Welch	University of British Columbia	CA
Larry Winter	University of Arizona	USA

Table of Contents

Uncertainties Using Genomic Information for Evidence-Based Decisions*

Pasky Pascual

U.S. Environmental Protection Agency
Washington, DC, USA
pascual.pasky@epa.gov

Abstract. For the first time, technology exists to monitor the biological state of an organism at multiple levels. It is now possible to detect which genes are activated or deactivated when exposed to a chemical compound; to measure how these changes in gene expression cause the concentrations of cell metabolites to increase or decrease; to record whether these changes influence the over-all health of the organism. By integrating all this information, it may be possible not only to explain how a person's genetic make-up might enhance her susceptibility to disease, but also to anticipate how drug therapy might affect that individual in a particularized manner.

But two related uncertainties obscure the path forward in using these advances to make regulatory decisions. These uncertainties relate to the unsettled notion of the term "evidence" — both from a scientific and legal perspective. From a scientific perspective, as models based on genomic information are developed using multiple datasets and multiple studies, the weight of scientific evidence will need to be established not only on long established protocols involving p-values, but will increasingly depend on still evolving Bayesian measures of evidentiary value. From a legal perspective, new legislation for the Food and Drug Administration has only recently made it possible to consider information beyond randomized, clinical trials when evaluating drug safety. More generally, regulatory agencies are mandated to issue laws based on a "rational basis," which courts have construed to mean that a rule must be based, at least partially, on the scientific evidence. It is far from certain how judges will evaluate the use of genomic information if and when these rules are challenged in court.

Keywords: genome, Bayesian model, scientific evidence, evidence-based decisions, regulatory decisions, systems biology, meta-analysis.

In 2000, in an event announcing that one of biology's long-standing challenges — the sequencing of the human genome — had finally been scaled, then US President Bill Clinton issued a bold prognostication: "It will revolutionize the

---

* The author is an environmental scientist and lawyer at the U.S. Environmental Protection Agency (EPA). However, this chapter does not represent the viewpoints of the EPA.

A. Dienstfrey and R.F. Boisvert (Eds.): WoCoUQ 2011, IFIP AICT 377, pp. 1–14, 2012.

diagnosis, prevention and treatment of most, if not all, human diseases" [3]. A decade later, while most biologists agree that mapping the human genome has revolutionized science, some also admit that it has increased — not diminished — the complexity of biological science by orders of magnitude. The complexity arises not because more genes have been discovered than had been previously anticipated. Indeed, before the Human Genome Project, biologists estimated that the genome might contain about 100,000 genes. The current estimate is that the human genome contains just a fraction of that — about 21,000 [7]. Rather, the challenge of interpreting genomic information lies in understanding the network of events through which genes are regulated.

The traditional model through which genes were thought to be expressed — that in response to environmental signals, one gene codes for one protein that may metabolize one or a few cellular functions — is insufficient to describe the full panoply of cellular behavior. The problem is that metabolic pathways, the series of cell-mediated chemical reactions necessary to maintain life, rarely proceed in a linear fashion. If a gene that triggers a series of reactions is deactivated, there still may be multiple other genes to ensure that the reactions continue to occur. In the words of Cell Biologist Tony Pawson, "When we started out, the idea was that signalling pathways were fairly simple and linear. Now, we appreciate that the signalling information in cells is organized through networks of information rather than simple discrete pathways. It's infinitely more complex" [7].

1   "Hairballs" as a Metaphor for Systems Biology

To do full justice to this complexity, Lander [11] suggests that the double helix — that icon of 20th century biology — should be replaced by the hairball as a metaphor for genomic science (see Fig. 1). A ubiquitous visualization tool for genomic data, the hairball consists of "nodes" (representing genes, proteins, or metabolites) and "edges" (which represent the associations among the nodes). A particular node may be the focus of a researcher's entire program. In 1977 for example, Andrew Schally, Roger Guillemin, and Rosalyn Sussman Yalow shared the Nobel Prize in Medicine for their investigations into a biologically significant "node" showing a connection between the nervous and endocrine systems [24]. Their work demonstrated that hormones secreted by an organism's hypothalamus could trigger the release of other hormones from its pituitary and gonadal glands. Elucidating biology's nodes — such as this so-called hypothalamus-pituitary-gonadal axis — is necessary to understand how an organism operates. But to the systems biologist intent on using genomic information in a quantitative way, the focal point of understanding is the hairball, i.e. the computational, systems-oriented model of how nodes relate to and function within a broader network of other nodes. And so, for example, Basu [2] in research funded by the US Environmental Protection Agency (EPA) proposes modeling how environmental toxicants disrupt fish reproduction and ultimately diminish fish populations by way of perturbations to the hypothalamus-pituitary-gonadal axis. That is, they will model how knowledge about the nodes describing the hypothalamus, pituitary,

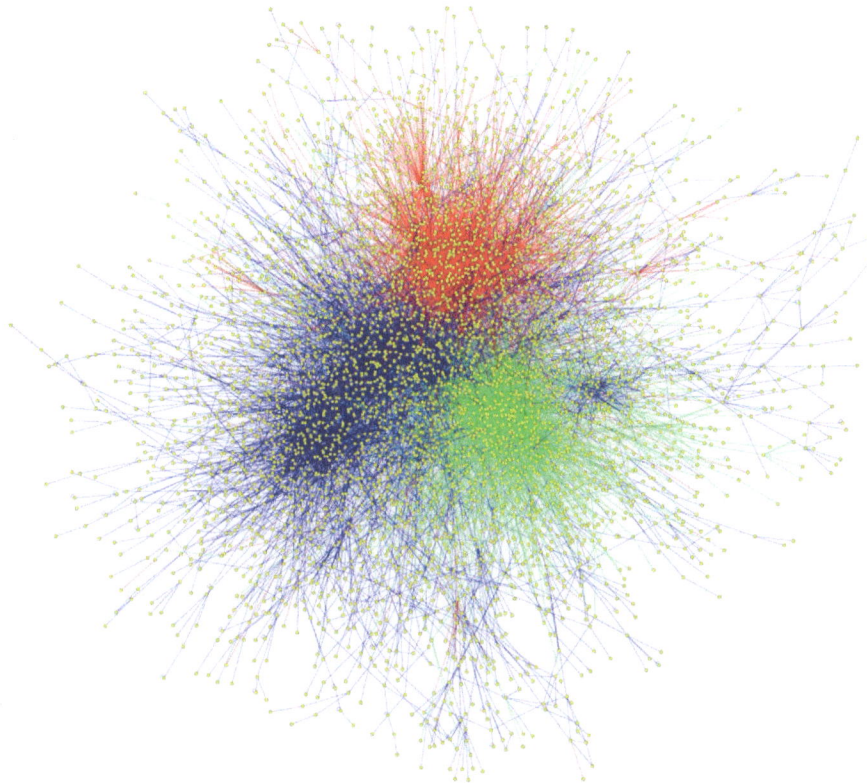

Fig. 1. The "hairball" of systems biology, consisting of "nodes" (representing genes, proteins, or metabolites) and "edges" (which represent the associations among the nodes). Original figure in color provided by Nicolas Simonis and Marc Vidal (see [5]).

and gonadal glands interact within a hairball to ultimately impact a resource protected by a regulatory agency's statutory mandate.

A report issued under the aegis of the National Academy of Sciences, *Toxicity Testing in the 21st Century: a Vision and a Strategy* [16], laid out a path for explaining the etiology of environmental disease by using the tools of genomic science. In that report, the Academy proposed that toxicity testing should become less reliant on whole animal tests and eventually rely instead on systems-oriented, computational models, which can be used to screen large numbers of chemicals, based on information from in vitro assays and in vivo biomarkers. Technology exists to monitor the biological state of an organism at multiple levels. It is now possible to detect which genes are activated or deactivated when exposed to a chemical compound; to measure how these changes in gene expression cause the concentrations of cell metabolites to increase or decrease; and to record whether these changes influence the over-all health of the organism. By integrating all this information, it may be possible not only to explain how a person's genetic make-up might enhance her susceptibility to disease, but to also anticipate how drug therapy might affect that individual in a particularized manner. One of the scientific leaders of the human genome project put it this way: "All biological science works by collecting the complexity and recognizing it is part of a limited repertoire of events. What's exciting about the genome is it's gotten us the big picture and allowed us to see the simplicity" [4].

2   Three Enabling Technologies for Genomic Information

Rusyn and Daston [20] highlight three interconnected, technological breakthroughs that have been accelerating developments in genomic science: continuing progress in computational power; advances in quickly and efficiently producing data streams with high information content; and novel biostatistical methods that take advantage of the previous two breakthroughs. More than 45 years ago, Intel's former Chief Executive Officer, Gordon Moore, first reported the observation that has come to be popularly referred to as "Moore's Law": the number of transistors that can be placed on an integrated circuit doubles every 18 months, even as the cost of producing these transistors has diminished over time [14]. In turn, the rapid increase in the cost-effectiveness of computing power has fueled the speed and economic efficiencies with which the genome can be sequenced. The National Institutes of Health's National Human Genome Research Institute has tracked data on the costs of sequencing a human-sized genome during the ten years since the genome was first mapped [15]. These costs have tracked and, since 2007, even exceeded the progress of Moore's Law (see Fig. 2). Similarly, computing power and the use of robotics have made it possible to test thousands of chemicals in plates containing hundreds of wells in order to evaluate a biological response — binding to a receptor site in a cell; producing a particular enzyme; transcribing a gene. These so-called "high-throughput technologies" have generated considerable data about an organism's reaction to chemical exposure.

By itself, this profusion of biological information would be nothing more than unrelated terabytes of data. Complemented with the appropriate analytical methods, the data can yield important insights into the human response to synthetic chemicals. The modeling objective for the systems biologist is the usual one for any modeler, which is to solve for (using the standard regression model):

$$Y = X\beta + \epsilon, \tag{1}$$

where Y is the n-vector of the categorical biological response in which the modeler is interested; X is the $[n \times p]$-matrix of predictors; β is the p-vector of parameters relating biological response to the predictors; and ϵ is the n-vector error term.

For modelers in genomic science, the high dimension of genomic information raises several challenges. Because high-throughput technologies can monitor for multiple biological and chemical attributes simultaneously, these modelers typically confront a situation in which the $[n \times p]$-matrix of predictors, X, is "short

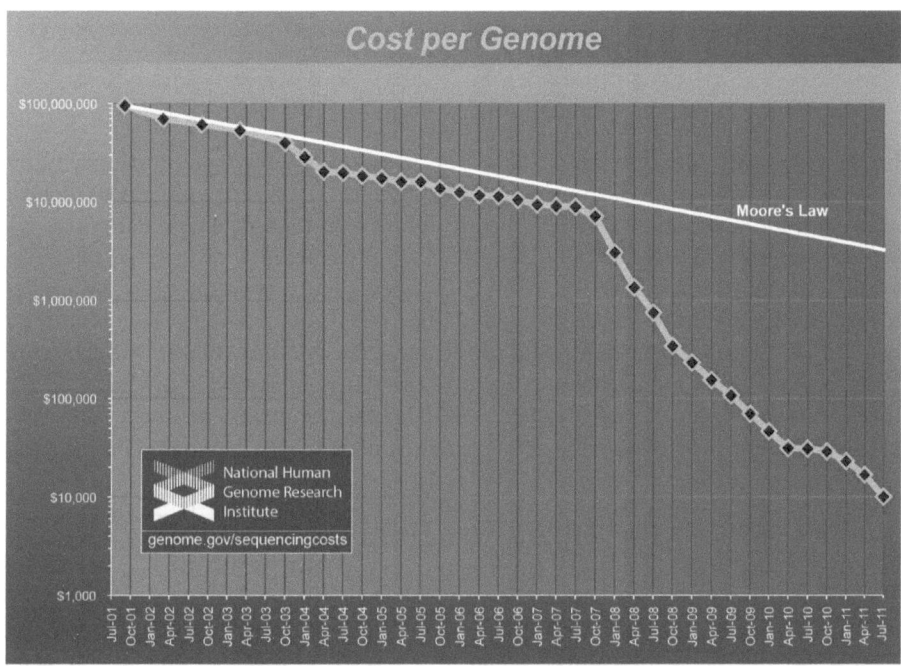

Fig. 2. Costs of sequencing the human genome. Figure from http://genome.gov/SequencingCosts .

and wide," i.e. the number of predictors far exceed the sample size, $p \gg n$ [25]. At the same time, the biological attributes monitored by high-throughput technologies may often be co-regulated by the same genes or may be involved in metabolic pathways that are correlated [10].

Fortunately for the modeler, the basic tenets of biology suggest that assuming an underlying structure can approximate biological data is not only analytically convenient, but also reasonable, plausible, and empirical. Natural selection, the key mechanism through which evolution selects biological traits that enable survival, imposes constraints on an organism's physical attributes. This is evidenced most clearly by cellular pathways that are conserved over long timescales and among widely disparate organisms [12]. The National Academy's report, *Toxicity Testing in the 21st Century*, defines a "toxicity pathway" as a cellular response that, when sufficiently perturbed, is expected to result in an adverse health effect [16]. Implicit in this definition is the notion that an organism's response to a toxic compound is the result of perturbation away from a stable, homeostatic system of cellular behavior that has evolved over time.

Bayesian methods are particularly well-suited for generating models of genomic information. The Bayesian approach is grounded in the view that because intractable uncertainties obscure any model's objective truth, one can only express the degree to which one believes in a model's truthfulness. If one can assume that these models conform to probability distributions and to certain axioms, then any initial, hypothesized model can accommodate emergent evidence according to the following relationship:

$$\text{Posterior model} \sim \text{Likelihood} \times \text{Prior model.}$$

Hierarchical Bayesian modeling, based on the notion that computational functions and probabilistic relationships can capture the underlying structure of data organized into discrete levels, conform to information about biological pathways that occur across multiple scales of biological information — from gene to cell to tissue to organ to the whole organism [13]. Additionally, several public databases are available that store data on genomic information, such as the Gene Ontology (GO) and the Kyoto Encyclopedia of Genes and Genomes (KEGG) [23]. Using hierarchical Bayes, the modeler can merge the datasets available at these repositories in order to improve statistical power, while accounting for sources of variability inherent in the experimental protocols used to generate each dataset [1].

3   Example: A Multinomial Probit Model for Genomic Information

As an example of how Bayesian methods can be used to develop computational models of biological information, Sha et al. [21] investigated the use of gene expression data in predicting rheumatoid arthritis, an autoimmune disease characterized by chronic inflammation and destruction of cartilage and bone in the joints. To glean useful insights from their data, the researchers used a multinomial probit (MNP) model. Like the more familiar multinomial logit model, the MNP is used to estimate how categorical, unordered response variables might be functionally related to explanatory variables. The MNP model is more appropriate in modeling genomic information because, unlike the multinomial logit, it allows for the possibility that the categories of response variables are not independent. The MNP model allows for dependence among these categories by estimating the variance-covariance matrix that quantifies any co-variability among them [27]. While this approach had long-standing theoretical appeal, applications of the MNP model were restricted by the computational complexities in fitting them. However, recent advances now implement a Markov Chain Monte Carlo (MCMC) method in order to estimate the MNP posterior model by taking random walks through the given data set [8].

In an MNP model, the response variable, Y_i , is modeled in terms of a latent variable $W_i = (W_{i1}, \ldots, W_{i,p-1})$, where

$$W_i = X_i \beta + \epsilon_i \quad \epsilon_i \sim \text{N}(0, \Sigma), \quad \text{for } i = 1, \ldots, n, \tag{2}$$

and Σ is a $p - 1 \times p - 1$ variance-covariance matrix. The response variable, Y_i, is then modeled using the latent variable W_i, as

$$Y_i(W_i) = \begin{cases} 0 \text{ if } \max(W_i) < 0 \\ j \text{ if } \max(W_i) = W_{ij} > 0 \end{cases} \quad \text{for } i = 1, \ldots, n \text{ and } j = 1, \ldots, p - 1,$$
$$\tag{3}$$

where max(W_i) is the largest element of the vector W_i and Y_i equal to 0 corresponds to an arbitrarily chosen base category.

In Sha et al.'s study [21], patients afflicted with rheumatoid arthritis were differentiated by whether they were in early or late stages of the disease, as measured by erythrocyte sedimentation, the rate of red blood cell sedimentation that is commonly used as an indicator of inflammation. As well, gene expression data for major functional categories were derived for these patients. Applying their MNP model, the investigators noted that genes regulating two sets of biological pathways were associated with patients afflicted with the late stages of rheumatoid arthritis — those regulating aspects of the cytoskeleton (i.e. the system of filaments that provide cells with their structure and shape) and those influencing cytokines (i.e., molecules that participate in regulating immune responses and inflammatory reactions).

It bears highlighting that in applying MCMC to estimate the MNP model, the quantification of uncertainty pervades the entire model estimation process. That is, the objective of model fitting is not merely to estimate the model parameters, but rather to estimate the entire probability distributions underlying the system being modeled.

Baragatti [1] extended this basic, single-level model to a hierarchical model with a categorical, binary response variable. Her study focused on the estrogen receptor status of a patient, a clinically measured indicator of breast cancer. The data were drawn from three different datasets and therefore, a hierarchical model of fixed and random effects were used. The former corresponded to gene expression measurements, while the latter corresponded to the variability introduced by using the different datasets.

4   Weight of Evidence and Meta-analysis

In their review of highly cited studies that have used in vitro and in vivo biological information in order to predict disease risk, Ioannidis and Panagiotou [9] suggest that the associations uncovered in these studies generally tend to be exaggerated, when compared to larger studies and subsequent meta-analysis. The authors attribute these false positives and spurious results partially to publication bias, i.e., the original researchers report only the data which indicate statistically significant results. To guard against misleading results, Zeggini and Ioannidis [26] propose the greater use of meta-analysis in genomic studies, for which a Bayesian framework provides an intuitive framework.

Once again, fixed and random effects models serve as a useful approach. When using these models in meta-analysis, one assumes that a common, fixed effect underlies every single study in the meta-analysis; i.e., if each study was infinitely large, there would be no heterogeneity between studies. However, no study is infinitely large and therefore one must assume that individual studies exert random effects. These random effects have some mean value and some measurement of variability stemming from between-study differences.

By meta-analyzing genomic studies, one increases sample size as well as the variation in genomic data, thereby enhancing the power to detect true

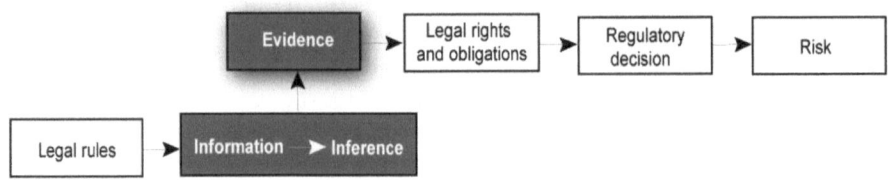

Fig. 3. Evidence is often the precursor to obligations, rights and responsibilities established by law, necessitating a decision that influences risk. But just as evidence leads to legal rights and obligations that constrain regulatory decisions, so too does the law constrain how evidence is established, sometimes in ways that are inconsistent with scientific best practices.

associations. The weight of evidence for genomic information accumulates over time. Previous, individual studies form the prior belief. With each additional study, estimates are updated to form the posterior belief in a way that takes into account all available evidence.

5   Linking Genomic Information to Regulatory Decisions with Evidence

While high-throughput technologies, the proliferation of genomic information, and evolving analytical techniques will continue to spur the scientific community's understanding of the cellular basis for disease, an important issue that remains uncertain is how these advances will be used to make regulatory decisions. The issue arises because, as a matter of administrative law, governmental agencies must issue regulations that have a "rational basis," which the courts have taken to mean that a regulation, among other things, must be based on the scientific evidence [17]. Some threshold of evidentiary burden must be satisfied before the evidence triggers legal obligations, rights or responsibilities that thereby necessitate a decision influencing risk (see Fig. 3). But just as scientific evidence constrains regulatory decisions, so too does the law constrain the way that evidence is established, sometimes in ways that are inconsistent with best scientific practices.

An example of how the law on scientific evidence can hamper the use of science for regulatory decisions is provided by the controversy surrounding the Food and Drug Administration's approval of the painkilling drug, Vioxx. Vioxx works by suppressing enzymes regulating the body's production of compounds associated with inflammation. Unfortunately, these compounds also play a role in maintaining the cardiovascular system [6]. Even before FDA's approval of Vioxx, there had been evidence indicating that inhibiting these enzymes may elevate blood pressure, may thicken artery walls, and may increase blood clots — all of which affect the risk of heart disease [19]. Two complications obfuscated the drug's risks. First, each single piece of evidence of risk — taken alone — did not dispositively evince a hazard [6]. Second, current research indicates that the response to Vioxx within a population is subject to genetic variation [22].

But in 1962, Congress mandated that FDA must assess whether a drug was effective for its intended use based on "substantial evidence" from "adequate and well-controlled investigations." The agency interpreted this statute to mean that a regulatory decision on drug effectiveness must be based on *randomized, replicated, controlled, clinical trials (RCTs)*. *Controlled* means a clinical trial is designed so that ideally, treatment and control groups are identical in every way but one, which is in the levels of treatment being tested. Ergo, any variability among these groups is attributed solely to the treatment. In reality, it is difficult to eliminate extraneous sources of variability. Therefore, one *randomizes* how treatments are assigned to the various groups so that, ideally, the effects of any extraneous sources of variability cancel out. Finally, to ensure that experimental results do not occur through sheer happenstance, one *replicates* or repeats the experiment several times. But as stated earlier, it was because of the variable response to Vioxx within the population that the risks of the drug were not fully appreciated. In order to maintain the homogeneous conditions necessitated by an RCT, the drug manufacturers left out data pertaining to those who would have been at greatest risk — an older demographic with previous history of heart disease. Given FDA's enshrinement of RCTs as the gold standard for substantial evidence supporting claims of drug safety, it is not difficult to see why false negatives — as in the Vioxx case — were inevitable.

Shortly after Vioxx was taken off the market, the National Academy of Science's Institute of Medicine (IOM) issued a report clearly stating what others had been saying for some time: that FDA's practices were unlikely to detect rare but serious drug risks [18]. Before drug approval by the FDA, RCTs simply do not have the statistical power to generate the information needed to assess risks that arise when the general population is exposed to a drug. After drug approval by the FDA, FDA did not possess the statutory authorities needed to implement a nation-wide system to continue gathering this information. The IOM report advocated assessing safety over a drug's life-cycle, in which data were to be continuously gathered from multiple sources for ongoing analyses.

In 2007, Congress passed the Food and Drug Administration Act, which corrected the FDA's over-reliance on RCTs and statistical p-value tests to evaluate drug safety. First, Congress directed FDA to establish a network of data systems to integrate any and all information that can be used to evaluate drug risks. Second, it provided FDA with new, extensive authorities to require continuous submission of risk information from drug companies.

Given the newness of the FDA Act of 2007, as well as the unprecedented use of genomic information to inform regulatory decisions, it remains to be seen how courts will rule when these decisions are challenged based on a lack of "rational basis."

6   Conclusion

For the first time, technology exists to monitor the biological state of an organism at multiple levels. It is now possible to detect which genes are activated or deactivated when exposed to a chemical compound; to measure how these changes in gene expression cause the concentrations of cell metabolites to increase or decrease; to record whether these changes influence the over-all health of the organism. By integrating all this information, it may be possible not only to explain how a person's genetic make-up might enhance her susceptibility to disease, but also to anticipate how drug therapy might affect that individual in a particularized manner.

But two related uncertainties obscure the path forward in using these advances to make regulatory decisions. These uncertainties relate to the unsettled notion of the term "evidence" — both from a scientific and legal perspective. From a scientific perspective, as models based on genomic information are developed using multiple datasets and multiple studies, the weight of scientific evidence will need to be established not only on long established protocols involving p-values, but will increasingly depend on still evolving Bayesian measures of evidentiary value. From a legal perspective, new legislation for the Food and Drug Administration has only recently made it possible to consider information beyond randomized, clinical trials when evaluating drug safety. More generally, regulatory agencies are mandated to issue laws based on a "rational basis," which courts have construed to mean that a rule must be based, at least partially, on the scientific evidence. It is far from certain how judges will evaluate the use of genomic information if and when these rules are challenged in court.

References

1. Baragatti, M.: Bayesian Variable Selection for Probit Mixed Models Applied to Gene Selection. Bayesian Analysis 6(2), 209–229 (2011)
2. Basu, N.: Proposal for EPA Grant. On file with author (2011)
3. Butler, D.: Science after the sequence. Nature 465, 1000–1001 (2010)
4. Cohen, J.: The Human Genome, a Decade Later. Technology Review (January/February 2011)
5. Ferrell Jr., J.E.: Q&A: Systems Biology. Journal of Biology 8(2), Article 2 (2009)
6. Grosser, T., Yu, Y., et al.: Emotion Recollected in Tranquility: Lessons Learned from the COX-2 saga. Annual Review of Medicine 61, 17–33 (2010)
7. Hayden, E.C.: Human Genome at Ten: Life is Complicated. Nature 464, 664–667 (2010)
8. Imai, K., van Dyk, D.A.: A Bayesian Analysis of the Multinomial Probit Model using Marginal Data Augmentation. Journal of Econometrics 124(2), 311–334 (2005)
9. Ioannidis, J.P.A., Panagiotou, O.A.: Comparison of Effect Sizes Associated With Biomarkers Reported in Highly Cited Individual Articles and in Subsequent Meta-analyses. JAMA-Journal of the American Medical Association 305(21), 2200–2210 (2011)
10. Kwon, D., Landi, M.T., et al.: An Efficient Stochastic Search for Bayesian Variable Selection with High-dimensional Correlated Predictors. Computational Statistics & Data Analysis 55(10), 2807–2818 (2011)
11. Lander, A.: The Edges of Understanding. BMC Biology 8(1), 40 (2010)
12. Lenormand, T., Roze, D., et al.: Stochasticity in Evolution. Trends in Ecology & Evolution 24(3), 157–165 (2009)

13. Lewin, A., Richardson, S.: Bayesian Methods for Microarray Data. In: Balding, G.J., Bishop, M., Cannings, C. (eds.) Handbook of Statistical Genetics, 3rd edn., pp. 267–295. Wiley (2007)
14. Matthews, J.N.A.: Moore Looks Beyond the Law. Physics Today 61, 20 (2008)
15. National Human Genome Research Institute (NHGRI): DNA Sequencing Costs, http://www.genome.gov/sequencingcosts/ (accessed November 11, 2011)
16. National Research Council (NRC): Toxicity Testing in the 21st Century: A Vision and a Strategy. National Academies Press, Washington, DC (2007)
17. Pascual, P.: Evidence-based Decisions for the Wiki World. International Journal of Metadata, Semantics and Ontologies 4(4), 287–294 (2009)
18. Pray, L.A., Robinson, S., et al.: Challenges for the FDA: the Future of Drug Safety: Workshop Summary. National Academies Press, Washington, DC (2007)
19. Ritter, J.M., Harding, I., et al.: Precaution, Cyclooxygenase Inhibition, and Cardiovascular Risk. Trends in Pharmacological Sciences 30(10), 503–508 (2009)
20. Rusyn, I., Daston, G.P.: Computational Toxicology: Realizing the Promise of the Toxicity Testing in the 21st Century. Environmental Health Perspectives 118(8), 1047 (2010)
21. Sha, N.J., Vannucci, M., et al.: Bayesian Variable Selection in Multinomial Probit Models to Identify Molecular Signatures of Disease Stage. Biometrics 60(3), 812–819 (2004)
22. St Germaine, C.G., Bogaty, P., et al.: Genetic Polymorphisms and the Cardiovascular Risk of Non-Steroidal Anti-Inflammatory Drugs. American Journal of Cardiology 105(12), 1740–1745 (2010)
23. Stingo, F.C., Chen, Y.A., et al.: Incorporating Biological Information into Linear Models: a Bayesian Approach to the Selection of Pathways and Genes. The Annals of Applied Statistics 5(3), 1978–2002 (2011)
24. Valentinuzzi, M.E.: Neuroendocrinology and its Quantitative Development: A Bioengineering View. Biomedical Engineering Online 9 (2010)
25. West, M.: Bayesian Factor Regression Models in the "Large p, small n" Paradigm. Bayesian Statistics 7, 723–732 (2003)
26. Zeggini, E., Ioannidis, J.P.A.: Meta-analysis in Genome-wide Association Studies. Pharmacogenomics 10(2), 191–201 (2009)
27. Zhang, X., Boscardin, W.J., et al.: Bayesian Analysis of Multivariate Nominal Measures using Multivariate Multinomial Probit Models. Computational Statistics & Data Analysis 52(7), 3697–3708 (2008)

Discussion

Speaker: Pasky Pascual

Brian Smith: You expressed the issue of injury-in-fact versus probability of the event as a concern with legal issues. Why is cost not a part of the issue, or why is it not discussed?

Pasky Pascual: The only reason why I did not specifically discuss the issue of cost was because of time constraints. As a matter of law, each major regulation that is issued must first undergo a cost-benefit analysis which is then submitted to the White House's Office of Management and Budget. So, when issuing a regulatory decision that is based on genomic information, an agency must also be able to estimate the monetary value of the costs and the benefits associated with a particular public health or environmental law.

Maurice Cox: My understanding of your main thesis is as follows. Scientific evidence is out there, usually in the form of data. You explain your assumptions and the statistical or computational model you are using. Then, if you have done your job properly, you should be able to convince the court. But, the court might question the validity of that data in terms of its reliability and consistency. I would welcome your comments.

Pasky Pascual: That's quite right. But part of the problem lies in the fact that the courts may not have the appropriate scientific training to evaluate the scientific evidence with which it is presented. As a matter of law, the courts will be deferential to agencies, particularly in areas that fall within agency's technical expertise and competence. But when a decision is challenged, the courts will subject the "rational basis," which includes the scientific basis, of an agency to a critical review. These reviews are necessarily *ad hoc* and depend on the particularities of the case. But few guidelines, if any, exist to assist the court in conducting this review.

Jeffrey Fong: Does the EPA have a policy statement on the minimum reliability of informatics data that is acceptable? If not, does the speaker have a personal opinion on this question?

Pasky Pascual: My personal opinion is that rather than have a standard score of reliability that then determines acceptability, I would find transparency of the informatics data and the analysis through which the data are used to derive inferences to be more useful. If I were to tell you, for example, that a particular dataset is 99% reliable, what would that mean? Perhaps it is unavoidable that people will demand some kind of seal of approval for a dataset or model that is used to make a decision, but I would want to make sure that the process through which this evaluation occurs is also communicated.

Tony O'Hagan: This is a conference on uncertainty quantification. You've talked a lot about the law. My understanding is that these two things don't go together. Lawyers hate uncertainty, unless they can use it as a weapon against someone foolish enough to admit uncertainty. They hate uncertainty quantification even more. For instance, you pointed out that the law would much prefer anecdotal evidence of actual harm to scientific reasoning of probabilistic harm. What do you feel can be done about this?

Pasky Pascual: I agree that the law tends to operate on binary terms — you comply with a rule or you don't; a drug is safe to market or it is not. So the legal decisions that ultimately get made based on scientific evidence do tend to eschew uncertainty. But at least within a regulatory context, these decisions do consider the uncertainties of science. For example, when EPA issues a regulation these days, it will generally conduct formal uncertainty analyses in order to better understand sources of uncertainty in the science. It may be something as simple as conducting Monte Carlo draws in order to derive a distribution of outputs from a model, rather than a single estimate. So quantification of uncertainty occurs at that phase of rule development. The decision itself may be binary — the Agency regulates or does not regulate a compound — but the analysis that enters into the decision is not. Moreover, when EPA does conduct formal uncertainty analyses when it proposes a rule, these analyses are generally discussed in the documents that accompany the issuance of a rule.

William Oberkampf: Given the strong aversion to uncertainty in the legal and judicial system, how will the EPA deal with more sophisticated uncertainty quantification methods in the future?

Pasky Pascual: My personal opinion is that uncertainty quantification is not going to go away. We will see more, rather than less, of it. It is in the best interest of regulatory agencies to be transparent in their analyses. Transparency is what leads to more defensible decisions. And part of analytical transparency is transparency about sources of uncertainties — both epistemic and alleatory.

Antonio Possolo: In relation with your stated goal of replacing in vivo animal experimentation with studies of differential gene expression: in 2005, colleagues and I published an article in Toxicological Sciences suggesting that studies of differential gene expression in vitro, using live rat and human liver cells, was an effective proxy for studies involving live animals, and also much more expeditious (days, including microarray processing and data analysis, versus the years that it takes for malignancy indications to express themselves), induced by PCBs. Why is it taking the EPA so long to put these and similar scientific, peer-reviewed results to widespread use?

Pasky Pascual: Part of it lies in the complexity of the organism. As we are realizing more and more, metabolic pathways rarely proceed in a linear fashion. For example, if we know that a gene that triggers a series of reactions is deactivated, there still may be other genes to ensure that the reactions will occur. So, the ways that a gene may relate to the manifestation of an observed harm is organized through networks of information rather than simple discrete pathways. And figuring what those networks are and how they operate is extremely hard, I think. Also, it's still not clear, to me anyway, what the evidentiary threshold has to be, before we can say — in a way that is legally defensible — that the behavior of this particular set of biomarkers are a reliable indicator that the likelihood of harm is increased to a level that warrants regulatory action.

Considerations of Uncertainty in Regulatory Decision Making

Mark A. Cunningham

Senior Risk Advisor, Office of Commissioner Apostolakis
United States Nuclear Regulatory Commission
Washington, DC 20555 USA
Mark.Cunningham@nrc.gov

Abstract. NRC's approach to ensuring the safety of nuclear power includes two complementary approaches, one more deterministic and one more probabilistic. These two approaches address the uncertainties in the underlying methods and data differently, with each approach having its strengths and limitations.

This paper provides some background on the historical evolution of deterministic and probabilistic methods in the regulation of nuclear power plants, describes the Commission's policy on the use of probabilistic methods to complement the more traditional deterministic approach, and identifies some example challenges as a staff group considers a strategic vision of how the agency should regulate in the future.

Keywords: risk assessment, nuclear safety, performance-based regulation, safety goals, probabilistic risk assessment.

1 Introduction

Nuclear power plants in the United States historically have been licensed using Part 50 of Title 10 of the Code of Federal Regulations [1]. Implementation of Part 50 has been achieved, for the most part, using deterministic methods and acceptance criteria. These deterministic methods and acceptance criteria were established originally in the 1960's and 1970's and were intentionally conservative in recognition of uncertainties in both routine operations and potential accident conditions. Concepts used included:

- A set of "design basis" accidents (DBAs) that was intended to envelope conditions from a credible set of events,
- A "single failure criterion," a qualitative approach to ensure that systems used to mitigate accidents were highly reliable,
- A "defense in depth" philosophy that introduced barriers between radioactive material and workers and the public, and
- Inclusion of safety margins, a traditional engineering approach for ensuring a robust design.

In 1975, NRC published its first probabilistic risk assessment (PRA) which examined two reactors designed using Part 50 from a different, more realistic,

A. Dienstfrey and R.F. Boisvert (Eds.): WoCoUQ 2011, IFIP AICT 377, pp. 15–26, 2012.

perspective [2]. This study considered a broader set of possible accidents (relative to the DBAs) and estimated their occurrence frequencies, evaluated system reliability quantitatively, estimated the potential public health consequences, and measured, in effect, the effectiveness of the included defense in depth and safety margins. Public health risk was estimated, including an estimate of the uncertainties in this risk. The value of PRA, highlighted by the investigations of the 1979 accident at Three Mile Island, led to the performance of a number of PRAs on other plants and, in the late 1980's, an examination by all operating nuclear power plants for potential vulnerabilities using PRA techniques [3].

2 Commission Policy on Use of Risk Assessment

By the mid-1990s, the nuclear industry had gained considerable experience with implementation of Part 50 and the results of PRAs. In 1995, NRC published a statement describing, among other things, the relationship between these two views of reactor safety [4]. This policy statement directed the NRC staff to increase the use of PRA, setting a course for staff activities that has resulted in a significant expansion in its use.

The Commission's policy statement summarizes the value of risk assessment as follows:

The NRC has generally regulated the use of nuclear material based on deterministic approaches. Deterministic approaches to regulation consider a set of challenges to safety and determine how those challenges should be mitigated. A probabilistic approach to regulation enhances and extends this traditional, deterministic approach, by: (1) allowing consideration of a broader set of potential challenges to safety, (2) providing a logical means for prioritizing these challenges based on risk significance, and (3) allowing consideration of a broader set of resources to defend against these challenges.

With this perspective on the value, and relative roles, of traditional and risk methods, the Commission established the following as its policy:

Increase use of PRA technology in all regulatory matters to the extent supported by the state-of-the-art in PRA methods and data and in a way that complements the deterministic approach and supports the traditional defense-in-depth philosophy.

Use PRA, where practical within the bounds of the state-of-the-art, to reduce unnecessary conservatism in current regulatory requirements, regulatory guides, license commitments, and staff positions and to support proposals for additional regulatory requirements in accordance with 10 CFR 50.109 (Backfit Rule).

PRAs used in regulatory decisions should be as realistic as practicable and supporting data should be publicly available.

Safety goals and subsidiary numerical objectives are to be used with appropriate consideration of uncertainties in making regulatory judgments on the need for new generic requirements.

NRC has made significant progress in implementing this policy in the past 15 years. In the late 1990's-early 2000's time frame, the NRC staff undertook a number of initiatives to better incorporate risk insights and performance considerations into its regulatory programs. In addition to regulatory changes, the NRC worked with other organizations (e.g., the American Society of Mechanical Engineers and the National Institute of Standards and Technology) to improve the technical infrastructure underlying risk assessments. These improvements included the development of consensus standards [20], the development of new methods [21], and performing research including developing better computational methods to validate these new assessment techniques [22]. These initiatives resulted in fundamental changes to how the NRC conducts its licensing, inspection and rulemaking programs.

NRC's Commission has also directed the NRC staff to solicit input from industry and other stakeholders on performance-based initiatives, including areas that are not amenable to risk-informed approaches, to supplement the NRC's traditional deterministic system of licensing and oversight. It should be noted that deterministic[1] and prescriptive[2] regulatory requirements were based mostly on experience, testing programs and expert judgment, considering factors such as safety margins and the principle of defense-in-depth. These requirements are viewed as being successful in establishing and maintaining adequate safety margins for NRC-licensed activities. The NRC has recognized that deterministic and prescriptive approaches can limit the flexibility of both the regulated industries and the NRC to respond to lessons learned from operating experience and support the adoption of improved designs or processes.

The NRC has as one of its primary safety goal strategies the use of sound science and state-of-the-art methods to establish, where appropriate, risk-informed and performance-based regulations. The NRC issued a paper [5] to define the terminology and expectations for evaluating and implementing the initiatives related to risk-informed, performance-based approaches. That paper defines a performance-based approach as follows:

---

[1] A deterministic approach to regulation establishes requirements for engineering margin and for quality assurance in design, manufacture, and construction. In addition, it assumes that adverse conditions can exist and establishes a specific set of design basis events and related acceptance criteria for specific systems, structures, and components based on historical information, engineering judgment, and desired safety margins. An example is a defined load on a structure (e.g., from wind, seismic events, or pipe rupture) and an engineering analysis to show that the structure maintains its integrity.

[2] A prescriptive requirement specifies particular features, actions, or programmatic elements to be included in the design or process, as the means for achieving a desired objective. An example is a requirement for specific equipment (e.g., pumps, valves, heat exchangers) needed to accomplish a particular function (e.g., remove a defined heat load).

A performance-based regulatory approach is one that establishes performance and results as the primary basis for regulatory decision-making, and incorporates the following attributes:

1. *measurable (or calculable) parameters (i.e., direct measurement of the physical parameter of interest or of related parameters that can be used to calculate the parameter of interest) exist to monitor system, including facility and licensee, performance,*
2. *objective criteria to assess performance are established based on risk insights, deterministic analyzes and/or performance history,*
3. *licensees have flexibility to determine how to meet the established performance criteria in ways that will encourage and reward improved outcomes; and*
4. *a framework exists in which the failure to meet a performance criterion, while undesirable, will not in and of itself constitute or result in an immediate safety concern.*[3]

Performance-based approaches can be pursued either independently or in combination with risk-informed approaches. After the paper's issuance, NRC continued to make progress on developing policies and guidance related to performance-based approaches and subsequently issued guidance documents [6] [7].

Perhaps the most significant programmatic adoption of risk-informed and performance-based considerations in the reactor area took place with implementation of the "reactor oversight process" (ROP) [8]. The ROP, intended to focus agency reactor inspection resources in the most risk-significant areas of reactor operation, replaced the previous program with explicit consideration of risk and performance considerations. The normal "baseline" inspection program is focused on the more risk-important areas of plant operations. In addition, events or conditions at plants are assessed for significance using probabilistic risk models. The results of such assessments are used to direct additional oversight to plants with more significant findings.

A more recent reactor initiative that adopts a risk-informed and performance-based approach relates to fire protection, in which standards from the National Fire Protection Association (NFPA-805) were incorporated into NRC's regulation in 10CFR50.48(c) [9]. This regulation provides deterministic requirements that are very similar to those in NRC's traditional fire protection regulations, and also includes performance-based methods for evaluating plant configurations that provide a comparable and equivalent level of safety intended by the conservative deterministic requirements. The performance-based methods allow engineering analyzes to demonstrate that the changes in overall plant risk that result from these plant configurations is acceptably small and that fire protection

---

[3] Using the previous example (footnote 2), a performance-based approach might provide additional flexibility to a licensee on plant equipment and configurations used to accomplish a safety function (e.g., removing a heat load), but the performance criteria could not be the actual loss of a safety function that would result in the release of radioactive materials.

defense-in-depth is maintained.[4] Defense-in-depth as applied to fire protection means that an appropriate balance is maintained between:

1. preventing fires from starting;
2. timely detection and extinguishing of fires that might occur; and
3. protection of systems, structures and components important to safety from a fire that is not promptly extinguished.

The adoption of NFPA 805, which is voluntary on the part of reactor licensees, provides a licensee with flexibility regarding how to implement its fire protection program while maintaining an acceptable level of fire safety.

In parallel, the NRC staff was incorporating risk insights into other regulatory areas. In the materials area, a staff document [12] was developed in the late 1990's to pull together into one place the various guidance documents written over the years for the wide variety of materials licensees. These documents allow license applicants to find the applicable regulations, guidance and acceptance criteria used in granting a materials license. Operational experience (performance) and risk insights guided the development of these documents. Over time the guidance has been revised to further incorporate risk insights, performance considerations and changing technology. A new revision to the series is under development to address security and other issues.

The materials inspection program was fundamentally revised in 2001 — both in terms of approach and frequency — in the Phase II Byproduct Material Review [13]. The inspection approach was modified to emphasize licensee knowledge and performance of NRC-licensed activities over document review. Inspectors now review a licensee's program against focus areas that reflect those attributes which are considered to be most risk-significant. If a licensee's performance against a given focus element during the inspection is considered to be acceptable, the inspector moves on to the next focus element. Performance concerns or questions lead an inspector to go deeper into that area. In addition, inspection frequencies were revised based on risk insights as well as licensee performance over time.

3   Developing a Strategic Vision

In early 2011, an NRC staff group was established [14] to, in effect, reflect on the past 15 years of experience and to develop a "strategic vision and options for

---

[4] Building upon the guidance in Regulatory Guide 1.174 [10], "An Approach for Using Probabilistic Risk Assessment in Risk-Informed Decisions on Plant-Specific Changes to the Licensing Basis," Regulatory Guide 1.205 [11], "Risk-Informed, Performance-Based Fire Protection for Existing Light-Water Nuclear Power Plants," states: "Prior NRC review and approval is not required for individual changes that result in a risk increase less than 1×10^{-7} per year for CDF (core damage frequency) and less than 1×10^{-8} per year for LERF (large early release frequency – a measure of potential offsite health consequences). The proposed change must also be consistent with the defense-in-depth philosophy and must maintain sufficient safety margins. The change may be implemented following completion of the plant change evaluation."

adopting a more comprehensive and holistic risk-informed, performance-based regulatory approach for reactors, materials, waste, fuel cycle, and transportation that would continue to ensure the safe and secure use of nuclear material." This group was established by Chairman Gregory Jaczko and is being headed by Commissioner George Apostolakis.

When this group was established, the agency was dealing with a number of challenging regulatory issues. In several instances, the technical understanding of the issue was relatively poor, with important uncertainties in key topics. Just after the group was established, the Fukushima nuclear power plants in Japan experienced a large earthquake and tsunami. The effects of these events, and their implications to the safety and regulation of US nuclear power plants, introduced additional challenges. Some of these challenges are discussed in more detail below.

3.1   Challenging Regulatory Topics

Performance of Emergency Core Cooling Systems. In 2004, NRC issued a communication [15] to all power reactor licensees requesting that each perform an evaluation to address:

> *The identified potential susceptibility of pressurized-water reactor (PWR) recirculation sump screens to debris blockage during design basis accidents requiring recirculation operation of [emergency core cooling system] ECCS or [containment spray system] CSS and on the potential for additional adverse effects due to debris blockage of flowpaths necessary for ECCS and CSS recirculation and containment drainage.*

Since that time, power reactor licensees have made modifications, as necessary, to address the 2004 issue. However, other related issues have also been identified, including the possible effects of chemical additives on the debris characteristics, potentially worsening the blockage potential, and the potential effects of debris entering the reactor core region and causing blockages there.

In this example, decision makers are provided technical information having uncertainties in several key technical areas, including the effect on debris accumulation, and possible cooling system blockage, of chemical interactions, as well the effect of debris potentially entering and blocking portions of the reactor core area. Experimental work is underway to provide additional data in both areas, but the applicability of the results to the different reactor types also complicates decision making.

NRC staff addressed the complex technical and regulatory issues in a paper [16] providing options to the NRC's Commission for decision. The paper recommended that the ongoing staff approach be continued, which included a determination "whether, given the conservatisms, nonconservatisms, and/or uncertainties in the various review areas, the licensee has demonstrated adequate strainer performance and therefore compliance with the regulations."

The schedule for this considered the risk of different types of accidents, such that it would "address any outstanding issues associated with more likely and risk-significant smaller loss of coolant accidents (LOCAs) (14 inches and below) in the short term, but would allow more time to address issues associated with the low-likelihood larger break LOCAs (above 14 inches). In this way, the more risk-significant issues would be closed quickly, and licensees would have the flexibility to reduce the impact (cost and dose) of addressing the less risk-significant LOCAs through planning, testing, or refined analyzes." The paper also recommended a second approach to be used in combination that "would provide more flexibility to licensees for addressing larger LOCAs than is currently permitted" using existing guidance and that would "likely reduce the scope of modifications needed to address [the issue] for some plants and would be consistent with agency policy regarding risk-informed regulation." NRC's Commission subsequently approved the recommended approach [17] with "comments and clarifications."

Earthquake Frequencies in the Central and Eastern United States. In 2010, NRC staff completed an assessment [18] of new information related to potential earthquakes in the central and eastern United States (that part of the United States east of the Rocky Mountains). In some cases, the new information indicated that estimated frequencies of earthquakes increased relative to previous estimates. Not surprisingly, these estimates had considerable uncertainty.

Since that time, NRC staff have been working to determine what, if any, actions need to be taken by power reactor licensees. A progressive screening approach is being considered which would include comparisons with deterministic information used in the initial design and, if necessary, two alternative seismic risk assessment methods.

One Impact of the Fukushima Accident. In March 2011, the Fukushima nuclear power station in Japan experienced a very large earthquake and subsequent tsunami. The resulting damage is described in a number of documents, and is summarized as follows in an NRC report [19]:

> As a result of the earthquake, all of the operating units appeared to experience a normal reactor trip within the capability of the safety design of the plants. The three operating units at Fukushima Dai-ichi automatically shut down, apparently inserting all control rods into the reactor. As a result of the earthquake, offsite power was lost to the entire facility. The emergency diesel generators started at all six units providing alternating current (ac) electrical power to critical systems at each unit, and the facility response to the seismic event appears to have been normal.
>
> Approximately 40 minutes following the earthquake and shutdown of the operating units, the first large tsunami wave inundated the site followed by multiple additional waves. The estimated height of the tsunami exceeded the site design protection from tsunamis by approximately 8 meters (27 feet). The tsunami resulted in extensive damage to site facilities

and a complete loss of ac electrical power at Units 1 through 5, a condition known as station blackout (SBO). Unit 6 retained the function of one of the diesel generators.

The operators were faced with a catastrophic, unprecedented emergency situation. They had to work in nearly total darkness with very limited instrumentation and control systems. The operators were able to successfully cross-tie the single operating Unit 6 air-cooled diesel generator to provide sufficient ac electrical power for Units 5 and 6 to place and maintain those units in a safe shutdown condition, eventually achieving and maintaining cold shutdown.

Despite the actions of the operators following the earthquake and tsunami, cooling was lost to the fuel in the Unit 1 reactor after several hours, the Unit 2 reactor after about 71 hours, and the Unit 3 reactor after about 36 hours, resulting in damage to the nuclear fuel shortly after the loss of cooling. Without ac power, the plants were likely relying on batteries and turbine-driven and diesel-driven pumps. The operators were likely implementing their severe accident management program to maintain core cooling functions well beyond the normal capacity of the station batteries. Without the response of offsite assistance, which appears to have been hampered by the devastation in the area, among other factors, each unit eventually lost the capability to further extend cooling of the reactor cores.

The current condition of the Unit 1, 2, and 3 reactors is relatively static, but those units have yet to achieve a stable, cold shutdown condition. Units 1, 2, 3, and 4 also experienced explosions further damaging the facilities and primary and secondary containment structures. The Unit 1, 2, and 3 explosions were caused by the buildup of hydrogen gas within primary containment produced during fuel damage in the reactor and subsequent movement of that hydrogen gas from the drywell into the secondary containment. The source of the explosive gases causing the Unit 4 explosion remains unclear. In addition, the operators were unable to monitor the condition of and restore normal cooling flow to the Unit 1, 2, 3, and 4 spent fuel pools.

From a decision making perspective, this accident raises issues with respect to the ability to predict the likelihood of very large earthquakes and to decide how large of an earthquake should be considered sufficiently likely that it must be considered in a nuclear power plant's design. In addition, this example introduces other important uncertainties. One key additional issue is the effectiveness of emergency response, relied upon in nuclear safety to help ensure that the potentially affected population near nuclear power plants would not be exposed to large amounts of radioactive material.

An NRC group provided recommendations to the NRC Commission [19] that included some that reflect on how to address issues with considerable uncertainty, and the relative role of deterministic and risk assessment methods. These recommendations are currently under evaluation.

3.2 Some Issues to Resolve

These example issues raise some important questions, including:

- What are the relative roles of traditional engineering approaches and risk assessment?
- Are there better ways to collect and analyze new information?
- Should performance based approaches be used to a greater extent to better reflect such new information?
- How should decision makers include consideration of very unlikely events that could result in very large consequences?

As noted above, an NRC staff group is now considering a strategic vision for a more comprehensive and holistic risk-informed, performance-based regulatory approach. This group is considering questions such as these; it expects to complete its report in the spring, 2012.

References

1. United States Nuclear Regulatory Commission (USNRC): Title 10 (Energy) Code of Federal Regulations, Part 50 (Domestic Licensing of Production and Utilization Facilities), http://www.nrc.gov/reading-rm/doc-collections/cfr/part050/
2. USNRC: NUREG 75/014, Reactor Safety Study: An Assessment of Accident Risks in US Commercial Nuclear Power Plants, http://www.nrc.gov/reading-rm/doc-collections/nuregs/staff/sr75-014/
3. USNRC: Generic Letter 88-20, Individual Plant Examination for Severe Accident Vulnerabilities, http://www.nrc.gov/reading-rm/doc-collections/gen-comm/gen-letters/1988/gl88020.html
4. USNRC: Use of Probabilistic Risk Assessment in Methods in Nuclear Regulatory Activities, Final Policy Statement, http://www.nrc.gov/reading-rm/doc-collections/commission/policy/60fr42622.pdf
5. USNRC: White Paper on Risk-Informed and Performance-Based Regulation, http://www.nrc.gov/reading-rm/doc-collections/commission/secys/1998/secy1998-144/1998-144scy.pdf
6. USNRC: SECY-2000-0191, High-Level Guidelines for Performance-Based Activities, http://www.nrc.gov/reading-rm/doc-collections/commission/secys/2000/secy2000-0191/2000-0191scy.pdf
7. USNRC: Guidance for Performance-Based Regulation, http://www.nrc.gov/reading-rm/doc-collections/nuregs/brochures/br0303/
8. USNRC: SECY-99-007, Recommendations for Reactor Oversight Process Improvements, http://www.nrc.gov/reading-rm/doc-collections/commission/secys/1999/secy1999-007/1999-007scy_attach.pdf
9. USNRC: Title 10 (Energy) Code of Federal Regulations, Part 50.48(c), National Fire Protection Standard NFPA 805, http://www.nrc.gov/reading-rm/doc-collections/cfr/part050/part050-0048.html
10. USNRC: Regulatory Guide 1.174, An Approach for Using Probabilistic Risk Assessment in Risk-Informed Decisions on Plant-Specific Changes to the Licensing Basis, http://pbadupws.nrc.gov/docs/ML1009/ML100910006.pdf

11. USNRC: Regulatory Guide 1.205, Risk-Informed, Performance-Based Fire Protection for Existing Light-Water Nuclear Power Plants,
http://pbadupws.nrc.gov/docs/ML0927/ML092730314.pdf
12. USNRC: NUREG-1556, Consolidated Guidance about Materials Licenses,
http://www.nrc.gov/reading-rm/doc-collections/nuregs/staff/sr1556/
13. USNRC: SECY-99-062, Nuclear Byproduct Material Risk Review,
http://www.nrc.gov/reading-rm/doc-collections/commission/secys/
1999/secy1999-062/1999-062scy.pdf
14. USNRC: Charter for Task Force for Assessment of Options for More Holistic Risk-Informed, Performance-Based Regulatory Approach. ADAMS document access system, accession number ML100270582
15. USNRC: Generic Letter 2004-02, Potential Impact of Debris Blockage on Emergency Recirculation during Design Basis Accidents at Pressurized Water Reactors,
http://www.nrc.gov/reading-rm/doc-collections/gen-comm/
gen-letters/2004/gl200402.pdf
16. USNRC: SECY-2010-0113, Closure Options for Generic Safety Issue-191, Assessment of Debris Accumulation on Pressurized Water Reactor Sump Performance,
http://www.nrc.gov/reading-rm/doc-collections/commission/secys/2010/
secy2010-0113/2010-0113scy.pdf
17. USNRC: Staff Requirements - SECY-2010-0113 - Closure Options for Generic Safety Issue-191, Assessment of Debris Accumulation on Pressurized Water Reactor Sump Performance, http://www.nrc.gov/reading-rm/doc-collections/
commission/srm/2010/2010-0113srm.pdf
18. USNRC: Safety/Risk Assessment Results for Generic Issue 199, "Implications of Updated Probabilistic Seismic Hazard Estimates in Central and Eastern United States on Existing Plants." ADAMS document access system, accession number ML100270582
19. USNRC: Recommendations for Enhance Reactor Safety in the 21st Century,
http://pbadupws.nrc.gov/docs/ML1118/ML111861807.pdf
20. ASME/ANS: ASME/ANS RA-Sa-2009, Standard for Level 1/Large Early Release Frequency Probabilistic Risk Assessment for Nuclear Power Plant Applications, Addendum A to RA-S-2008. ASME, New York (2009)
21. USNRC and Electric Power Research Institute: NUREG/CR-6850, EPRI/NRC-RES Fire PRA Methodology for Nuclear Power Facilities,
http://www.nrc.gov/reading-rm/doc-collections/nuregs/
contract/cr6850v1.pdf
22. USNRC: NUREG-1824, Verification and Validation of Selected Fire Models for Nuclear Power Plant Applications,
http://www.nrc.gov/reading-rm/doc-collections/nuregs/staff/sr1824/

Discussion

Speaker: Mark Cunningham

Philip Starhill: Presumably the landscape of low probability events is much more diverse than that of higher probability events. How does that impact the utility of deterministic modeling of low probability events?

Mark Cunningham: The estimation of very low probability events, such as very large earthquakes or floods, or the failure probability of very large pipes used in nuclear power plants, is a particularly challenging aspect of nuclear safety analyzes. Historically, nuclear safety organizations such as NRC have made conservative deterministic statements on what magnitudes of earthquakes or sizes of pipe breaks needed to be considered in plant designs. These statements included, of course, some consideration on the credibility of the specified earthquake or pipe break, which was tied to some perception at the time of their probability. Modern risk assessment methods supplement these statements by including more quantitative probability estimates. Expert interpretation of available information, and the translation of such interpretations in statements of probability, is used for low probability events. This approach, especially when a number of experts are used, has been found to be an acceptable method when sufficient data are not otherwise available.

Van Snyder: A numeric limit is established for risk of death within a specified distance due to nuclear power plant operations. Has there been a quantification, and limit, of total system risk to the public at large, and a comparison of that total system risk to total system risk from alternatives, especially coal?

Mark Cunningham: In 1986, NRC established "safety goals" for the operation of nuclear power plants. These goals were intended to ensure that public risk from nuclear power plant operations was a small fraction of other risks, including the risks from other forms of electric power generation. Risk assessments performed by NRC indicate that nuclear power plants meet the established safety goals. NRC has not made comparisons between, for example, risks from electric power generation from coal. Some such studies were performed, however, by staff at Oak Ridge National Laboratory a number of years ago.

William Kahan: If I remember correctly, burning coal releases Radon; and coal ash is very weakly radioactive, but there is a vast amount of it to be stored. When the totals of radioactivity liberated by coal mining are added up, do they bring the coal industry under the purview of the Nuclear Regulatory Commission? Should they?

Mark Cunningham: The coal industry is not subject to regulatory review by NRC. Since its establishment in 1975 by Congress, the scope of NRC's statutory authority has been discussed by Congress to possibly include, for example, Department of Energy facilities using nuclear materials. To my knowledge, these discussions have not considered the coal industry.

Jeffrey Fong: My question is on stability of operation. After deregulation, electricity is being treated as a commodity purchased every six minutes. This creates instability and pressure on operators. Does NRC have new regulations to address this type of operational instability?

Mark Cunningham: When deregulation was beginning to occur, NRC recognized that nuclear power plant operations could be impacted by deregulation of the electrical energy market, and reassessed its regulatory approach. To my knowledge, no new regulations were established specifically to address the possible effects of deregulation.

Jeffrey Fong: Do risk assessment standards include quantitative estimates of human errors during all phases of nuclear reactor operations?

Mark Cunningham: Yes, risk assessments do include quantitative consideration of human errors.

An Industrial Viewpoint on Uncertainty Quantification in Simulation: Stakes, Methods, Tools, Examples

Alberto Pasanisi[1] and Anne Dutfoy[2]

[1] EDF R&D. Industrial Risk Management Dept.
6 quai Watier, 78401 Chatou, France
[2] EDF R&D. Industrial Risk Management Dept.
1 avenue du Général de Gaulle, 92141 Clamart, France
{alberto.pasanisi,anne.dutfoy}@edf.fr

Abstract. Simulation is nowadays a major tool in R&D and engineering studies. In industrial practice, in both design and operating stages, the behavior of a complex system is described and forecast by a computer model, which is, most of time, deterministic. Yet, engineers coping with quantitative predictions using deterministic models deal actually with several sources of uncertainties affecting the inputs (and occasionally the model itself) which are transferred to the outputs. Therefore, uncertainty quantification in simulation has garnered increased importance in recent years. In this paper we present an industrial viewpoint of this practice. After a reminder of the main stakes related to uncertainty quantification and probabilistic computing, we will focus on the specific methodology and software tools which have been developed for treating this problem at EDF R&D. We conclude with examples illustrating applied studies recently performed by EDF R&D engineers arising from different physical domains.

Keywords: Simulation, Computer Experiments, Risk, Uncertainty, Reliability, Sensitivity Analysis.

1 Introduction

Computer simulation is undoubtedly a fundamental issue in modern engineering. Whatever the purpose of the study, computer models help analysts to forecast the behavior of the system under investigation in conditions which cannot be reproduced in physical experiments such as accident scenarios, or when physical experiments are theoretically possible but at a very high cost.

The need for simulating and forecasting gained dramatic momentum in recent decades due to the growth of computers' power and vice versa. Since the very first large scale numerical experiments carried out in the 40's, the development of computers (and computer science) has gone pairwise with the desire to simulate more and more deeply, more and more precisely, physical, industrial, biological, economic systems. A profound change in science and engineering has resulted

A. Dienstfrey and R.F. Boisvert (Eds.): WoCoUQ 2011, IFIP AICT 377, pp. 27–45, 2012.

in which the role of the computer has been compared to the one of the steam engine in the first industrial revolution [1]. Together with formulating theories and performing physical experiments, computer simulation has been labeled a "third way to Science" [2] allowing researchers to explore problems which were unaffordable in a not so distant past. In turn this has raised epistemic issues.

Some scholars identify a shift from a scientific culture of calculation, linear, logical, aiming to simplifying and unpacking results, to a culture of simulation, empirical, opaque, aiming at providing high-dimension results, under nice forms of graphs, or even movies, (see [3] for an interesting discussion). A quite negative perspective of computer models, seen as opaque boxes one can play with to obtain whatever desired result, has also emerged which directly attacks the credibility of computational models as tools for guiding decisions [4].

Our viewpoint is pragmatic. We believe that computer simulation can serve as a major tool in daily engineers' work and is furthermore capable to assist in understanding, forecasting, and guiding decision making. The possibility to simulate more and more complex phenomena, taking into account the effect of more and more input parameters is better viewed as an opportunity as opposed to a threat. However, at the same time, we are aware that quantitative uncertainty assessment of results is a fundamental issue for assuring the credibility of computer model based studies and remains a challenge as well.

Besides technical and theoretical difficulties, in industrial practice a key difficulty is to bridge the cultural gap between a traditional engineering deterministic viewpoint and the probabilistic and statistical approach which considers the result of a model as an *uncertain* variable.

Even if the fundamentals of these topics are rooted in probabilistic and statistic literature from decades back, in recent years there has been a significant increase in interest on the part of industries and academia in uncertainty quantification (UQ) applied to computer models. A casual survey of recent papers reveals the variety of disciplinary fields involved: e.g. nuclear waste disposal [5], [6] (which was also one of the first contexts in which UQ on large computer models was applied), water quality modeling [7], avalanche forecasting [8], welding simulation [9], buildings performance simulation [10], galaxies formation [11], climate modeling [13], fires simulation [14] to name a few.

The remainder of the paper is organized as follows. Section 2 presents the common framework for uncertainty assessment as it is currently used in the industrial practice by EDF (Électricité de France) and other major stakeholder companies and industrial research institutions, for example: European Aeronautic Defence and Space Company (EADS), French Atomic Energy and Alternative Energies Commission (CEA), and Dassault Aviation. Uncertainty analysis requires a multi-disciplinary approach, and this framework is an useful tool to establish dialog between experts in the disciplinary field of the model application and those with a more probabilistic or statistical background. It also allows to put into evidence, at the very early stage of the study, what is the really relevant expected outcome and, consequently, to choose the most proper and effective mathematical tools to obtain it.

Section 3 presents *Open TURNS*, an open source software package, jointly developed by the three companies EDF, EADS and Phiméca, which implements the methodology described in Section 2. Together with proper and recent mathematical algorithms, an interesting feature of Open TURNS is that it can be linked in an effective and rather easy way to any external computer code, which is simply seen as a deterministic function of a random input. By a mathematical viewpoint, that corresponds to linking a deterministic model to a probabilistic model of its inputs.

Section 4 shows some practical examples of applied studies carried on at EDF. These studies concern different problems and involve a great variety of physical problems. The wide range covered by those examples, due to the diversity of EDF's business areas, illustrates the motivation for both a general approach to the problem, as well as the need for specific and powerful mathematical and software tools.

Section 5 very briefly sketches some conclusions and perspectives.

2   The Common Framework for Uncertainty Management

2.1   An Iterative Methodology in Four Steps

In the last decade, thanks to the numerous multi-disciplinary challenges it has to cope with, EDF established a global methodology for uncertainty treatment for models and simulations. This framework has subsequently been accepted and improved by other industrial and research institutions.

The EDF focus is on so-called parametric uncertainties, i.e. uncertainties characterizing dispersion of input parameters of a model, where a model could be a complex numerical code which requires an approximated resolution or an analytical expression. Our analysis does not explicitly address uncertainties attached to the computer model itself, arising from simplifying assumptions for the model of the physical phenomenon under investigation, nor numerical uncertainties due to its practical implementation as a computer code. The methodology is based on the probabilistic paradigm, i.e. uncertainties are represented by associated probability distribution functions (pdf). Even if some perspective works are carried at EDF R&D on extra-probabilistic approaches [15], they are currently considered not yet sufficiently mature for engineering application.

The common framework of uncertainty management is a four step process the genericity of which facilitates application across a broad variety of disciplinary fields: (i) *Step A "Uncertainty Specification"* defines the structure of the UQ study by selecting the random parameters, the outcomes of interest and the features of the output's pdf which are relevant for the analysis; (ii) *Step B "Uncertainty Quantification"* defines the probabilistic modeling of the random parameters; (iii) *Step C : Uncertainty Propagation* evaluates the criteria defined on Step A; (iv) *Step C' "Uncertainty Importance Ranking"* determines which uncertainty sources have the greatest impact on the outcome (sensitivity analysis).

In practice, the process is often iterative: a too large uncertainty tainting the final outcome and/or the ranking step could motivate corrective feedback R&D actions for reducing uncertainties where possible, e.g. setting up new experiments to improve the probabilistic modeling of model's inputs.

The following sections review the principal methods used in each step of the framework.

2.2   Step A - The Uncertainty Problem Specification

This step first involves selection of the input parameters to be represented as random variables. The remaining parameters are considered as fixed either because they are supposed to be known with a negligible uncertainty or (as it is typical in safety studies) because they are given values, generally conservative, which are characteristic of a given accident scenario. In the following we will denote by X the vector of the random input parameters, and $Z = G(X)$ the random outcome of interest of the deterministic model $G(\cdot)$.

Step A requires also to select the relevant features of the outputs' pdf, depending on the stakes which motivated the study (the so-called *quantities of interest*). In most cases they formalize, in a simplified yet explicitly normative manner, some decision criteria. For instance, during the design stage of a system, the analyst is often asked to provide the mean and the standard deviation (or the range) of a given performance indicator of the system–e.g., fuel consumption–in order to check its general conception. Whereas in operating stages, one must often verify that the system satisfies (or not) regulatory requirements for licensing or certification. Therefore, depending on the context of the study, the decision criteria may be: (i) a *min/max* criterion, i.e. the range of the outcomes given the variability of the inputs; (ii) a *central dispersion* criterion, i.e. central tendency and dispersion measures; (iii) a *threshold exceedance* criteria, i.e. the probability for a state variable of the system to be greater than a threshold safety value.

A rudimentary analysis of the computer code is also necessary: does it require a high CPU time for a single run, does it provide a precise evaluation of its gradient with respect to the probabilistic input parameters are typical questions.

Depending on these considerations, the uncertainty quantification methodology proceeds through different algorithms.

2.3   Step B - The Input's Uncertainty Quantification

The methods used for the probabilistic modeling of the inputs depend on the nature and the amount of available information.

In case of scarce information, the analyst first needs to interview experts. The literature proposes numerous protocols (e.g. [16]) that can assist in obtaining unbiased and relevant information which may then be translated into a pdf. In addition, a commonly used approach consists in applying the Maximum Entropy Principle, that leads to the pdf maximizing the lack of information (modeled by the Shannon entropy [17]), given the available expertise on the variable to be modeled. Whatever the chosen model, it is critical to validate it. One means for this consists of establishing a dialog with the expert to clearly identify key features for comparison with the established pdf–e.g. mean and quantiles, or alternative shape and scale parameters.

When data sets are available, the analyst can use the traditional statistical inference tools following a parametric or non parametric approach. The kernel smoothing technique is useful to model distributions which do not present usual shapes, for example, multimodal distributions. Then, the model is validated by a numerical fitting test, adapted to the objective of the analysis: for example the Kolmogorov test is used in the central zone or Anderson-Darling if one is more interested in tail fitting.

The EDF framework requires that the random input parameters $X_1, \ldots, X_m$ be represented as a random vector X with a multivariate pdf, the dependence structure of which must be explicit. A common way is to define the multivariate pdf $p(X)$ by its univariate marginal distributions $p_1(X_1), \ldots, p_m(X_m)$ and its copula $\mathcal{C}$, the later encoding the dependence structure [18]. In practice, inference on copula parameters can present problems and, as shown in [19], Kendall's τ or Spearman's ρ coefficients are not sufficient to fully determine the structure. Mismodeling the dependence structure is potentially dangerous as it can lead to an error of several orders of magnitude in the estimate of a threshold exceedance probability [20]. Our recommendation is that the copula inference be performed using the same techniques (e.g. Maximum Likelihood Estimation) as those for the univariate marginals.

2.4   Step C - The Uncertainty Propagation

Once quantified, uncertainties are propagated to the model outcomes. The propagation algorithms depend on the decision criteria and on the model characteristics specified in Step A.

In case of a *min/max* analysis, the range of the outcome is determined either as a result of an optimization algorithm or by sampling techniques. The input sample may come from a deterministic scheme –e.g., factorial, axial or composite grid–or randomly generated from the distribution accorded to the input vector. The choice of the method is informed by the CPU time required for model execution, $G(\cdot)$.

In case of a *central dispersion* analysis, the mean value and the variance of the outcome can be evaluated using Monte Carlo sampling, which also provides confidence intervals of the estimated values. If a high CPU time forbids such a sampling method, it is possible to evaluate the mean of the outcomes using a Taylor variance decomposition method that requires the additional evaluation of the partial derivatives of the model $G(\cdot)$. No confidence interval is estimated to quantify the quality of the Taylor approximation.

Finally, in case of a *threshold exceedance* criteria $\mathbb{P}[G(x) \geq z^*]$, the most widespread techniques are the simulation-based, such as Monte Carlo method and its variants that reduce the variance of the probability estimator: LHS, importance sampling, directional sampling, ... All of the simulation techniques provide confidence intervals. More sophisticated sampling methods exist to evaluate rare events (e.g. particle sampling [21]). In case of high CPU runtime, alternatives (FORM and SORM methods) exist to estimate the exceedance probabilities, based on an isoprobabilistic transformations such as the Generalized Nataf transformation [20], [22] in case of elliptical copula of the input random vector, and the Rosenblatt one [23], [24] in the other cases. These transformations are designed to map the input random vector into a standard space of spherical distributions. In that space, the integral defining the exceedance probability:

$$\int_{\mathcal{D}_f} p(x)dx, \text{ where: } \mathcal{D}_f := \{x; G(x) \geq z^*\}$$

is approximated using geometrical considerations [25,26,27]. These popular techniques provide approximations of very low exceedance probabilities with very few calls to the model. But no confidence interval is estimated to validate the geometrical approximations.

Finally, an alternative technique is to use a given budget of model runs to build a surrogate model $\tilde{G}(\cdot)$ which requires negligible CPU time for subsequent runs. Monte Carlo is then performed on on $\tilde{G}(\cdot)$ instead of $G(\cdot)$. Many techniques are provided in the Open TURNS package, among them the polynomial chaos expansion (PCE) [28] and the kriging approach [29].

The analyst is invited to mix methods and optimize an evaluation strategy with respect to a calculus budget. The validation comes from the confrontation of results obtained by different methods.

2.5   Step C' - Uncertainty Importance Ranking

The ranking of the uncertainty sources is based on the evaluation of some importance factors, correlation coefficients and sensitivity factors, the choice of which varies according to the quantities of interest specified in Step A. See [30] for an introductory overview of the problem.

In central dispersion studies, the Sobol's indices explain the variability of the outcomes by the variability of the input parameters or sets of parameters. Their evaluation by Monte Carlo sampling is costly and surrogate modeling approach (in particular PCE) are of great help. More simple correlation based indices (SRC, SRRC, PCC, PRCC indicators) could also be useful in practice.

In a threshold exceedance study, importance factors could be defined as particular Sobol indices after the linearization of the model around a specific point in the standard space. They quantify the impact of the global input uncertainty on the estimated exceedance probability.

According to the nature of the highest impact uncertainty source, feedback actions will differ: epistemic uncertainty requires some additional work to increase knowledge; reducible stochastic uncertainty requires some variability reducing actions; and irreducible stochastic uncertainty requires some modifications of the system in order to protect it against that unavoidable highest impact variability. These actions do not have the same consequences from the economical point of view, and do not address time equivalent issues.

3   The Open TURNS Software

3.1   An Open Source Software

Open TURNS [31] is an open source software package designed to implement the uncertainty framework sketched above. The package is distributed under LGPL and FDL licenses for the code source and its documentation respectively.

Running under the Windows and Linux environment, Open TURNS is a C++ library proposing a Python textual interface. It can be linked to any code communicating through input - output files (thanks to generic wrapping files) or to any Python-written functions. It also proposes standard interface for complex wrappings (distributed wrappers, binary data).

Gradients of the external code are taken into account when available and otherwise can be approximated automatically by finite differences schemes. In addition to its more than 40 continuous/discrete univariate/multivariate distributions, Open TURNS proposes several dependence models based on copulas: independent, empirical, Clayton, Frank, Normal, Gumbel, Sklar copulas. It offers a great variety of definitions of a multivariate distribution: list of univariate marginals and the copula, linear combination of probability density functions or random variables. Uncertainty propagation step is accomplished through numerous simulation algorithms. Open TURNS implements the innovative Generalized Nataf transformation and the Rosenblatt one for the FORM/SORM methods. For ranking analysis, Open TURNS implements the Sobol indices, in addition to the usual statistical correlation coefficients.

Open TURNS has a rich documentation suite comprising more than 1000 pages, dispatched within 8 documents covering all the aspects of the platform: scientific guidelines (Reference Guide), end-user guides (Use Cases Guide, User Manual and Example Guide) and some software documentations (Architecture Guide, Wrapper Guide, Contribution Guide and Windows port Guide).

Open TURNS implements select high performance computing capabilities such as the parallelisation of algorithms manipulating large data set (up to 10^8 scalars) using the threading building blocks technology (TBB). It also provides a generic parallel implementation of the evaluation of models over large data set using either pthreads or TBB.

3.2   Some Innovative Aspects

Open TURNS is innovative in several aspects. Its input data model is based on the multivariate cumulative distribution function (cdf). This enables the usual sampling approach, as would be appropriate for statistical manipulation of large data sets, but also facilitates analytical approaches. If possible, the exact final cdf is determined (thanks to characteristic functions implemented for each distribution, the Poisson summation formula, the Cauchy integral formula, ...). Furthermore, different sophisticated analytical treatments may be explored: aggregation of copulas, composition of functions from $\mathbb{R}^n$ into $\mathbb{R}^p$, extraction of copula and marginals from any distribution.

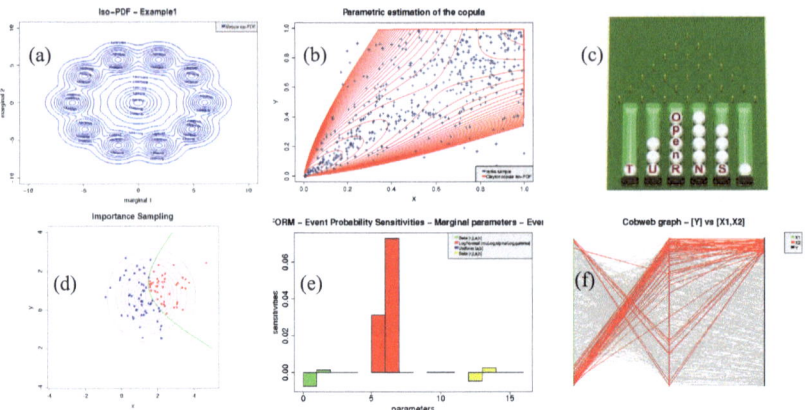

Fig. 1. Some Open TURNS snapshots: (a) the Open TURNS logo inspired by Galton's box experience, (b) modeling a multi-modal random vector of $\mathbb{R}^2$, (c) copula's fitting, (d) importance sampling in the standard space around the FORM design point, (e) FORM importance factors, (f) cobweb plots

Open TURNS implements up-to-date and efficient sampling algorithms. The Mersenne Twister Algorithm is used to generate uniform random variables [32], the Ziggurat method for normal variables [33], the Sequential Rejection Method for binomial variables and the Tsang & Marsaglia method for Gamma variables [34]. Exact Kolmogorov statistics are evaluated with the Marsaglia Method and the Non Central Student and Non Central χ^2 distribution with the Benton & Krishnamoorthy method [35].

Open TURNS is the repository of recent results of PhD research performed at EDF R&D. In 2011, sparse PCE based on the LARS method [36] was implemented. In a future release the ADS (Adaptive Directional Stratification, [37]) accelerated Monte Carlo sampling technique will be made available for Open TURNS users.

4 Examples of Applied Studies

We sketch examples from applied studies excerpted from recent works performed at EDF R&D. Despite being representative of real industrial problems, these examples are provided for demonstration purposes only and the results cannot be used to draw any general conclusion about EDF risk assessment studies.

4.1 Flood Risk Assessment after the Failure of an Earth Dam

Considering that EDF is a major hydro-power operator (operating more than 200 dams and 400 power stations) and the role played by sea and river water in the nuclear power generation, it follows that hydraulic simulation is an important topic of interest for EDF R&D. In particular, most EDF studies are concerned with flood risk. As an example, a recent study [38] investigates the flood risk assessment of a valley in the event that a dominating earth dam fails. Unlike concrete dams, which generally collapse and empty instantaneously, earth dam failure is assumed to be progressive and characterized by a so called *failure hydrograph*, i.e. a function $Q = H(t)$, describing the emptying discharge Q as a function of time t. Due to the complexity of the physics involved during the failure process, the precise shape of a hydrograph is not well known. Oft-used *ansatz* in these studies are that (i) the hydrograph has a triangular shape, (ii) the reservoir volume W at the beginning of the failure ($t = 0$) is known and (iii) the reservoir will completely empty during the observation period $[0, T_{obs}]$, i.e. $\int_0^{T_{obs}} H(t) \cdot dt = W$. Under these assumptions, the failure hydrograph is completely determined by the peak discharge Q_{max} and the time T_m at which the maximum discharge occurs.

The hydraulic modeling of the flood through the underlying valley is implemented by the MASCARET software [39] (resolution of 1D shallow water De St. Venant's equations) jointly developed by EDF R&D and CETMEF (Centre d'Etudes Techniques Maritimes et Fluviales). The geometrical features of the valley, here modeled as a 200 km long 1D channel (length, slope, section shape) are supposed to be known. On the other hand, the hydraulic friction parameter K_s (Strickler's coefficient) is uncertain and modeled as a random variable.

Three random variables are propagated trough the hydraulic model: Q_{max}, T_m and K_s. The output variables of interest are the maximum water level Z_{max} reached by the wave front in the most dangerous points of the valley and the corresponding arrival time T_f. The two most dangerous points (located downstream a section narrowing, which gives raise to an hydraulic jump) have been previously identified by physical consideration. They are located 11 km (Point 1) and 60 km (Point 2) downstream from the dam, respectively.

The uncertainty propagation has been performed by first building a polynomial response surface, then Monte Carlo sampling. A sensitivity analysis has also been made to find out the most influential variables on Z_{max} and T_f in different points of the valley. The most interesting results of the study are: (i) the quantiles (95%, 99% and 99.9%) of Z_{max} in Points 1 and 2 and (ii) the Spearman ranks correlation coefficients between Z_{max} (T_f, respectively) and the three input random variables. As an example of results the 99% quantiles of Z_{max} in Points 1 and 2 are respectively 675.6 and 516.5 m above mean see level (amsl). The analysis of Spearman's coefficients is particularly interesting. The most influential variable with respect to Z_{max} evaluation is the peak discharge Q_{max}. On the other hand, as far as T_f is concerned, it can be noticed that for the abscissas located close to the dam the most influential variable is T_m, but as one moves more and more downstream, the influence of Q_{max} and K_s raises. 90 km downstream from the dam, the friction coefficient becomes the dominant variable in the response evaluation.

This kind of study is valuable for supporting public powers in preparing the Emergency Response Plans in case of dam failure (e.g. planning evacuation of the most exposed areas). The sensitivity analysis is important to let the decision maker be aware of the weight of the hypotheses taken on the input variable and to possibly guide further study to reduce the uncertainties tainting the influential variables.

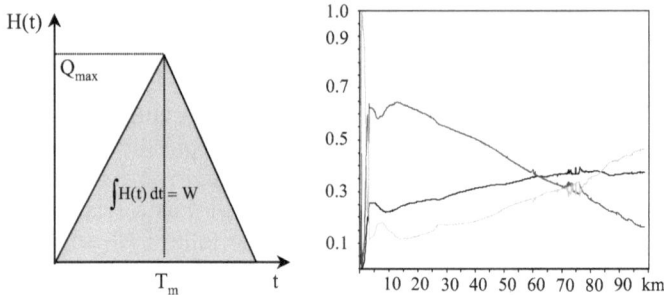

Fig. 2. Flood risk example [38]: hypothesis on the failure hydrograph (left) and sensitivity analysis by Spearman's ranks correlation coefficients (right) of T_f with respect to Q_{max}, T_m and K_s along the valley, up to 100 km downstream the dam

4.2   Globe Valve Sensitivity Analysis

EDF R&D has a deep history and experience in the application of uncertainty analysis methods in the field of solid mechanics. Many applied studies have been performed over the years, concerning for instance cooling towers, containment structures, thermal fatigue problems, lift-off assessment of fuel rods etc. We will focus on an application concerning reliability and sensitivity analysis of globe valves [40]. This investigation was one of the industrial case studies proposed by EDF in the context of the program, *Open Source Platform for Uncertainty Treatment in Simulation* (OPUS), funded by the French National Research Agency (ANR) between April 2008 and September 2011 [41].

Industrial globe valves are used for isolating a piping part inside a fluid circuit. This study is concerned with the mechanical behavior of the valve under water pressure. For this exemplary study, the variables of interest are the maximum displacement of the rod and the contact pressures. The tightness performance of the valve is assured if these variables stand below stated threshold values. The numerical model has been implemented thanks to the *Code_Aster* software, developed by EDF R&D and distributed under GPL license [42].

We will focus here on the sensitivity analysis of the maximum rod displacement Z. The problem has six uncertain input variables X_i, $i = 1 \ldots 6$: packings, glands, beams, steel rod Youngs modulus, hydraulic load and clearance. The goal of the study is the evaluation of Sobol's indices, which quantifies the contribution of each input X_i (or combinations of inputs, e.g. X_i and X_j) to the variance of the output $\mathbb{V}[Z]$:

$$S_i[Z] = \frac{\mathbb{V}\left[\mathbb{E}[Z|X_i]\right]}{\mathbb{V}[Z]}, \quad S_{ij}[Z] = \frac{\mathbb{V}\left[\mathbb{E}[Z|X_i, X_j],\right]}{\mathbb{V}[Z]} - S_i[Z] - S_j[Z] \quad \ldots$$

In practice, the Monte Carlo evaluation of the variances of the conditional expectations above is unfeasible due to computational resource constraints. One path to resolve this problem is by implementation of a polynomial chaos expansion (PCE). This technique consists in replacing the random output of the

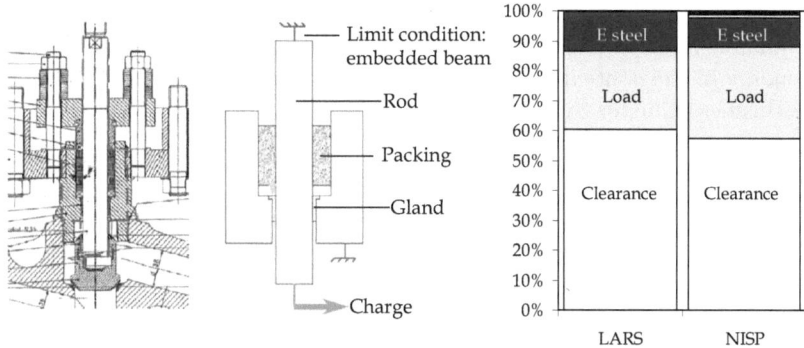

Fig. 3. Globe valve example [40]: cross sectional view (left), mechanical scheme (center) and Sobol's first order indices (right)

physical model by a decomposition onto a basis of orthonormal polynomials. The problem is reduced to the estimation of a finite set of coefficients, under the basis of a given number of previous runs of the physical model. As shown for instance in [43], once the coefficients have been determined, the evaluation of Sobol's indices is straightforward due to the orthonormality of the polynomials. PCE is particularly suited for this kind of problem.

Different methods have been tested for estimating the PCE coefficients. We have found that the LARS method [36] (cf. Section 3.2) and the NISP library (*Non Intrusive Spectral Projection*, developed by CEA [44]) return similar results, cf. Fig. 3.

5   Conclusion

Throughout this paper we have attempted to present an industrial perspective on UQ as we see it in our current practice. Of course, we do not pretend to provide an exhaustive nor prescriptive vision of this large problem.

Our approach is strictly non-intrusive and the problem is primarily viewed as a propagation of uncertainties from inputs to outputs of a numerical code. Some further steps for improving this methodology include: (i) systematically accounting for uncertainties tainting the computer model itself (the works carried by the MUCM [45], based on the Bayesian analysis of numerical codes [46,47] will be indeed of great help), (ii) linking of the common methodology sketched in Section 2 with decision theory, (iii) dealing with high dimension stochastic problems ($m \approx 100$) and (iv) treatment of functional inputs and outputs.

There can be no doubt that UQ is currently deeply rooted into EDF R&D practice. Our motivation for further work takes inspiration by our belief that industrial studies benefit from the consolidated practice of a common methodological framework and the Open TURNS software.

Acknowledgements. We gratefully thank the colleagues who worked on the industrial studies, sketched in Section 4 and in particular A. Arnaud,

M. Berveiller, G. Blatman and N. Goutal, as well as E. Remy and B. Iooss for their valuable advices. This work was partially supported by the French Ministry of Economy in the context of the CSDL project (*Complex Systems Design Lab*) of the Business Cluster System@tic Paris-Région.

References

1. Schweber, S., Wachter, M.: Complex Systems, Modelling and Simulation. Stud. Hist. Phil. Mod. Phys. 31, 583–609 (2000)
2. Heymann, M.: Understanding and misunderstanding computer simulation: The case of atmospheric and climate science - An introduction. Stud. Hist. Phil. Mod. Phys. 41, 193–200 (2010)
3. Sundberg, M.: Cultures of simulations vs. cultures of calculations? The development of simulation practices in meteorology and astrophysics. Stud. Hist. Phil. Mod. Phys. 41, 273–281 (2010)
4. Pilkey, O.H., Pilkey-Jarvis, L.: Useless Arithmetic: Why Environmental Scientists Can't Predict the Future. Columbia University Press, New York (2007)
5. Helton, J.C., Sallaberry, C.J.: Conceptual basis for the definition and calculation of expected dose in performance assessments for the proposed high-level radioactive waste repository at Yucca Mountain, Nevada. Reliab. Eng. Syst. Safe. 94, 677–698 (2009)
6. Helton, J.C., Hansen, C.W., Sallaberry, C.J.: Uncertainty and sensitivity analysis in performance assessment for the proposed repository for high-level radioactive waste at Yucca Mountain, Nevada. Reliability Engineering & System Safety (2011) (in press), doi:10.1016/j.ress.2011.07.002
7. Zheng, Y., Wang, W., Han, F., Ping, J.: Uncertainty assessment for watershed water quality modeling: A Probabilistic Collocation Method based approach. Adv. Water Resourc. 34, 887–898 (2011)
8. Eckert, N., Naaim, M., Parent, E.: Long-term avalanche hazard assessment with a Bayesian depth averaged propagation model. J. Glaciol. 56, 563–586 (2010)
9. Asserin, O., Loredo, A., Petelet, M., Iooss, B.: Global sensitivity analysis in welding simulations - What are the material data you really need? Fin. El. Analys. Des. 47, 1004–1016 (2011)
10. Hopfe, C.J., Hensen, J.L.M.: Uncertainty analysis in building performance simulation for design support. Energ. Buildings 43, 2798–2805 (2011)
11. Vernon, I., Goldstein, M., Bower, R.G.: Galaxy Formation: a Bayesian Uncertainty Analysis. Bayes. Anal. 5, 619–670 (2010)
12. Crucifix, M., Rougier, J.: A Bayesian prediction of the next glacial inception. Eur. Phys. J. Special Topics 174, 11–31 (2009)
13. Antoniadis, A., Helbert, C., Prieur, C., Viry, L.: Spatio-temporal prediction for West African monsoon. Environmetrics 23, 24–36 (2012)
14. Allard, A., Fischer, N., Didieux, F., Guillaume, E., Iooss, B.: Evaluation of the most influent input variables on quantities of interest in a fire simulation. J. Soc. Franc. Stat. 152, 103–117 (2011)
15. Baraldi, P., Pedroni, N., Zio, E., Ferrario, E., Pasanisi, A., Couplet, M.: Monte Carlo and fuzzy interval propagation of hybrid uncertainties on a risk model for the design of a flood protection dike. In: Berenguer, C., Grall, A., Guedes Soares, C. (eds.) Advances in Safety, Reliability and Risk Management: ESREL 2011. CRC Press, Leiden (2011)

16. O'Hagan, A., Buck, C.E., Daneshkhah, A., Eiser, J.R., Garthwaite, P.H., Jenkinson, D.J., Oakley, J.E., Rakow, T.: Uncertain judgements: eliciting expert probabilities. John Wiley & Sons, Chichester (2006)
17. Shannon, C.: A mathematical theory of communication. Bell Syst. Tech. J. 27, 379–423, 623–656 (1948)
18. Embrechts, P., Lindskog, F., McNeil, A.: Modelling dependence with copulas and applications to risk management. In: Rachev, S.T. (ed.) Handbook of Heavy Tailed Distributions in Finance, pp. 329–384. Elsevier, Amsterdam (2003)
19. Dutfoy, A., Lebrun, R.: A practical approach to dependence modeling using copulas. Proc. Inst. Mech. Eng. O J. Risk Reliab. 223, 347–361 (2009)
20. Lebrun, R., Dutfoy, A.: An innovating analysis of the Nataf transformation from the viewpoint of copula. Probabilist. Eng. Mech. 24, 312–320 (2009)
21. Del Moral, P.: Feynman-Kac Formulae - Genealogical and interacting particle systems with applications. Springer, New York (2004)
22. Lebrun, R., Dutfoy, A.: A generalization of the Nataf transformation to distributions with elliptical copula. Probabilist. Eng. Mech. 24, 172–178 (2009)
23. Lebrun, R., Dutfoy, A.: Do Rosenblatt and Nataf isoprobabilistic transformations really differ? Probabilist. Eng. Mech. 24, 577–584 (2009)
24. Rosenblatt, M.: Remarks on a multivariate transformation. Ann. Math. Statist. 23, 470–472 (1952)
25. Dolinski, K.: First-order second-moment approximation in reliability of structural systems: critical review and alternative approach. Struct. Saf. 1, 211–231 (1983)
26. Hasofer, A.M., Lind, N.C.: Exact and invariant second moment code format. J. Eng. Mech. 100, 111–121 (1974)
27. Tvedt, L.: Second order probability by an exact integral. In: Thoft-Christensen, P. (ed.) 2nd IFIP Working Conference on Reliability and Optimization on Structural Systems, pp. 377–384. Springer, Berlin (1988)
28. Ghanem, R.G., Spanos, P.D.: Stochastic Finite Elements: A Spectral Approach, Revised Edition. Dover, Mineola (2003)
29. Sacks, J., Welch, W.J., Mitchell, T.J., Wynn, H.P.: Design and Analysis of Computer Experiments. Stat. Sci. 4, 409–435 (1989)
30. Saltelli, A., Tarantola, S., Campolongo, F., Ratto, M.: Sensitivity analysis in practice: a guide to assessing scientific models. John Wiley & Sons, Chichester (2004)
31. Open TURNS, Open Treatment of Uncertainties, Risk's aNd Statistics, an open source source platform, http://www.openturns.org
32. Saito, M., Matsumoto, M.: SIMD-oriented Fast Mersenne Twister: a 128-bit Pseudorandom Number Generator. In: Keller, A., Heinrich, S., Niederreiter, H. (eds.) Monte Carlo and Quasi-Monte Carlo Methods 2006, pp. 607–622. Springer, Berlin (2006)
33. Doornik, J.A.: An Improved Ziggurat Method to Generate Normal Random Samples. Working paper, Department of Economics, University of Oxford (2005)
34. Marsaglia, G., Tsang, W.W.: The Ziggurat method for generating random variables. J. Stat. Softw. 5, 1–7 (2000)
35. Benton, D., Krishnamoorthy, K.: Computing discrete mixtures of continuous distributions: noncentral chisquare, noncentral t and the distribution of the square of the sample multiple correlation coefficient. Comput. Stat. Data An. 43, 249–267 (2003)
36. Blatman, G., Sudret, B.: Efficient computation of global sensitivity indices using sparse polynomial chaos expansions. Reliab. Eng. Syst. Safe. 95, 1216–1229 (2010)

37. Munoz Zuniga, M., Garnier, J., Remy, E., De Rocquigny, E.: Analysis of adaptive directional stratification for the controlled estimation of rare event probabilities. Stat. Comput. 22, 809–821 (2012)
38. Arnaud, A., Goutal, N., De Rocquigny, E.: Influence des incertitudes sur les hydrogrammes de vidange de retenue en cas de rupture progressive d'un barrage en enrochements sur les zones inondées en aval. In: SimHydro 2010 Conference, Sophia Antipolis (2010) (in French)
39. Goutal, N., Maurel, F.: A finite volume solver for 1D shallow-water equations applied to an actual river. Int. J. Numer. Meth. Fl. 38, 1–19 (2002)
40. Berveiller, M., Blatman, G.: Sensitivity and reliability analysis of a globe valve using an adaptive sparse polynomial chaos expansion. In: 11th International Conference on Applications of Statistics and Probability in Civil Engineering, Zurich (2011)
41. OPUS Contributors: Final Report of the ANR OPUS project (2011), http://www.opus-project.eu
42. EDF R&D: Code_Aster, Analysis of Structures and Thermomechanics for Studies & Research, http://www.code-aster.org
43. Sudret, B.: Global sensitivity analysis using polynomial chaos expansions. Reliab. Eng. Syst. Safe. 93, 964–979 (2008)
44. Crestaux, T., Martinez, J.M., Le Maitre, O.: Polynomial chaos expansions for sensitivitiy analysis. Reliab. Eng. Syst. Safe. 94, 1161–1172 (2009)
45. MUCM: Managing Uncertainty in Complex Models, http://www.mucm.ac.uk/
46. O'Hagan, A.: Bayesian analysis of computer code outputs: a tutorial. Reliab. Eng. Syst. Safe. 91, 1290–1300 (2006)
47. Goldstein, M.: External Bayesian analysis for computer simulators. In: Bernardo, J.M., Bayarri, M.J., Berger, J.O., Dawid, A.P., Heckerman, D., Smith, A.F.M., West, M. (eds.) Bayesian Statistics 9, pp. 201–228. Oxford University Press (2011)

Discussion

Speaker: Alberto Pasanisi

Michael Goldstein: I consider the notion of a unified formulation across applications as natural and valuable. I wonder how the model discrepency, i.e., the difference between the model and the system that the model tries to represent, is represented within the unified formulation.

Alberto Pasanisi: That is a crucial question and I thank Michael for giving me the opportunity to discuss our viewpoint. Actually, our general formulation of the problem and the methodological framework going with do not cope explicitly with model discrepancy. According to this scheme, UQ is mainly seen as the propagation of parametric uncertainties from inputs X to output Z of a deterministic computer model $Z = G(X)$. In some studies, generally concerning inverse problems (e.g. [1]), we did explicitly account for an additive Gaussian error in the observation model, so that $Z = G(X) + \epsilon$, but in most cases we simply consider the model, provided by the experts of given specific application fields as a black-box admitted *as is*. That is a purely pragmatic viewpoint, as in most cases the phases of verification & validation of the numerical code and parametric uncertainties propagation are made in separate times by different teams. In addition, a quite shared viewpoint in Uncertainty Analysis practitioners' community is that if the analyst does not trust enough the computer model, he/she must first improve it, before carrying an uncertainty and a (following) risk analysis [2].

Even if I acknowledge that the simplified framework I sketched makes thing easier in the engineering practice, I am aware of the limits of such a scheme and I hope that the use of more extended approaches quantifying both parametric and model's uncertainties will soon become a more standard practice in our studies. And I think that the work you carried with A. O'Hagan and your colleagues of MUCM will be of great help.

William Oberkampf: Model form uncertainty and, in many cases, model parametric uncertainty is epistemic (lack of knowledge) uncertainty. Epistemic uncertainty may be characterized as a probability distribution, but this represents incertitude as a random variable; which it is not. A more proper representation is to characterize incertitude as an interval-valued quantity, i.e. no knowledge structure over the range of the interval. This type of uncertainty analysis requires the use of a broader framework usually referred to as imprecise probability theory. Has EDF investigated the use of imprecise probability distributions in its uncertainty quantification?

Alberto Pasanisi: This question concerns a very important topic, namely, in a slightly reformulated way, "is the probabilistic assumption too informative for coping with purely epistemic uncertainties?" Actually, in our current practice, we use probabilistic modeling for both epistemic and stochastic uncertainty; indeed the Bayesian paradigm seems to give us a reasonable setting for coping with both sources of uncertainties in a decisional framework. Nevertheless, we have also carried more perspective works involving Dempster-Shäfer Theory [3,4] or a *hybrid* framework [5] combining probabilistic and possibilistic modeling of uncertainties. The results about extra-probabilistic modeling are encouraging but, as far as we are concerned, some issues (e.g. modeling dependencies) still need to be investigated before these methods could be widely applied in engineering practice.

Jeffrey Fong: Is the failure envelope boundary line the result of a deterministic analysis, or, stochastic with a 95% lower limit? If the former, then the failure envelope is not uncertainty-quantification-based at all. Please clarify.

Alberto Pasanisi: As we do not tackle explicitly model's uncertainty in our current framework, the failure domain $\mathcal{D}_f := \{x; G(x) \geq z^*\}$ is deterministic: it is just the set of values of the input vector X that produce values of the output Z corresponding to failure conditions. So, the probability of failure is the probability for the random input X to belong to the failure domain $\mathcal{D}_f$.

Maurice Cox: In your framework for uncertainty management you referred to various principles. Such principles are set out in the GUM (Guide to the Expression of Uncertainty in Measurement) and Supplements to the GUM. Does EDF use these documents, or is this a parallel development by EDF?

Alberto Pasanisi: Yes, absolutely. Actually, our framework is largely inspired by the GUM (and its supplements) which is a reference document in EDF's practice. The GUM is widely used by EDF's engineers and technicians working in R&D and Engineering Departments, and in power plants.

Pasky Pascual: Maybe I misunderstood, but you seemed to suggest that Bayesian inference is a way to address the issue of imprecise probabilities. But doesn't Bayes assume well-described pdfs or at least probability distributions that can be (somewhat) estimated?

Alberto Pasanisi: I think that it was a misunderstanding. My idea was that Bayesian setting allows to take into account an additional level of uncertainty tainting the probabilistic distributions of inputs, and that this framework fits comfortably most industrial requirements. Of course, imprecise probabilities constitute a different way to address the problems.

Mark Campanelli: How does your framework do sensitivity analysis? In particular, is sensitivity analysis done prior to uncertainty analysis in order to determine which parameters can be treated as fixed? Furthermore, can these sensitivity anayses incorporate dependencies between random variables, and if so, how?

Alberto Pasanisi: According to our schematic framework, sensitivity analysis (SA) is performed at the same time than uncertainty propagation. That happens, for instance, when putting into practice advanced method of SA, as Sobol's decomposition of the output's variance. Nevertheless, in particular when the number of inputs is large it is recommended to perform, prior to uncertainty analysis a first SA with less costly techniques (e.g. screening [6] or Morris method [7]) which, even if based on simplified assumptions, provides a first selection of influent variables. Then fixed values are given to less influent variables, thus reducing the stochastic dimension of the problem. Cf. also [8] for a pragmatic approach to the choice of the SA method, depending on the context of the study. The second part of the question is much more tricky and concerns more advanced research works than industrial R&D practice. In engineering studies, the most pragmatic way for coping with this problem is to gather dependent variables in groups and evaluate sensitivity indices for each one of these groups. Moving to more advanced works, you can see [9] for an introduction to the problem. That is a quite active research topic. Recent interesting papers concerned with this problem proposes several different approaches. In the linear case, Xu & Gertner [10] distinguish two indices, quantifying the contribution of a parameter due to its correlation with the other parameters and the contribution due to its own effect. In the non-linear case, Li et al. [11] propose a technique for the covariance decomposition and three types of Sobol's indices. Other works propose indices based on the distance [12] between the actual pdf of the output and conditional densities. These indices can be used in presence of correlated inputs, but their interpretation could not be easy. Finally techniques based on the copulas' formalism are proposed by Kucherenko et al. [13].

Wayne King: Could you describe EDF needs in life extension and life prediction as it relates to reactors?

Alberto Pasanisi: That is an interesting and actually very wide question. I will give hereby some elements, focused on the key topic "running safely nuclear power units in the long-term" [14]. EDF operates today 58 pressurized water reactor units with three different power levels: 900 MW (34 units), 1300 MW (20 units) and 1500 MW (4 units). Nuclear generation represents about 85% of EDF power generation. EDF nuclear power plants were designed for operating during 40 years at least. The lifespan of some components is supposed to be greater than 40 years, while others must be replaced before, e.g. transformers should be replaced after 25-30 years. As another example, the EDF Board has approved in September 2011 an order for 44 steam generators, for its 1300 MW units.

The mean age of EDF power units is around 29 years: most of nuclear reactors operating worldwide are contemporary or older than EDF's ones. Nowadays, according to international studies lead on both lifespan and maintenance policies of several nuclear units in different countries, the target of 60 years lifespan is considered to be granted, by a technical point of view. Since several years, EDF sets the technical conditions to operate its nuclear unit well beyond 40 years: components' refurbishing and replacing programs currently run and will continue in the next years. The extension of the lifespan will have to meet the compliance with specific safety objectives which will be fixed by the French Regulation Authority (ASN). By its side, the strategy of EDF for assuring a safe and high performance running of its units in the long term is based on the five pillars below:

- The ten-yearly inspection of each unit. Under the control of the Regulation Authority, it is actually a complete check-up of the unit's facilities. With a duration of about 90 days, this exceptional shutdown period allows to realize a huge number of controls and maintenance works. As a matter of facts, the safety level of each unit is then reinforced every ten years. After the end of the inspection, the Regulation Authority gives its verdict about the continuation of the unit's operation for ten more years.
- The modification and the refurbishing of unit's devices. These operations are lead regularly and can take place during ten-yearly inspection or other shutdown periods. Thanks to the technical similarity of EDF's units, these works take profit of technological advances and the enhanced feedback of the whole nuclear fleet.
- The survey and the anticipation of equipment's ageing. EDF has set an ambitious maintenance policy of its components. Depending on the features and the role played by each equipment, the maintenance policy could be scheduled, condition-based or corrective. Other actions as long-term partnership with subcontractors to avoid technical obsolescence or shortage problems complete this set of actions. As a key figure, EDF spends yearly about 2 billions of euros for the maintenance, refurbishing and modification of nuclear units' equipment.
- The preservation and the renewal of human skills. Nuclear units' operation needs workers with very specific and high skills. As a significant part of technicians and engineers working in nuclear units will retire in the very next years, this question is crucial and challenging and several actions have already been set by HR to cope with (e.g. recruitment campaigns, tutoring, partnerships with universities and engineering schools).
- The improvement of technical and technology knowledge. With its large staff and budget (around 2000 people and 486 millions of euros in 2010 respectively [15]), EDF R&D has a key role in this action. R&D activities cover all disciplinary fields involved in nuclear process. As an example, two recently acquired equipments of EDF R&D: the TITAN Transmission Electron Microscope and the IBM Blue Gene super-computer witness the ambition to enhance more and more the knowledge of components and materials of nuclear units, by both physical and computer experiments.

References

1. Celeux, G., Grimaud, A., Lefebvre, Y., De Rocquigny, E.: Identifying intrinsic variability in multivariate systems through linearised inverse methods. INRIA Research Report RR-6400 (2007)
2. Aven, T.: Some reflections on uncertainty analysis and management. Reliab. Eng. Syst. Safe. 95, 195–201 (2010)
3. Limbourg, P., De Rocquigny, E.: Uncertainty analysis using evidence theory – confronting level-1 and level-2 approaches with data availability and computational constraints. Reliab. Eng. Syst. Safe. 95, 550–564 (2010)

4. Le Duy, T.D., Vasseur, D., Couplet, M., Dieulle, L., Bérenguer, C.: A study on up-dating belief functions for parameter uncertainty representation in Nuclear Proba-bilistic Risk Assessment. In: 7th International Symposium on Imprecise Probabil-ity: Theories and Applications, Innsbruck (2011)
5. Baraldi, P., Pedroni, N., Zio, E., Ferrario, E., Pasanisi, A., Couplet, M.: Monte Carlo and fuzzy interval propagation of hybrid uncertainties on a risk model for the design of a flood protection dike. In: Berenguer, C., Grall, A., Guedes Soares, C. (eds.) Advances in Safety, Reliability and Risk Management: ESREL 2011. CRC Press, Leiden (2011)
6. Campolongo, F., Tarantola, S., Saltelli, A.: Tackling quantitatively large dimen-sionality problems. Comput. Phys. Commun. 117, 75–85 (1999)
7. Morris, M.: Factorial sampling plans for preliminary computational experiments. Technometrics 33, 161–174 (1991)
8. De Rocquigny, E., Devictor, N., Tarantola, S. (eds.): Uncertainty in Industrial Practice. Wiley, Chichester (2008)
9. Kurowicka, D., Cooke, R.: Uncertainty analysis with high dimensional dependence modelling. Wiley, Chichester (2006)
10. Xu, C., Gertner, G.Z.: Uncertainty and sensitivity analysis for models with corre-lated parameters. Reliab. Eng. Syst. Safe. 93, 1563–1573 (2008)
11. Li, G., Rabitz, H., Yelvington, P.E., Oluwole, O.O., Bacon, F., Kolb, C.E., Schoen-dorf, J.: Global Sensitivity Analysis for Systems with Independent and/or Corre-lated Inputs. J. Phys. Chem. 114, 6022–6032 (2010)
12. Borgonovo, E.: A new uncertainty importance measure. Reliab. Eng. Sys. Safe. 92, 771–784 (2007)
13. Kucherenko, S., Munoz Zuniga, M., Tarantola, S., Annoni, P.: Metamodelling and Global Sensitivity Analysis of Models with Dependent Variables. In: AIP Conf. Proc., vol. 1389, pp. 1913–1916 (2011)
14. EDF Group: Exploiter les centrales nucléaires dans la durée (in French). Informa-tion Note (2011), http://energie.edf.com/nucleaire/publications/notes-d-information-46655.html
15. EDF Group: 2010 Activity and Sustainable Development Report (2011), http://www.edf.com/html/RA2010/en/

Living with Uncertainty

Patrick Gaffney

Bergen Software Services International (BSSI)
Bergen, Norway
pat@bssi-tt.com

Abstract. This paper describes 12 years of experience in developing simulation software for automotive companies. By building software from scratch, using boundary integral methods and other techniques, it has been possible to tailor the software to address specific issues that arise in painting processes applied to vehicles and to provide engineers with results for real-time optimization and manufacturing analysis. The title provides the focus and the paper describes how living under the shadow of uncertainty has made us more innovative and more resourceful in solving problems that we never really expected to encounter when we started on this journey in 1999.

Keywords: electrocoat simulation, computational science, simulating painting processes.

1 Prologue

- On the one hand, this is a story of doing software business with over 14 automotive manufacturers spanning the USA, Europe, and Asia during the period 1999 2011. The story is true, I have withheld the names of the guilty and innocent, and the story is ongoing.
- On the other hand, it is the story of how little influence we, as a community of computational scientists and numerical analysts, have had on engineers, those doing the daily business of manufacturing, wherever they may be.
- But mainly, the story is about VALIDATION the process by which a manufacturer thinks they can assess the quality of software.

In the attempt to assess the quality of software, manufacturers often forget that it takes a level of skill to understand how to use software properly. Without a clear understanding of the requirements of software and, more importantly, an understanding of the imperfections in the manufacturers own processes, it is impossible to come to a valid conclusion given the present scheme of VALIDATION undertaken by almost all automotive manufacturers.

This paper provides evidence to support these claims and gives some simple recommendations that make it easier for the manufacturer to make wiser decisions.

A. Dienstfrey and R.F. Boisvert (Eds.): WoCoUQ 2011, IFIP AICT 377, pp. 46–59, 2012.
© IFIP International Federation for Information Processing 2012

2   Electrodeposition

The software used to exemplify our thesis is one simulator from a family of software simulators called Virtual Paint Operations, or VPO^{TM} Software for short. This family contains software predictors for the processes involved in electro-coating a vehicle, namely drainage, electrocoating, void detection, drag-out, and baking. Everything said in this paper applies to all members of the VPO^{TM} Family but for the purposes of clarity the paper focuses on the simulator for predicting the electrodeposition of paint on a vehicle body, frame, or part. Automotive manufacturers and the coatings industry call this type of paint **e-coat** and it is the first coat of paint applied to protect a vehicle from salt spray induced corrosion and stone chipping.

The application of e-coat is through a process called electrodeposition that resembles the electro-plating technique that most people are familiar with. In this case, a vehicle is immersed in a tank containing the e-coat material and an electric potential (voltage) is applied to anodes positioned along the sides, and possibly the top and bottom of the tank. The resulting electrical current transports paint solids in the e-coat from the tank to the surface of the vehicle. There, and as a result of the hydrolysis of water that also occurs, the paint coagulates and adheres to the surface. As the paint adheres to the exterior surfaces of the vehicle, an electrical resistance builds and since electricity always seeks the path of least resistance the electrical current automatically flows to areas of the vehicle not previously coated. To access recessed and other areas of the body that are difficult to reach, engineers construct pathways, by manually placing holes in the body of the vehicle to allow current to flow through.

Areas of the vehicle with an inadequate coating of e-coat are liable to corrode. Therefore, the correct identification and subsequent choice of pathways are vital to the vehicles quality, its corrosion protection, and the reduction of the vehicles warrantee costs. Relying on manual experience and testing to insert a suitable configuration of holes, of the right size, number, and placement, is not a good idea, especially when software simulation guarantees success at far less cost to the manufacturer.

An *adequate coating of e-coat* usually means exterior surfaces of the vehicle have a thickness of paint, called film build, which for many manufacturers is around 20 microns while interior surfaces can have less thick film build, usually around 10 microns.

It is this difference in film build requirements between exterior and interior surfaces that contributes to the cost of applying e-coat because to ensure interior surfaces have an *adequate* film build often requires the exterior surfaces to have more than adequate coverage. One goal of any manufacturer is to avoid excessive coverage on exterior surfaces while maintaining adequate coverage on both exterior and interior surfaces.

Software experiments are an ideal way of investigating how to achieve this goal because, unlike physical testing, they can guarantee success. Physical testing has limitations —the amount of material wastage to perform the tests is bad for theenvironment, it is financially expensive, and because the number of physical tests is limited, they are unsure of success. There are no such limitations when using software but what is required is an understanding of the physical process and the knowledge of how to use correctly a software tool that provides accurate predictions of this process.

3   Validation

Since electrical resistance increases with paint thickness, one way of modeling electrodeposition is to model resistivity correctly. This is the approach taken by the simulator of this paper. A paint model, whose parameters are all measurable and determined by a simple laboratory experiment, encapsulates the behavior of e-coat over time, and allows accurate predictions at any instant the vehicle is in the tank.

Verification of the simulator is the responsibility of the software developer and in the present case; verification is a continual process of daily life, beginning in 2003 and continuing today.

Validation is the process of comparing results from the simulator against an experiment performed by the manufacturer. Validation should be a contract between the customer and the developer, specifying precise conditions of the experiment to ensure comparisons are fair. This specificity often falls short because of uncertainties in the way the manufacturer conducted the experiment.

A manufacturer usually requests an experiment with the following components for validating electrodeposition software.

1. The manufacturer applies e-coat to a vehicle and then bakes the vehicle.
2. Either the manufacturer or a third-party cut up the vehicle (a process called teardown) and the manufacturer measures the thickness of electrocoat at measurement points chosen by the manufacturer.
3. The manufacturer compares the measured thickness of electrocoat to the predicted values obtained by the simulator.

On the face of it, these three components seem simple enough until one delves into the details of how and when the manufacturer conducts the experiment.

4   The Vehicle to E-Coat

Western manufacturers usually intend to conduct item 1 of the experiment using a **production** vehicle, in other words, one that contains all of the reinforcements and other structural entities that are in the actual vehicle that a customer purchases. However, a manufacturers good intentions often go awry because the description of the vehicle that is passed to the software developer misses some important material, such as reinforcements, bolts, separators, and non-conductive material, the presence and position of all of which are necessary if the simulator is to predict accurately the electrodeposition behavior.

On the other hand, Asian manufacturers will often have a specific point in their design schedule for when they will perform a teardown. This teardown is for reasons other than validating electrodeposition software and can occur at a time when the vehicle does not contain all of its reinforcements or other structural entities. These manufacturers are reluctant, and often refuse, to perform a second teardown for validating specific software. Regardless of the reasons for this attitude, it behooves the software developer to accommodate the manufacturers wishes and in this case provide an alternative benchmark by which to compare the software predictions.

5   The Time Elapsed

Another uncertainty the developer needs to address is the time elapsed between item 1 above and the comparison mentioned in item 3. For Asian manufacturers this time can be on the order of years, because of their rigid teardown schedule, and this means it is doubtful the e-coat used in step 1 still exists. Without a sample of this e-coat, it is impossible to produce a paint model that encapsulates the behavior of the e-coat used to paint the teardown vehicle.

The manufacturers paint vendor has been known to suggest that the laboratory experiment required to determine the model parameters is performed using e-coat that has *"the same characteristics"* as that used to paint the teardown vehicle. This is tantamount to using fresh e-coat paint and is definitely not acceptable for validation because the resistivity of such paint tends to be much lower than the e-coat drawn from the manufacturers tank. Low resistivity implies high film builds and the use of fresh paint will therefore result in distorted and incorrect predictions from any simulator that uses properties of the e-coat paint.

When a sample of the e-coat used to paint the aged teardown vehicle does not exist then **validation** is inappropriate because it is impossible to reproduce in software the precise conditions under which the vehicle was painted.

Provided the CAD file of the teardown vehicle exists, it is possible to model the vehicle and if the e-coat used to paint the vehicle does not exist then one can use a sample of the e-coat used to paint the existing vehicles of the same model and type. In this way it is possible to evaluate *trends* in film build coverage over the 3-D model of the teardown vehicle.

6   The Measurement Points

The manufacturer selects points on the vehicle surface and takes measurements of the film thickness at them using one of the methods described in the following sub-section. However, recording the accurate position of these measurement points is not always easy for the manufacturer to remember and this is a potential major source of error when it comes to obtaining predictions from the simulator at the same measurement points. Very often, the person running the simulator receives a photograph or image showing the positions and then has to interpret them for input to the simulator. Ironically, to produce a predicted result at an arbitrary point on the surface of a 3-D computational model requires a substantial amount of programming and technical effort, which is of little use when the manufacturer *forgets* the position of a measurement point.

7   Measuring Film Build

Whichever geometry the manufacturer uses as a benchmark against which comparisons are made of the software predictions, it is necessary to consider how the manufacturer measures the thickness of paint noted in item 2 above. There are essentially two methods in common use by well known reputable manufacturers. The first method consists of a special instrument designed specifically for this purpose and is usually considered to be accurate. The second method however, is manual and involves:

1. Measuring the thickness of the vehicle part.
2. Using sand paper to remove the paint layer and reveal the phosphate layer.
3. Measuring the thickness of the vehicle part again.
4. Computing the difference between the two measurements in 1 and 3.

Needless to say, item 2 of the second method is undesirable when considering using the measured values as sufficiently accurate to be part of a validation process. A small undue pressure on the sand paper can remove more of the paint layer and thereby result in an incorrect measurement of film thickness. However, this is unlikely to happen because usually the person performing the removal has acquired a deft touch from experience over many years. Passing on this experience to a new trainee will be problematical though.

8   Measured Film Build

An automobile manufacturer usually records measured film build by writing on the surface of the teardown vehicle the thickness value at the measurement point. A continuous sequence of voids usually indicates a problem in one or more stages of the electrocoating process as opposed to a problem with the e-coat itself.

The phosphate process, that the vehicle undergoes prior to electrocoating, cleans all surfaces of the vehicle from residual stamping oils and other contaminants, and deposits on the surface a protective layer of zinc and iron phosphate crystals. The color of adequate phosphate crystal deposits are dullish gray and rough and these facts help identify problem areas. For example, in the present case, by examining a color photograph of the teardown or better still, the actual teardown vehicle, the shiny color of the metal surrounding the voids indicates that the phosphate system has not been effective in removing stamping oils, and therefore the e-coat could not get to this surface of the vehicle causing the voids to occur.

An engineer rarely divulges information about areas of the vehicle with phosphate problems or grease problems and therefore the simulator has no way of knowing that contamination has compromised e-coat coverage in these areas of the vehicle. Consequently, predictions from the simulator cannot be correct in these areas of contamination.

Surprisingly, engineers rarely understand this fact and tend to expect the simulator to still produce accurate predictions in these areas even though they know the areas are contaminated. This is a great pity, for in the spirit of a validation *contract*, it is possible for the contractual partner —in this case the software developer —to provide **variations** of the simulator to take account of *known* uncertainties such as contamination from bad phosphating or grease problems, or voids due to air pockets. For example, in the case of air pockets, a variation of the simulator discussed in this paper can identify their location and predict electrodeposition in their presence, which makes it unnecessary for the engineer to know in advance where the air pockets may occur. The same is true of grease spots where new research at BSSI may lead also to the identification of grease spots a priori.

9   Towards a Turnkey HPC Solution

The attitude inherent in the last section is prevalent among engineers and is due in large part from a combination of their bad experiences with software and its providers, their lack of understanding of the power of mathematics, and their lack of technical experience with software simulation.

The latter is especially true in countries with a low standard of education in computational science. Unfortunately, there are few signs that this situation will improve. Many Universities no longer teach the basic numerical methods that are required to use software correctly, and employer-based training schemes have little understanding of non-business or technical courses. Therefore, to optimize the chances of a manufacturer using software simulation to improve their processes, it behooves the developer to write software so that the level of technical skill required to use it is minimal.

The software must include and convey the required expertise and knowledge in a way that makes the software easier to use and *control* by non-expert users. This means the developer must write the software to take account of imperfections in the input to the software and must convey enough information to the user in a way that allows them to either correct the input or to abandon the software run. If done correctly then, the software supplements the present lack of knowledge and expertise of an individual user in a way that opens up for the use of the software by a much wider community, which is a beneficial outcome for the manufacturer since a broader user space allows workers to migrate from one responsibility to another without the company having to sacrifice the use of the software.

Software produced in this way will not fail because invalid input is trapped and either fixed automatically or reported to the user in a way that will allow a correction.

Towards the goal of providing software that is significantly easier to use, the designers of all the VPO^{TM} Software made the following three conscious decisions:

1. A single user interface provides access to all VPO^{TM} Software.
2. The computational mesh of the vehicle is the same for all VPO^{TM} Software.
3. The system provides Software Generators to specify the input of tanks (VPO^{TM} Tank Generator) and bake ovens (VPO^{TM} Oven Generator) and to produce their respective 3-D models automatically.

Item 3 enables the manufacturer to specify dip tanks, e-coat tanks, and bake ovens without the need of CAD files for these structures, up to date versions of which often do not exist.

Item 2 ensures that only one mesh is required to model all the processes addressed by VPO^{TM} Software and thus the manufacturer only has to maintain a single mesh for modeling painting processes. It is also the only item for which a CAD file is required and the odds that this exists for a so-called Body In White (BIW) are greatly improved.

Item 1 provides access to the VPO^{TM} Software components from a single interface and this makes it easier for users to invoke the individual related tools for the topic of painting. A single user interface also makes it easier to *combine* the use of tools for modeling related processes, for example, to determine the effects on drainage from hole configurations used by electrocoating and vice versa.

10   Coping with Imperfections in the Input

The input to the simulator discussed in this paper requires the manufacturer to specify three datasets describing

(a) The e-coat material
(b) The e-coat tank, and
(c) The vehicle.

11   The E-Coat Material

Characterization of the e-coat material is determined from a laboratory experiment, usually performed by the automobile manufacturers paint vendor, according to a set of rules specified by BSSI. As mentioned previously, a sample of e-coat drawn from the operational e-coat tank is crucial for the correct determination of the model parameters that simulate the e-coat used to paint the vehicle. This is the main contender for error in specifying the e-coat material input for the simulator, as shown by the following items.

1. The automobile manufacturers paint vendor asserts that material used for the laboratory experiment has the same characteristics as that in the operational tank, and this is blatantly impossible if the material is fresh or it has not been drawn from the tank.
2. The paint vendor does not take adequate precautions when transporting the e-coat from the operational tank, especially if it is possible for the e-coat to be frozen during transit.

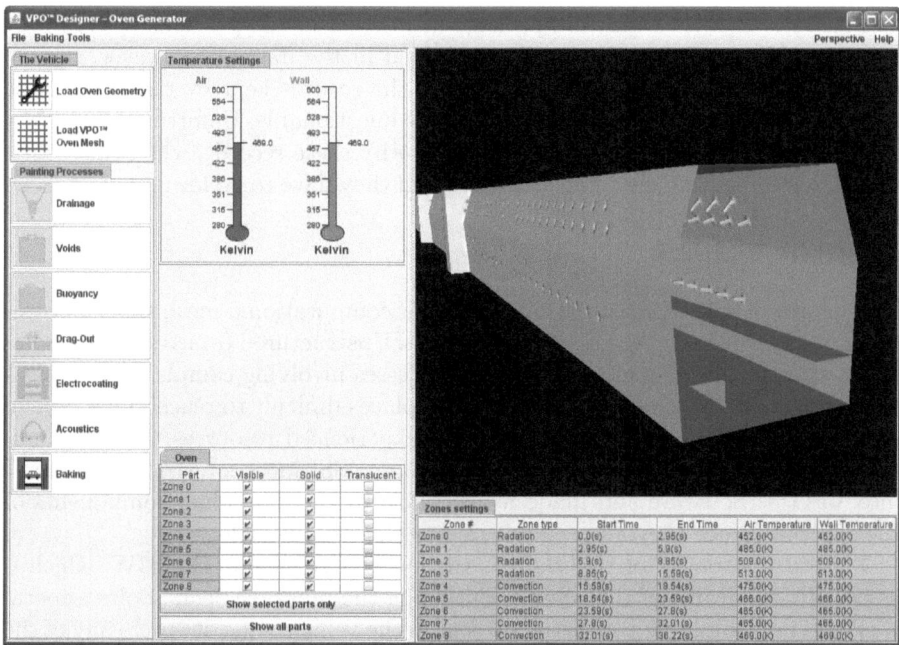

Fig. 1. A single user interface to all tools makes it easier to use the software. In this picture, invoking the Baking tool gives access to the Oven Generator.

BSSI has had to contend with these and other related issues from different automobile manufacturers at one time or another. Since the manufacturer or the paint vendor does not divulge these issues, it was necessary for BSSI to develop techniques, independent of the simulator, to analyze the laboratory results ahead of calculating the paint model parameters. These techniques, which, among others, identify the issues of 1 and 2 above, now form part of an *Input Verification* package that all input to the simulator must pass through successfully before running a simulation.

12   The E-Coat Tank

The dataset that describes the e-coat tank includes information about the anodes contained in the tank: their type, the voltages applied to them, and the time the vehicle spends in front of them.

Initially, many automobile manufacturers thought they had to provide information about the *shape* of the tank and they instigated special laser measurements to determine this information for their tanks. However, BSSI has removed the burden from the manufacturer of providing this information because a built in *Tank Generator* includes the information from the provided dataset (b) to generate automatically a 3-D model of the tank suitable for the simulator.

Thus, the *Input Verification* package mentioned in the previous section only has to verify the anode information provided in the dataset. However, the veracity of this information relies heavily on the records kept by the automobile manufacturer and in some cases these can be questionable. Unfortunately, only a physical inspection of the e-coat tank can verify these records, when they exist, and this is something that BSSI has accepted they have to do for manufacturers.

13   The Vehicle

Every form of software simulation requires a computational mesh that captures the physical features of the simulated process. Unstructured (or irregular) meshes are usually required to model physical processes involving complex 3-D geometries that contain recessed areas and other places difficult to reach. This type of mesh is difficult to generate and requires experienced resources that most automotive companies do not possess. Therefore, BSSI incorporated a *Mesh-Suite* into VPO^{TM} Software and made it accessible to the individual components of the software, since they all use the same mesh of the vehicle.

The starting point for this *Suite* is the native CAD file of the BIW, which is essentially the frame of the vehicle containing the reinforcements, bolts, separators, and non-conductive material present in the vehicle when it enters any of the tanks or the bake oven. For the electrodeposition simulator described in this paper, dataset (c) is this native CAD file. Unfortunately, for one reason or another, the automotive company rarely provides a CAD file of the BIW because one or more of the listed items is usually missing. The omission of any one of them will affect the predictions of e-coat but omitting one or more reinforcements will have the most serious effect.

The present validation process defined by most automobile manufacturers makes it very difficult to ascertain when the supplied CAD file differs from the vehicle used to provide the benchmark measurements of film build. However, in some cases where a manufacturer is willing to share these measurements with the software developer it is possible for BSSI, using techniques it has developed, to determine where differences occur in the geometry of the vehicle and then together, the manufacturer and BSSI, can evaluate the situation and agree on how to proceed. This is the ideal way of working but is all too often rare.

14   Observations

The standard validation process, used by most automobile manufacturers, can never work because a painted vehicle is never free from imperfections. Unless the manufacturer is prepared to accept this fact and to point out these imperfections BEFORE making simulations, predictions can never be totally correct. Identifying imperfections makes it possible to use a simulator, modified to take imperfections into account.

In an ideal world, validation is a contract meant to assist both the manufacturer and the developer. Unfortunately, for one reason or another, most manufacturers, but not all, view the validation process as adversarial. Until this attitude changes, manufacturers will not reap the benefits of using software to minimize physical experiments.

To optimize the chances of a manufacturer using software simulation to improve their processes, the developer must write software so that the level of technical skill required to use it is minimal. Since the use of the software requires the INPUT to be correct in order for the OUTPUT to be likewise (garbage in garbage out), and since the input requires three pieces of information that may be contaminated with imperfections, the software must be written to recognize this fact, inform the user, and supply enough information for corrections to be made. To enhance this process and to foster a good working relationship, BSSI has instigated the following new validation procedure, which is a modification of the standard one.

At the manufacturers premises anywhere worldwide, BSSI will inspect the vehicle while it is painted and baked, noting:

- Areas where grease or phosphate problems may have occurred prior to applying e-coat
- Any other imperfections in the vehicle or in the operations
- How the manufacturer takes measurements and transfers them to BSSI.

Of course the implementation of this new procedure will run into trouble when the manufacturer wishes to use a teardown vehicle significantly older than the inspected ones. However, provided the CAD file of the teardown vehicle exists, it is possible to model it. Depending on the age of the teardown, the e-coat used to paint it will most likely not exist and therefore validation is not appropriate. Rather, the only recourse is to evaluate trends in film build coverage over the 3-D model of the teardown vehicle.

15   Conclusions

This paper gives a brief summary of the discoveries made by a naive software developer when trying to interest automotive companies in the use of advanced mathematical simulators of painting processes.

Naiveté on the part of the developer is from two perspectives: (1) the assumption (unjustified) that the manufacturer's people were knowledgeable in the basics of modeling and understood the physical process and (2) the reality that as a community of numerical analysts and computational scientists we have failed to communicate with the very people we are in business to serve.

In the first case, the manufacturers technical people, almost universally, are deeply skeptical, for whatever reason. In the second case, we have not been successful to the extent that nowadays people who use software have little or no concept of the importance of **input** to that software —the old adage of *garbage-in garbage-out* is just that: a standalone saying that almost nobody thinks about when using software.

Input to the software comes from the manufacturer and it has been our unfortunate experience that almost no single one of them has got all aspects of the input correct when requesting a validation.

One way of avoiding this problem is for the software to diagnose invalid input, correct it automatically where possible, and if not then convey to the user what they need to do to correct the input. Another way is to remove the responsibility from the user of providing aspects of the input that they are incapable of providing correctly. The mesh of the vehicle is one example of both these situations.

For instance, the generation of a computational mesh requires that the elements of the mesh (for example, triangles) are of a size, and placement that accurately reproduces the simulated physical process. For example, to model accurately the flow of paint between two plates, the number of triangles and their size on both plates will depend on the gap between them. It should be intuitively obvious that, if one is to obtain an *accurate mesh* that provides accurate predictions from the simulator then many smaller triangles will be required on each plate for a gap of 1 mm than for a gap of 3 mm or greater. In this context, an accurate mesh is a mesh that has converged in space and mesh builders unfortunately often overlook this aspect of mesh generation. It is not uncommon to find the same mesh spacing used for gaps of 1, 2, 4, 8, and 10 mm and accompanying wonderment that the predicted results look nothing like what they should be!

One way to avoid this problem is for the simulation system to generate the converged mesh automatically without input from the user. Another way is for the simulation system to monitor the mesh input by the user and determine if it has converged in space or not. If it has not converged then, before continuing with the run, the system informs the user that the mesh is too crude to get accurate predictions and suggests where more refinement may be necessary.

I am definitely not advocating an artificial intelligence or expert systems approach to using the type of computational simulator addressed in this paper. What I am advocating is much simpler: **get the input right and the code will work**.

Since we know exactly the three pieces of information needed for the input then we know what getting it right means. The three pieces of information come from three different sets of people:

1. CAD people know CAD but not meshing - therefore let them provide the CAD and the simulation system will do the meshing properly. The Mesh-Suite mentioned above without input from the user produces an irregular 3-D mesh that is both mathematically correct and converged in space.
2. Paint vendors know paint - therefore let them supply what is required and the simulation system will monitor and analyze their data to see that it is correct.
3. The tank operators - they may delegate the operation of the tank to the paint vendors, who, in turn, may not record the settings of that tank. Unfortunately, only a physical inspection of the e-coat tank can verify these records, when they exist, and this is something that BSSI, as a developer, have accepted to do for manufacturers.

In the paper, I have tried to make it clear that to perform computational simulation successfully the software must minimize the information required of the operator or engineer because they cannot easily get this information. Instead, the software must be written in a way that either avoids the user having to provide any information or that provides questions that the user is able to answer.

As mentioned in the prologue, this process is ongoing. One tool that is essential to ensure accuracy of code changes and to assist with parallel programming implementations is Brian Smiths Test Harness (TH) [1]. This tool is an indispensible part of working towards a turnkey HPC solution.

16   Afterword

> *"We are the heirs to a tradition that has left science and society out of step."* – Jacob Bronowski, 1951

Over 60 years ago, the eminent scientist and philosopher Jacob Bronowski made these remarks, in his seminal work *The Common Sense of Science* [2]. His words are as true today as they were then.

Moving to today, I believe the further remark is unfortunately, no less true:

> *"We are the heirs to a tradition that has left computational science and society out of step."* – 2011

I believe both statements are true for the same reasons: a lack of understanding between both parties in each case, borne of a tradition where the practitioners of Science and Computational Science tend to work in an environment populated by the elite. This fact is even more apt today where computational scientists have allowed the gap between themselves and their users to grow.

We cannot hope to stem or reverse this situation and therefore we have to accommodate our knowledge and expertise in our software so that society at large could find it as easy to use as their other *Apps* on their so-called *smart-phones*.

References

1. Smith, B.T.: Measuring Uncertainty in Scientific Computation Using Numerica 21's Test Harness. In: Dienstfrey, A., Boisvert, R.F. (eds.) WoCoUQ 2011. IFIP AICT, vol. 377, pp. 165–179. Springer, Heidelberg (2012)
2. Bronowski, J.: The Common Sense of Science. Harvard University Press, Cambridge (1978)

Discussion

Speaker: Patrick Gaffney

Felipe Montes: You mention a series of problems that affect the validation procedures of the simulation code. Those problems appear also in real life experiments (not simulations). So how do car companies reconcile and argue against simulation while real experimental results show the same problems as the model simulation?

Patrick Gaffney: Experiments regarding electrodeposition (the subject of the talk), baking, drainage, or void detection are limited to the extent that experimenting with each of these on full vehicles in real operational situations would require an interruption to the production line and that is a no-no.

Automotive Manufacturers who conduct real life experiments for electrodeposition, usually perform them under controlled laboratory conditions using simplified geometries and they do not exhibit the problems discussed in the talk and presented in the paper. It is only when scaling up to real world operating circumstances that things go wrong, and they do so primarily because one obtains the input for the code from several disjoint entities, many of whom are not privy to the reasons, nor do they understand, why recording and monitoring of conditions are necessary.

William Welch: How are OEMs using the simulator once verified? For example, are they optimizing hole positions? This looks like a high-dimensional optimization.

Patrick Gaffney: Once verified, OEMs use the software for a variety of purposes, primarily at the Design and Re-Work stages, the latter for reducing the costs of re-tooling.

In the Design stage, OEMs are using the software to determine configurations of pathways (not necessarily holes) to enhance the flow of electric current and hence e-coat coverage. For holes and other pathways we have provided software that makes the task of running different scenarios easier.

Similarly, for optimizing hole or pathway positions in a vehicle part, the combination of software tools together with 3-D animation to see how current flows, is sufficient. High-dimensional optimization is not involved nor is it necessary if one needs an answer quickly.

Brian Smith: Your model is that the cars are manufactured to be consistent from one vehicle to the next–e.g., the holes are in the same place, the caps are the same shape. Can you conceive of assessing the "vehicle" so that your software can adjust the painting parameters to the "new" conditions?

Patrick Gaffney: Excellent question! Yes, we can absolutely conceive of this situation, especially with the prospect of using fast GPUs to run the models "on-the-fly" and this is something we are presently investigating.

William Kahan: Can universities, that offer "computational science" minors with engineering (but these courses are mostly about how to use software packages) be realistically expected to teach the kind of fundamentals about numerical methods for PDE's, whose lack you bemoaned to students whose syllabi are already packed tight?

Patrick Gaffney: A "computational science" minor that only teaches how to use software packages is itself performing a disservice because I would contend that in order to use a software package correctly a user should have some basic knowledge of numerics, at least they should understand how important it is to get the input correct.

This is not going to happen in the West, and therefore one cannot rely upon Universities to fulfill this role.

However, the OEM has a responsibility for ensuring proper training of its staff and technical people. For those OEMs, I would suggest that they include a very basic course that exemplifies why the INPUT to the software is so important to get right.

In our experience, it would be beneficial if engineers and those leaving University had knowledge and experience with the following basic items.

- Numbers
- Precision and significant figures
- Floating point arithmetic
- IEEE Standard
- The processes of rounding and truncation
- Rounding error

- Differences and divided differences
- Iteration methods
- Convergence
- Continuity
- Limiting processes

- Basic numerical integration

Anyone wishing more details should contact me directly.

Uncertainty and Sensitivity Analysis: From Regulatory Requirements to Conceptual Structure and Computational Implementation

Jon C. Helton[1] and Cédric J. Sallaberry[2]

[1] Department of Mathematics and Statistics, Arizona State University, Tempe, AZ 85287 USA
[2] Sandia National Laboratories, Albuquerque, NM 87185 USA
{jchelto,cnsalla}@sandia.gov

Abstract. An approach to the conversion of regulatory requirements into a conceptual and computational structure that permits meaningful uncertainty and sensitivity analyses is descibed. This approach is predicated on the description of the desired analysis in terms of three basic entities: (i) a probability space characterizing aleatory uncertainty, (ii) a probability space characterizing epistemic uncertainty, and (iii) a model that predicts system behavior. The presented approach is illustrated with results from the 2008 performance assessment for the proposed repository for high-level radioactive waste at Yucca Mountain, Nevada.

Keywords: Aleatory uncertainty, Epistemic uncertainty, Performance assessment, Regulatory requirements, Sensitivity analysis, Uncertainty analysis.

1   Introduction

An approach to the conversion of regulatory requirements into a conceptual and computational structure that permits meaningful uncertainty and sensitivity analyses is descibed. This approach is predicated on the description of the desired analysis in terms of three basic entities: (i) a probability space characterizing aleatory uncertainty, (ii) a probability space characterizing epistemic uncertainty, and (iii) a model that predicts system behavior. The presented approach is illustrated with results from the 2008 performance assessment (PA) for the proposed repository for high-level radioactive waste at Yucca Mountain (YM), Nevada, carried out by the U.S. Department of Energy (DOE) to assess compliance with regulations promulgated by the U.S. Nuclear Regulatory Commission (NRC) [1-3].

2   Example: DOE's Licensing Requirements for YM Repository

The NRC's licensing requirements for the YM repository provide a good example of the challenges that are present in the conversion of regulatory requirements into the conceptual structure and associated computational implementation of an analysis that establishes compliance (or noncompliance) with those requirements [4; 5].

A. Dienstfrey and R.F. Boisvert (Eds.): WoCoUQ 2011, IFIP AICT 377, pp. 60–77, 2012.

The following two radiation protection requirements for a reasonably maximally exposed individual (RMEI) are at the core of the NRC's requirements for the YM repository ([6], p. 10829): "(a) DOE must demonstrate, using performance assessment, that there is a reasonable expectation that the reasonably maximally exposed individual receives no more than the following annual dose from releases from the undisturbed Yucca Mountain disposal system: (1) 0.15 mSv (15 mrem) for 10,000 years following disposal; and (2) 1.0 mSv (100 mrem) after 10,000 years, but within the period of geologic stability. (b) DOE's performance assessment must include all potential environmental pathways of radionuclide transport and exposure." In addition, the following elaboration on the preceding dose requirements for the RMEI is also given ([6], p. 10829): "Compliance is based upon the arithmetic mean of the projected doses from DOE's performance assessments for the period within 1 million years after disposal".

The preceding dose requirements indicate (i) that dose results must be determined for long time periods into the future and also for many different potential modes of exposure and (ii) that some type of averaging process is to be used to determine the dose values to which the regulatory requirements apply. The indicated averaging process (i.e., "arithmetic mean of projected doses") is vague and thus particularly challenging to the design of an analysis to assess compliance with the indicated bounds on (mean) dose. However, of necessity, implementation of this averaging process requires some form of a probabilistic representation of uncertainty.

Additional detail on what is desired in assessing compliance with the indicated dose requirements is provided by the NRC in the following definition for PA ([7], p. 55794): "Performance assessment means an analysis that: (1) Identifies the features, events, processes (except human intrusion), and sequences of events and processes (except human intrusion) that might affect the Yucca Mountain disposal system and their probabilities of occurring during 10,000 years after disposal, (2) Examines the effects of those features, events, processes, and sequences of events and processes upon the performance of the Yucca Mountain disposal system; and (3) Estimates the dose incurred by the reasonably maximally exposed individual, including the associated uncertainties, as a result of releases caused by all significant features, events, processes, and sequences of events and processes, weighted by their probability of occurrence."

The preceding definition makes very clear that a PA used to assess regulatory compliance for the YM repository must (i) consider what could happen in the future, (ii) assign probabilities to what could happen in the future, (iii) model the effects of what could happen in the future, (iv) consider the effects of uncertainties, and (v) weight potential doses by the probability of the occurrence of such doses. Of particular interest and importance to the design of an analysis to assess compliance is the indicated distinction between "uncertainty" and "probability of occurrence". This is a distinction between what is often called epistemic uncertainty and aleatory uncertainty [8; 9]. Specifically, epistemic uncertainty derives from a lack of knowledge about the appropriate value to use for a quantity that is assumed to have a fixed value in the context of a particular analysis, and aleatory uncertainty derives from an inherent randomness in the properties or behavior of the system under study.

The NRC further emphasizes the importance of an appropriate treatment of uncertainty in assessing regulatory compliance for the YM repository in the following definition of reasonable expectation ([7], p. 55813): "Reasonable expectation means that the Commission is satisfied that compliance will be achieved based upon the full record before it. Characteristics of reasonable expectation include that it: (1) Requires less than absolute proof because absolute proof is impossible to attain for disposal due to the uncertainty of projecting long-term performance; (2) Accounts for the inherently greater uncertainties in making long-term projections of the performance of the Yucca Mountain disposal system; (3) Does not exclude important parameters from assessments and analyses simply because they are difficult to precisely quantify to a high degree of confidence; and (4) Focuses performance assessments and analyses on the full range of defensible and reasonable parameter distributions rather than only upon extreme physical situations and parameter values." As the preceding definition makes clear, the NRC intends that a thorough treatment of uncertainty is to be an important part of assessing compliance with licensing requirements for the YM repository.

Similar requirements to the NRC's requirements for the YM repository, either by explicit statement or implication, underlie requirements for analyses of other complex systems, including (i) the NRC's safety goals for nuclear power stations [10], (ii) the U.S. Environmental Protection Agency's certification requirements for the Waste Isolation Pilot Plant [11; 12], and (iii) the National Nuclear Security Administration's mandate for the quantification of margins and uncertainties in assessments of the nation's nuclear stockpile [13-15]. Three recurrent ideas run through all of these examples: (i) the occurrence of future events (i.e., aleatory uncertainty), (ii) prediction of the consequences of future events (i.e., the modeling of physical processes), and (iii) lack of knowledge with respect to appropriate models and associated model parameters (i.e., epistemic uncertainty). The challenge in each case is to define a conceptual model and an associated computational implementation that appropriately incorporates these ideas into analyses supporting compliance determinations.

3   Conceptual Structure and Computational Implementation

The needed conceptual structure and path to computational implementation is provided by viewing the analysis of a complex system as being composed of three basic entities: (i) a probability space $(\mathcal{A}, \mathbb{A}, p_A)$ characterizing aleatory uncertainty, (ii) a probability space $(\mathcal{E}, \mathbb{E}, p_E)$ characterizing epistemic uncertainty, and (iii) a model that predicts system behavior (i.e., a function $f(t|\mathbf{a}, \mathbf{e})$, or more typically a vector function $\mathbf{f}(t|\mathbf{a}, \mathbf{e})$, that defines system behavior at time t conditional on elements $\mathbf{a}$ and $\mathbf{e}$ of the sample spaces $\mathcal{A}$ and $\mathcal{E}$ for aleatory and epistemic uncertainty). In the context of the three recurrent ideas indicated at the end of the preceding section, the probability space $(\mathcal{A}, \mathbb{A}, p_A)$ defines future events and their probability of occurrence; the functions $f(t|\mathbf{a}, \mathbf{e})$ and $\mathbf{f}(t|\mathbf{a}, \mathbf{e})$ predict the consequences of future events; and the probability space $(\mathcal{E}, \mathbb{E}, p_E)$ defines "state of knowledge uncertainty" with respect to the appropriate values to use for analysis inputs and characterizes this uncertainty with probability.

In turn, this conceptual structure leads to an analysis in which (i) uncertainty in analysis results is defined by integrals involving the function $f(t|\mathbf{a}, \mathbf{e})$ and the two indicated probability spaces and (ii) sensitivity analysis results are defined by the relationships between epistemically uncertain analysis inputs (i.e., elements e_j of $\mathbf{e}$) and analysis results defined by the function $f(t|\mathbf{a}, \mathbf{e})$ and also by various integrals of this function. Computationally, this leads to an analysis in which (i) high-dimensional integrals must be evaluated to obtain uncertainty analysis results and (ii) mappings between high-dimensional spaces must be generated and explored to obtain sensitivity analysis results. In general, $f(t|\mathbf{a}, \mathbf{e})$ is just one component of a high dimensional function $\mathbf{f}(t|\mathbf{a}, \mathbf{e})$. It is also possible for $f(t|\mathbf{a}, \mathbf{e})$ and $\mathbf{f}(t|\mathbf{a}, \mathbf{e})$ to be functions of spatial coordinates as well as time.

In general, the elements $\mathbf{a}$ of $\mathcal{A}$ are vectors

$$\mathbf{a} = \left[a_1, a_2, ..., a_m \right] \tag{1}$$

that define one possible occurrence in the universe under consideration. In practice, the uncertainty structure formally associated with the set $\mathbb{A}$ and the probability measure p_A is defined by defining probability distributions for the individual elements a_i of $\mathbf{a}$. Formally, this corresponds to defining a density function $d_{Ai}(a_i)$ on a set $\mathcal{A}_i$ characterizing aleatory for each element a_i of $\mathbf{a}$ (or some other uncertainty structure such as a cumulative distribution function (CDF) or a complementary CDF (CCDF) when convenient). Collectively, the sets $\mathcal{A}_i$ and density functions $d_{Ai}(a_i)$, or other appropriate uncertainty characterizations, define the set $\mathcal{A}$ and a density function $d_A(\mathbf{a})$ for $\mathbf{a}$ on $\mathcal{A}$, and thus, in effect, define the probability space $(\mathcal{A}, \mathbb{A}, p_A)$.

Similarly, the elements $\mathbf{e}$ of $\mathcal{E}$ are vectors

$$\mathbf{e} = \left[\mathbf{e}_A, \mathbf{e}_M \right] = \left[e_1, e_2, ..., e_n \right] \tag{2}$$

that define one possible set of epistemically uncertainty analysis inputs, where the vector $\mathbf{e}_A$ contains uncertain quantities used in the characterization of aleatory uncertainty and the vector $\mathbf{e}_M$ contains uncertain quantities used in the modeling of physical processes. As in the characterization of aleatory uncertainty, the uncertainty structure formally associated with the set $\mathbb{E}$ and the probability measure p_E is defined by defining probability distributions for the individual elements e_i of $\mathbf{e}$. Formally, this corresponds to defining a density function $d_{Ei}(e_i)$ (or some other uncertainty structure such as a CDF or CCDF when convenient) on a set $\mathcal{E}_i$ characterizing epistemic uncertainty for each element e_i of $\mathbf{e}$. Collectively, the sets $\mathcal{E}_i$ and density functions $d_{Ei}(e_i)$, or other appropriate uncertainty characterizations, define the set $\mathcal{E}$ and a density function $d_E(\mathbf{e})$ for $\mathbf{e}$ on $\mathcal{E}$, and thus, in effect, define the probability space $(\mathcal{E}, \mathbb{E}, p_E)$. In practice, the distributions for the individual elements of $\mathbf{e}$ are often obtained through an extensive expert review process (e.g., [16]).

The model, or system of models, that predict analysis results can be represented by

$$y(t/\mathbf{a},\mathbf{e}_M) = f(t/\mathbf{a},\mathbf{e}_M) \text{ for a single result} \tag{3}$$

and

$$\mathbf{y}(t/\mathbf{a},\mathbf{e}_M) = \left[y_1(t/\mathbf{a},\mathbf{e}_M), y_2(t/\mathbf{a},\mathbf{e}_M),... \right] = \mathbf{f}(t/\mathbf{a},\mathbf{e}_M) \text{ for multiple results}, \tag{4}$$

where t represents time. In practice, $f(t|\mathbf{a}, \mathbf{e})$ and $\mathbf{f}(t|\mathbf{a}, \mathbf{e})$ are very complex computer models and may produce results with a spatial as well as a temporal dependency.

In concept, the probability space $(\mathcal{A}, \mathbb{A}, p_A)$ and the function $y(t|\mathbf{a}, \mathbf{e}) = f(t|\mathbf{a}, \mathbf{e})$ are sufficient to determine the expected value $E_A[y(t|\mathbf{a}, \mathbf{e})]$ of $y(t|\mathbf{a}, \mathbf{e})$ over aleatory uncertainty conditional on the values for uncertain analysis inputs defined by an element $\mathbf{e} = [\mathbf{e}_A, \mathbf{e}_M]$ of $\mathcal{E}$ (i.e., risk in the terminology of many analyses and expected dose in the terminology of the NRC's regulations for the YM repository). Specifically,

$$E_A\left[y(t|\mathbf{a},\mathbf{e}_M)|\mathbf{e}_A \right] = \int_A y(t|\mathbf{a},\mathbf{e}_M)d_A(\mathbf{a}|\mathbf{e}_A)d\mathcal{A}$$

$$\cong \begin{cases} \displaystyle\sum_{j=1}^{nS} y(t|\mathbf{a}_j,\mathbf{e}_M)/nS & \begin{cases} \text{for random sampling with } \mathbf{a}_j, j=1,2,...,nS, \\ \text{sampled from } \mathcal{A} \text{ consistent with } d_A(\mathbf{a}|\mathbf{e}_A) \end{cases} \\ \displaystyle\sum_{j=1}^{nS} y(t|\mathbf{a}_j,\mathbf{e}_M)p_A(\mathcal{A}_j|\mathbf{e}_A) & \begin{cases} \text{for stratified sampling with } \mathbf{a}_j \in \mathcal{A}_j \\ \text{and } \mathcal{A}_j, j=1,2,...,nS, \text{ partitioning } \mathcal{A} \end{cases} \end{cases} \tag{5}$$

with the inclusion of "$|\mathbf{e}_A$" in $d_A(\mathbf{a}|\mathbf{e}_A)$ and $p_A(\mathcal{A}_j|\mathbf{e}_A)$ indicating that the distribution (i.e., probability space) for $\mathbf{a}$ is dependent on epistemically uncertain quantities that are elements of $\mathbf{e}_A$. Similarly, the probabilities that define CDFs and CCDFs that show the effects of aleatory uncertainty conditional on a specific element $\mathbf{e} = [\mathbf{e}_A, \mathbf{e}_M]$ of $\mathcal{E}$ are defined by

$$p_A\left[y(t|\mathbf{a},\mathbf{e}_M) \le y|\mathbf{e}_A \right] = \int_A \underline{\delta}_y\left[y(t|\mathbf{a},\mathbf{e}_M) \right]d_A(\mathbf{a}|\mathbf{e}_A)d\mathcal{A} \tag{6}$$

and

$$p_A\left[y < y(t|\mathbf{a},\mathbf{e}_M)|\mathbf{e}_A \right] = \int_A \overline{\delta}_y\left[y(t|\mathbf{a},\mathbf{e}_M) \right]d_A(\mathbf{a}|\mathbf{e}_A)d\mathcal{A}, \tag{7}$$

respectively, where

$$\underline{\delta}_y\left[y(t|\mathbf{a},\mathbf{e}_M) \right] = \begin{cases} 1 \text{ for } y(t|\mathbf{a},\mathbf{e}_M) \le y \\ 0 \text{ for } y < y(t|\mathbf{a},\mathbf{e}_M) \end{cases} \tag{8}$$

and

$$\overline{\delta}_y\left[y(t|\mathbf{a},\mathbf{e}_M) \right] = 1 - \underline{\delta}_y\left[y(t|\mathbf{a},\mathbf{e}_M) \right] = \begin{cases} 0 \text{ for } y(t|\mathbf{a},\mathbf{e}_M) \le y \\ 1 \text{ for } y < y(t|\mathbf{a},\mathbf{e}_M). \end{cases} \tag{9}$$

The integrals in Eqs. (6) and (7) can be approximated with procedures analogous to the sampling-based procedures indicated in Eq. (5).

The integrals in Eqs. (5)-(7) must be evaluated with multiple values of $\mathbf{e} = [\mathbf{e}_A, \mathbf{e}_M]$ in order to determine the effects of epistemic uncertainty. As illustrated in Sect. 4, the indicated multiple values for $\mathbf{e} = [\mathbf{e}_A, \mathbf{e}_M]$ are often obtained with a Latin hypercube sample (LHS)

$$\mathbf{e}_k = \left[\mathbf{e}_{Ak}, \mathbf{e}_{Mk} \right] = \left[e_{1k}, e_{2k}, ..., e_{nk} \right], k = 1, 2, ..., nLHS, \qquad (10)$$

of size $nLHS$ from the sample space $\mathcal{E}$ for epistemic uncertainty due to the efficient stratification properties of Latin hypercube sampling [17; 18]. This sample provides the basis for both (i) the numerical estimation of the effects of epistemic uncertainty and (ii) the implementation of a variety sensitivity analysis procedures [19-21].

Just as expected values, CDFs and CCDFs related to aleatory can be defined as indicated in Eqs. (5)-(7), similar quantities can be defined that summarize the effects of epistemic uncertainty. Several possibilities exist: (i) epistemic uncertainty in a result $y(t|\mathbf{a}, \mathbf{e}_M)$ conditional on a specific realization $\mathbf{a}$ of aleatory uncertainty, (ii) epistemic uncertainty in an expected value over aleatory uncertainty, and (iii) epistemic uncertainty in the cumulative probability $p_A[y(t|\mathbf{a}, \mathbf{e}_M) \leq y|\mathbf{e}_A]$ or exceedance (i.e., complementary cumulative) probability $p_A[y < y(t|\mathbf{a}, \mathbf{e}_M) |\mathbf{e}_A]$ for a specific value y of an analysis result.

For a result $y(t|\mathbf{a}, \mathbf{e}_M)$ conditional on a specific realization $\mathbf{a}$ of aleatory uncertainty, the expected value, cumulative probability and exceedance probability over epistemic uncertainty are given by

$$E_E\left[y(t|\mathbf{a},\mathbf{e}_M) \right] = \int_{\mathcal{E}M} y(t|\mathbf{a},\mathbf{e}_M) d_E(\mathbf{e}_M) d\mathcal{E}M \cong \sum_{k=1}^{nLHS} y(t|\mathbf{a},\mathbf{e}_{Mk}) / nLHS, \qquad (11)$$

$$p_E\left[y(t|\mathbf{a},\mathbf{e}_M) \leq y \right] = \int_{\mathcal{E}M} \delta_y \left[y(t|\mathbf{a},\mathbf{e}_M) \right] d_E(\mathbf{e}_M) d\mathcal{E}M \\ \cong \sum_{k=1}^{nLHS} \delta_y \left[y(t|\mathbf{a},\mathbf{e}_{Mk}) \right] / nLHS, \qquad (12)$$

and

$$p_E\left[y < y(t|\mathbf{a},\mathbf{e}_M) \right] = \int_{\mathcal{E}M} \overline{\delta}_y \left[y(t|\mathbf{a},\mathbf{e}_M) \right] d_E(\mathbf{e}_M) d\mathcal{E}M \\ \cong \sum_{k=1}^{nLHS} \overline{\delta}_y \left[y(t|\mathbf{a},\mathbf{e}_{Mk}) \right] / nLHS, \qquad (13)$$

respectively, where (i) $\mathcal{E}M$ corresponds to the subspace of $\mathcal{E}$ that contains only the vectors $\mathbf{e}_M$ and (ii) the vectors $\mathbf{e}_{Mk}$ are part of the LHS in Eq. (10).

For an expected result $E_A[y(t|\mathbf{a}, \mathbf{e}_M)|\mathbf{e}_A]$ over aleatory uncertainty, the expected value, cumulative probability and exceedance probability over epistemic uncertainty are defined analogously to the corresponding results in Eqs. (11)-(13). Specifically,

$$E_E\left\{E_A\left[y(t\mid\mathbf{a},\mathbf{e}_M)\mid\mathbf{e}_A\right]\right\}=\int_{\mathcal{E}}\left[\int_{\mathcal{A}}y(t\mid\mathbf{a},\mathbf{e}_M)d_A(\mathbf{a}\mid\mathbf{e}_A)d\mathcal{A}\right]d_E(\mathbf{e})d\mathcal{E}$$
$$\cong\sum_{k=1}^{nLHS}\int_{\mathcal{A}}y(t\mid\mathbf{a},\mathbf{e}_{Mk})d_A(\mathbf{a}\mid\mathbf{e}_{Ak})d\mathcal{A}/nLHS,\qquad(14)$$

$$p_E\left\{E_A\left[y(t\mid\mathbf{a},\mathbf{e}_M)\mid\mathbf{e}_A\right]\le\overline{y}\right\}=\int_{\mathcal{E}}\delta_{\overline{y}}\left[\int_{\mathcal{A}}y(t\mid\mathbf{a},\mathbf{e}_M)d_A(\mathbf{a}\mid\mathbf{e}_A)d\mathcal{A}\right]d_E(\mathbf{e})d\mathcal{E}$$
$$\cong\sum_{k=1}^{nLHS}\delta_{\overline{y}}\left[\int_{\mathcal{A}}y(t\mid\mathbf{a},\mathbf{e}_{Mk})d_A(\mathbf{a}\mid\mathbf{e}_{Ak})d\mathcal{A}\right]/nLHS,\qquad(15)$$

and

$$p_E\left\{\overline{y}<E_A\left[y(t\mid\mathbf{a},\mathbf{e}_M)\mid\mathbf{e}_A\right]\right\}=\int_{\mathcal{E}}\overline{\delta}_{\overline{y}}\left[\int_{\mathcal{A}}y(t\mid\mathbf{a},\mathbf{e}_M)d_A(\mathbf{a}\mid\mathbf{e}_A)d\mathcal{A}\right]d_E(\mathbf{e})d\mathcal{E}$$
$$\cong\sum_{k=1}^{nLHS}\overline{\delta}_{\overline{y}}\left[\int_{\mathcal{A}}y(t\mid\mathbf{a},\mathbf{e}_{Mk})d_A(\mathbf{a}\mid\mathbf{e}_{Ak})d\mathcal{A}\right]/nLHS,\qquad(16)$$

where, in general, the inner integrals over $\mathcal{A}$ will have to be evaluated with some appropriate integration procedure as indicated in Eq. (5).

For a cumulative probability $p_A[y(t\mid\mathbf{a},\mathbf{e}_M)\le y\mid\mathbf{e}_A]$ or an exceedance (i.e., complementary cumulative) probability $p_A[y<y(t\mid\mathbf{a},\mathbf{e}_M)\mid\mathbf{e}_A]$ over aleatory uncertainty for a specific value y of an analysis result, the expected value, cumulative probability and exceedance probability over epistemic uncertainty are defined analogously to the corresponding results in Eqs.(14)-(16). For example, the expected value and cumulative probability for $p_A[y<y(t\mid\mathbf{a},\mathbf{e}_M)\mid\mathbf{e}_A]$ that derive from epistemic uncertainty are defined by

$$E_E\left\{p_A\left[y<y(t\mid\mathbf{a},\mathbf{e}_M)\mid\mathbf{e}_A\right]\right\}=\int_{\mathcal{E}}\left\{\int_{\mathcal{A}}\overline{\delta}_y\left[y(t\mid\mathbf{a},\mathbf{e}_M)\right]d_A(\mathbf{a}\mid\mathbf{e}_A)d\mathcal{A}\right\}d_E(\mathbf{e})d\mathcal{E}$$
$$\cong\sum_{k=1}^{nLHS}\int_{\mathcal{A}}\overline{\delta}_y\left[y(t\mid\mathbf{a},\mathbf{e}_{Mk})\right]d_A(\mathbf{a}\mid\mathbf{e}_{Ak})d\mathcal{A}/nLHS\qquad(17)$$

and

$$p_E\left\{p_A\left[y<y(t\mid\mathbf{a},\mathbf{e}_M)\mid\mathbf{e}_A\right]\le p\right\}$$
$$=\int_{\mathcal{E}}\delta_p\left\{\int_{\mathcal{A}}\overline{\delta}_y\left[y(t\mid\mathbf{a},\mathbf{e}_M)\right]d_A(\mathbf{a}\mid\mathbf{e}_A)d\mathcal{A}\right\}d_E(\mathbf{e})d\mathcal{E}\qquad(18)$$
$$\cong\sum_{k=1}^{nLHS}\delta_p\left\{\int_{\mathcal{A}}\overline{\delta}_y\left[y(t\mid\mathbf{a},\mathbf{e}_{Mk})\right]d_A(\mathbf{a}\mid\mathbf{e}_{Ak})d\mathcal{A}\right\}/nLHS,$$

respectively.

The conceptual structure and computational procedures described in this section are illustrated in the next section with results from the 2008 YM PA [1].

4   Example: 2008 PA for YM Repository

The individual elements of the sample space $\mathcal{A}$ for aleatory uncertainty in the 2008 YM PA are vectors of the form

$$\mathbf{a} = \left[nEW, nED, nII, nIE, nSG, nSF, \mathbf{a}_{EW}, \mathbf{a}_{ED}, \mathbf{a}_{II}, \mathbf{a}_{IE}, \mathbf{a}_{SG}, \mathbf{a}_{SF} \right],$$ (19)

where, for a time interval $[a, b]$ (e.g., $[0, 10^4$ yr] or $[0, 10^6$ yr]), nEW = number of early waste package (WP) failures, nED = number of early drip shield (DS) failures, nII = number of igneous intrusive (II) events, nIE = number of igneous eruptive (IE) events, nSG = number of seismic ground (SG) motion events, nSF = number of seismic fault (SF) displacement events, $\mathbf{a}_{EW}$ = vector defining the nEW early WP failures, $\mathbf{a}_{ED}$ = vector defining the nED early DS failures, $\mathbf{a}_{II}$ = vector defining the nII igneous intrusive events, $\mathbf{a}_{IE}$ = vector defining the nIE igneous eruptive events, $\mathbf{a}_{SG}$ = vector defining the nSG seismic ground motion events, and $\mathbf{a}_{SF}$ = vector defining the nSF seismic fault displacement events. The definition of the probability space $(\mathcal{A}, \mathbb{A}, p_A)$ for aleatory uncertainty was completed by defining probability distributions for the individual elements of $\mathbf{a}$ (see [1], App. J). In the 2008 YM PA, elements of the sample space $\mathcal{A}$ are referred to as scenarios, and elements of the set $\mathbb{A}$ are referred to as scenario classes. With this usage, scenarios and scenario classes correspond to what are called elementary events and events, respectively, in the usual terminology of probability theory.

The individual elements of the sample space $\mathcal{E}$ for epistemic uncertainty in the 2008 YM PA are vectors of the form

$$\mathbf{e} = \left[\mathbf{e}_A, \mathbf{e}_M \right] = \left[e_1, e_2, ..., e_{392} \right],$$ (20)

where, as examples, the following quantities are elements of $\mathbf{e}$: *DSNFMASS* = scale factor used to characterize uncertainty in radionuclide content of defense spent nuclear fuel; *IGRATE* = frequency of intersection of the repository footprint by a volcanic event (yr^{-1}); *MICTC99* = groundwater biosphere dose conversion factor (BDCF) for ^{99}Tc in modern interglacial climate; *SCCTHRP* = residual stress threshold for stress corrosion cracking nucleation of Alloy 22 (as a percentage of yield strength in MPa); *SZGWSPDM* = logarithm of scale factor used to characterize uncertainty in groundwater specific discharge (dimensionless); and *WDGCA22* = temperature dependent slope term of Alloy 22 general corrosion rate (K). Distributions characterizing epistemic uncertainty were assigned to the individual elements of $\mathbf{e}$ and, in effect, defined the probability space $(\mathcal{E}, \mathbb{E}, p_E)$ for epistemic uncertainty. A complete listing of the 392 elements of $\mathbf{e}$ and sources of additional information on these variables and the development of their distributions are given in Table K3-3 of Ref. [1].

A very complex system of models was used to predict a large number of time-dependent results related to evolution of the repository, including (i) the release of radionuclides from WPs, (ii) the transport of radionuclides away from the engineered component of the repository, and (iii) human exposure to released radionuclides (see [1], Table K3-4, for a listing of the analysis results selected for study). An overview description of these models and extensive sources of additional information are available in Ref. [1]. As an example, a high-level overview of the models used in the analysis of seismic ground motion events is given in Fig. 1. The models indicated in Fig. 1 correspond to part of what is very simplistically represented by $\mathbf{f}(t|\mathbf{a}, \mathbf{e})$ in Eq. (4).

Owing to its central role in the NRC's regulatory requirements for the YM repository,

$$D\left(t \mid \mathbf{a}, \mathbf{e}_M\right) = \text{dose (mrem/yr) to the RMEI at time } t \text{ (yr) conditional on } \mathbf{a} \in \mathcal{A}$$
$$\text{and } \mathbf{e} = \left[\mathbf{e}_A, \mathbf{e}_M\right] \in \mathcal{E} \tag{21}$$

will be used as an example to illustrate results of the form indicated in Sect. 3. It is in the generation of such results where the challenge of bringing conceptual structure and computational implementation together arises. Bluntly put, it is not possible to evaluate integrals of the form indicated in Sect. 3 for $D(t|\mathbf{a}, \mathbf{e}_M)$ defined by modeling systems of the complexity shown in Fig. 1 without a carefully designed computational strategy that makes efficient use of what will almost always be a limited number of detailed, mechanistic calculations. Such a strategy will be analysis specific and designed to take advantage of particular properties of the models in use.

A core quantity in the NRC's regulatory requirements for the YM repository is the expected value $E_A[D(t|\mathbf{a}, \mathbf{e}_M)|\mathbf{e}_A]$ of $D(t|\mathbf{a}, \mathbf{e}_M)$ over aleatory uncertainty, with $E_A[D(t|\mathbf{a}, \mathbf{e}_M)|\mathbf{e}_A]$ being an example of the expected value formally defined in Eq. (5). Specifically,

$$E_A\left[D\left(t \mid \mathbf{a}, \mathbf{e}_M\right) \mid \mathbf{e}_A\right] = \int_{\mathcal{A}} D\left(t \mid \mathbf{a}, \mathbf{e}_M\right) d_A\left(\mathbf{a} \mid \mathbf{e}_A\right) d\mathcal{A}$$
$$\cong \int_{\mathcal{A}} \left[\sum_{C \in \mathcal{C}} D_C\left(t \mid \mathbf{a}, \mathbf{e}_M\right)\right] d_A\left(\mathbf{a} \mid \mathbf{e}_A\right) d\mathcal{A}, \mathcal{C} = \{N, EW, ED, II, IE, SG, SF\}$$
$$= \sum_{C \in \mathcal{C}} \int_{\mathcal{A}} D_C\left(t \mid \mathbf{a}, \mathbf{e}_M\right) d_A\left(\mathbf{a} \mid \mathbf{e}_A\right) d\mathcal{A}$$
$$= \sum_{C \in \mathcal{C}} E_A\left[D_C\left(t \mid \mathbf{a}, \mathbf{e}_M\right) \mid \mathbf{e}_A\right], \tag{22}$$

where the subscripts contained in the set $\mathcal{C}$ are used to denote doses to the RMEI from individual scenario classes with N designating the scenario class in which no disruptions occur and the remaining subscripts designating scenario classes associated with the correspondingly subscripted aleatory occurrences in Eq. (19). Results associated with dose $D_C(t|\mathbf{a}, \mathbf{e}_M)$ for scenario class C in Eq. (22) are assumed to be calculated with only those elements of $\mathbf{a}$ and $\mathbf{e} = [\mathbf{e}_A, \mathbf{e}_M]$ that are related to scenario class C. The approximation to $E_A[D(t|\mathbf{a}, \mathbf{e}_M)|\mathbf{e}_A]$ in Eq. (22) can be justified on the basis of tradeoffs between the effects of high probability-low consequence scenario classes and low probability-high consequence scenario classes.

Epistemic uncertainty is propagated in the 2008 YM PA with an LHS

$$\mathbf{e}_k = \left[\mathbf{e}_{Ak}, \mathbf{e}_{Mk}\right] = \left[e_{1k}, e_{2k}, ..., e_{392k}\right], k = 1, 2, ..., nLHS = 300, \tag{23}$$

from the sample space $\mathcal{E}$ for epistemic uncertainty. In turn, the approximation

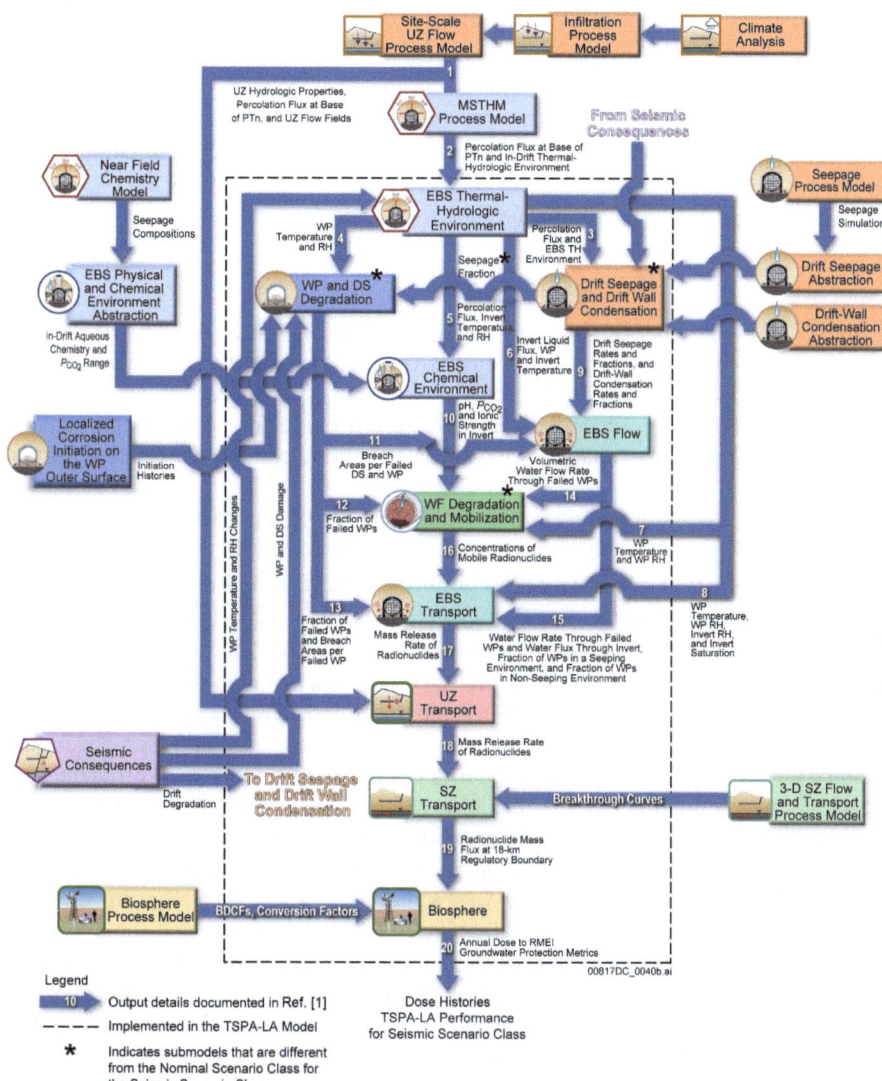

Fig. 1. Information transfer between the model components and submodels for the seismic scenario class in the 2008 YM PA ([1], Fig. 6.1.4-6)

$$\hat{E}_A\left[D\left(t\mid\mathbf{a},\mathbf{e}_{Mk}\right)\mid\mathbf{e}_{Ak}\right] \cong \sum_{C\in C}\hat{E}_A\left[D_C\left(t\mid\mathbf{a},\mathbf{e}_{Mk}\right)\mid\mathbf{e}_{Ak}\right] \qquad (24)$$

results for each element $\mathbf{e}_k$ of the LHS in Eq. (23), where $\hat{E}_A[D(t\mid\mathbf{a},\mathbf{e}_{Mk})\mid\mathbf{e}_{Ak}]$ and $\hat{E}_A[D_C(t\mid\mathbf{a},\mathbf{e}_{Mk})\mid\mathbf{e}_{Ak}]$ denote approximations to $E_A[D(t\mid\mathbf{a},\mathbf{e}_{Mk})\mid\mathbf{e}_{Ak}]$ and $E_A[D_C(t\mid\mathbf{a},\mathbf{e}_{Mk})\mid\mathbf{e}_{Ak}]$, respectively. Because of the occurrence of the same elements of $\mathbf{e}_M$ in the evaluation of $D_C(t\mid\mathbf{a},\mathbf{e}_M)$ for different values of C (i.e., for different scenario classes), it is essential that the doses $D_C(t|\mathbf{a}, \mathbf{e}_M)$ in Eq. (24) be evaluated for the same elements of the LHS in Eq. (23) for the indicated approximation to $E_A[D(t|\mathbf{a}, \mathbf{e}_{Mk})|\mathbf{e}_{Ak}]$ to be valid.

As an example, analysis results are presented for seismic ground motion events occurring in the time interval $[0, 2\times10^4$ yr] (i.e., for what is called the seismic ground motion scenario class in the 2008 YM PA). This restriction reduces the elements of the sample space $\mathcal{A}$ for aleatory uncertainty to

$$\mathbf{a} = \left[nSG, t_1, v_1, A_1, t_2, v_2, A_2, ..., t_{nSG}, v_{nSG}, A_{nSG} \right], \tag{25}$$

where (i) nSG = number of seismic ground motion events in 20,000 yr, (ii) t_i = time (yr) of event i, (iii) v_i = peak ground velocity (m/s) for event i, (iv) A_i = damaged area (m^2) on individual WPs for peak ground velocity v_i, (v) the occurrence of seismic ground motion events is characterized by a hazard curve for peak ground velocity, and (vi) damaged area is characterized by distributions conditional on peak ground velocity.

To evaluate results of the form defined in Sect. 3 for the seismic ground motion scenario class, it is necessary to integrate the function $D_{SG}(t|\mathbf{a}, \mathbf{e}_M)$ over the sample space $\mathcal{A}$ for aleatory uncertainty with $\mathbf{a}$ defined as indicated in Eq. (25). In full detail, $D_{SG}(t|\mathbf{a}, \mathbf{e}_M)$ is defined by the model system shown in Fig. 1. Evaluation of this system is too computationally demanding to permit its evaluation 1000's of times for each element $\mathbf{e}_k = [\mathbf{e}_{Ak}, \mathbf{e}_{Mk}]$ of the LHS in Eq. (23). This is a common situation in analyses of complex systems, where very detailed physical models are developed which then turn out to be too computationally demanding to be naively used in the propagation of aleatory uncertainty. In such situations, it is necessary to find ways to efficiently use the results of a limited number of model evaluations to predict outcomes for a large number of different possible realizations of aleatory uncertainty.

For the seismic ground motion scenario class and the time interval $[0, 2\times10^4$ yr], the needed computational efficiency was achieved by evaluating $D_{SG}(t|\mathbf{a}, \mathbf{e}_{Mk})$ at a sequence of times (i.e., 100, 1000, 3000, 6000, 12000, 18000 yrs) and for a sequence of damaged areas areas (i.e., $10^{-8+s}(32.6$ m$^2)$ for $s = 1, 2, ..., 5$ with 32.6 m^2 corresponding to the surface area of a WP) at each time (Fig. 2a). This required $6\times5 = 30$ evaluations of the system indicated in Fig. 1 for each LHS element in Eq.(23). Once obtained, these evaluations can be used with appropriate interpolation and additive procedures to evaluate $D_{SG}(t|\mathbf{a}, \mathbf{e}_{Mk})$ for different values of $\mathbf{a}$ for each LHS element $\mathbf{e}_k = [\mathbf{e}_{Ak}, \mathbf{e}_{Mk}]$.

The individual CCDFs in Fig. 2b are defined by probabilities of the form shown in Eq. (7) with (i) $D_{SG}(t|\mathbf{a}, \mathbf{e}_{Mk})$ and $\mathbf{e}_{Ak}$ replacing $y(t|\mathbf{a}, \mathbf{e}_M)$ and $\mathbf{e}_A$ and (ii) $t = 10^4$ yr. Numerically, the integrals that define exceedance probabilities for the individual CCDFs are approximated with (i) random sampling from the possible values for $\mathbf{a}$ as indicated in Eq. (5) and (ii) estimated values $\hat{D}_{SG}(t|\mathbf{a}_j, \mathbf{e}_{Mk})$ for $D_{SG}(t|\mathbf{a}, \mathbf{e}_{Mk})$ constructed from results of the form shown in Fig. 2a. Specifically,

$$\hat{p}_A\left[y < D_{SG}\left(t|\mathbf{a}, \mathbf{e}_{Mk}\right)|\mathbf{e}_{Ak} \right] = \sum_{j=1}^{nS} \bar{\delta}_y\left[\hat{D}_{SG}\left(t|\mathbf{a}_j, \mathbf{e}_{Mk}\right) \right]/nS, \tag{26}$$

with the $\mathbf{a}_j$, $j = 1, 2, ..., nS$, sampled in consistency with the density function $d_A(\mathbf{a}|\mathbf{e}_{Ak})$ for vectors of the form shown in Eq. (25). The mean and quantile curves in Fig. 2b are (i) defined and approximated as indicated in Eqs. (17) and (18) and (ii) provide a summary of the epistemic uncertainty present in the estimation of exceedance probabilities (i.e., $p_A[y < D_{SG}(10^4|\mathbf{a}, \mathbf{e}_M)|\mathbf{e}_A])$ for $D_{SG}(10^4|\mathbf{a}, \mathbf{e}_M)$.

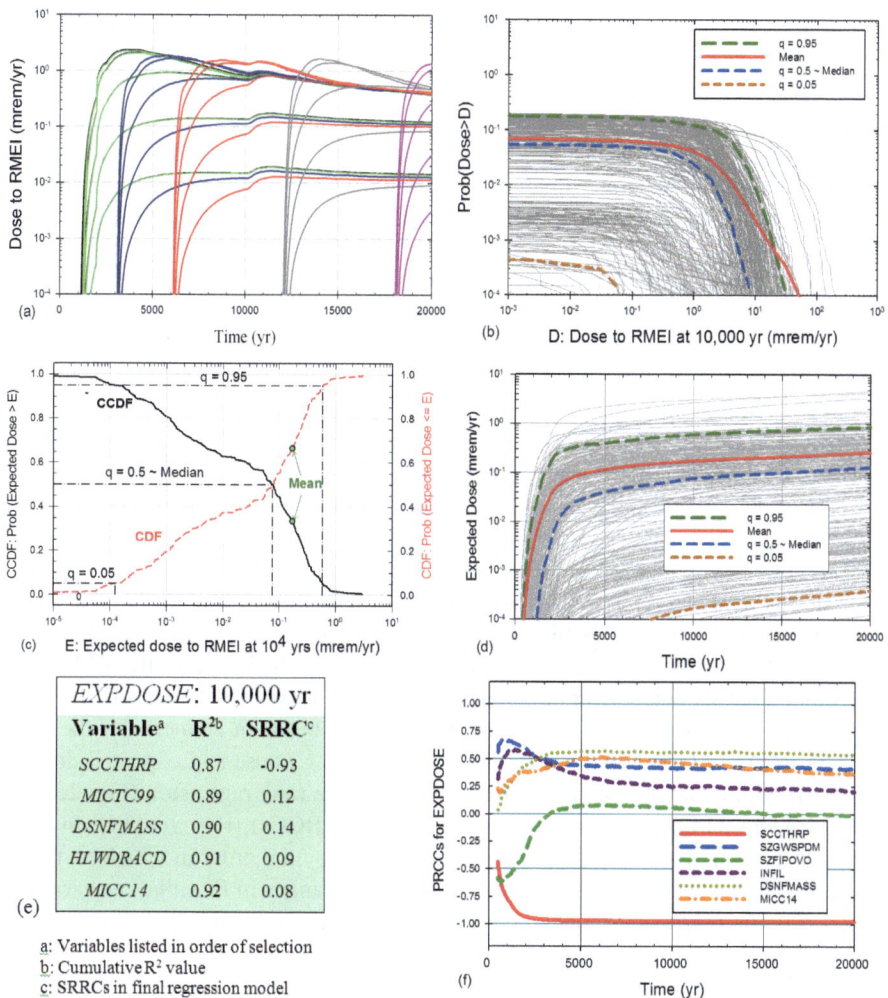

Fig. 2. Example results for dose (mrem/yr) to RMEI for seismic ground motion scenario class: (a) dose for seismic events occurring at different times and causing different damaged areas on WPs ([1], Fig. J8.3-3a), (b) CCDFs for dose at 10,000 yr ([1], Fig. J8.3-10a), (c) CCDF for expected dose at 10,000 yr ([1], Fig. J8.3-5c), (d) time-dependent expected dose ([1], Fig. J8.3-6), (e) stepwise rank regression for expected dose at 10,000 yr ([1], Fig. K7.7.1-2a), and (f) time-dependent PRCCs for expected dose ([1], Fig. K7.7.1-1c)

As indicated in Eq. (5), the expected value $E_A[D_{SG}(t|\mathbf{a}, \mathbf{e}_{Mk})|\mathbf{e}_{Ak}]$ of $D_{SG}(t|\mathbf{a}, \mathbf{e}_{Mk})$ over aleatory uncertainty can also be defined and estimated, with the estimate $\hat{E}_A[D_{SG}(t|\mathbf{a}, \mathbf{e}_{Mk})|\mathbf{e}_{Ak}]$ obtained as shown in Eq. (26) with removal of the indicator function $\bar{\delta}_y$. The expected values $E_A[D_{SG}(t|\mathbf{a}, \mathbf{e}_{Mk})|\mathbf{e}_{Ak}]$ and their corresponding estimates are the result of reducing each CCDF in Fig. 2b to a single number. As indicated in Eqs. (14)-(16) and illustrated in Fig. 2c, epistemic uncertainty associated $E_A[D_{SG}(t|\mathbf{a}, \mathbf{e}_M)|\mathbf{e}_A]$ can be summarized by (i) an expected (mean) value $E_E\{E_A[D_{SG}(t|\mathbf{a}, \mathbf{e}_M)|\mathbf{e}_A]\}$ over epistemic uncertainty as defined in Eq. (14), (ii) a CDF as defined by the cumulative probabilities in Eq. (15), or (iii) a CCDF as defined by the complementary cumulative probabilities in Eq. (16). The indicated mean defined in Eq. (14) and illustrated in Fig. 2c is the outcome of reducing all the information in Fig. 2b to a single number. The approximation process for a CDF also provides the basis for obtaining specific quantile values (e.g., $q = 0.05, 0.5 \sim$ median, 0.95) as indicated in Fig. 2c.

In the NRC's regulatory requirements for the YM repository, bounds apply over time to expected dose to the RMEI. Thus, the analysis results of greatest interest are expected dose and the uncertainty in expected dose as a function of time (Fig. 2d). Specifically, expected doses for individual LHS elements correspond to the lighter lines in Fig. 2d, and quantile and mean values for expected dose that summarize the effects of epistemic uncertainty correspond to the darker dashed and solid lines. The results on Fig. 2d at 10,000 years correspond to the results shown in more detail in Fig. 2c. For reasons of computational efficiency, the individual expected dose curves in Fig. 2d were estimated with a quadrature procedure as described in Sect. J8.3 of Ref. [1] rather than with a sampling-based procedure as illustrated in Fig. 2b.

Sensitivity analysis is an important component of the 2008 YM PA and contributes to an establishment of "reasonable expectation" by supporting a detailed examination of the operation of the models that predict dose and expected dose to the RMEI and many other analysis outcomes of interest. Specifically, sensitivity analysis in the 2008 YM PA was based on an exploration of the mapping between elements of the LHS indicated in Eq. (23) and analysis results of interest (e.g., dose and expected dose to the RMEI) with a variety techniques including stepwise rank regression (Fig. 2e) and time-dependent partial rank correlation coefficients (PRCCs) (Fig. 2f) [19]. In stepwise rank regression, variable importance is indicated by the order in which variables are selected in the stepwise process, the incremental changes in R^2 values as variables are added to the regression model, and the values of the standardized rank regression coefficients (SRRCs) in the regression model. The indicated results with R^2 values and SRRCs provide a measure of the amount of epistemic uncertainty in the dependent variable under consideration that derives from the epistemic uncertainty in individual analysis inputs (i.e., elements of $\mathbf{e}$); in contrast, PRCCs provide a measure of the strength of the monotonic relationship between individual epistemically uncertain analysis inputs and the dependent variable under consideration after removal of the monotonic effects of all other epistemically uncertain analysis inputs. Definitions for selected variables appearing Figs. 2e and 2f are given after Eq. (20).

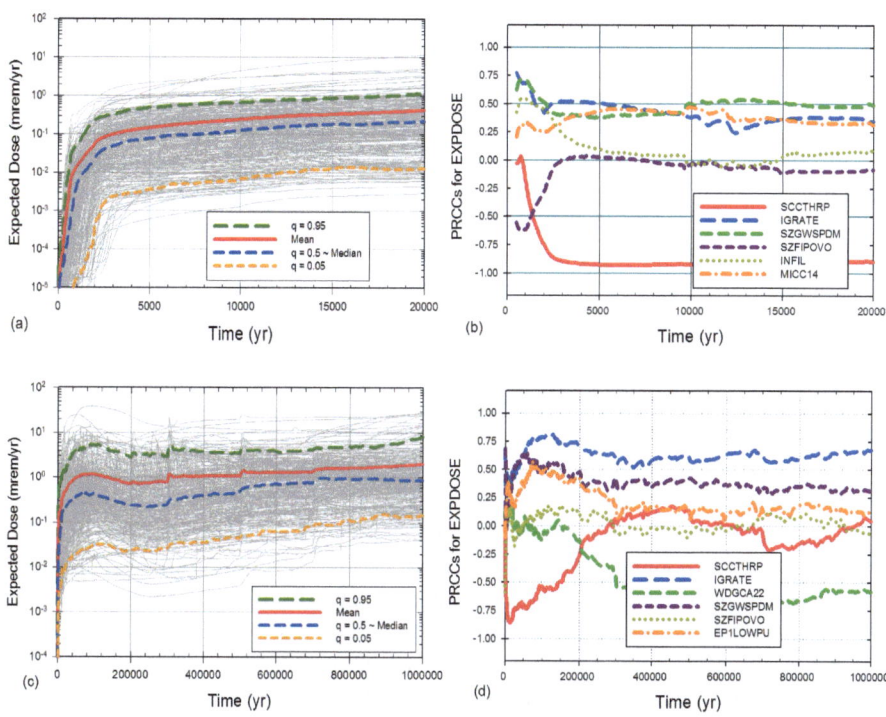

Fig. 3. Expected dose to RMEI: (a, b) Expected dose and associated PRCCs for $[0, 2 \times 10^4$ yr] ([1], Fig. K8.1-1[a]) and (c, d) Expected dose and associated PRCCs for $[0, 10^6$ yr] ([1], Fig. K8.2-1[a])

Additional detailed information on uncertainty and sensitivity analysis for the seismic ground motion scenario class in the 2008 YM PA is available in Sects. J8 and K7 of Ref. [1]. Included in this material are extensive examples of results conditional on individual realizations of aleatory uncertainty as defined in Eqs. (11)-(13).

As indicated in Eq. (22), expected dose results are obtained for individual scenario classes and then summed to determine expected dose over all scenario classes (i.e., over all futures as described by the vectors defined in Eq. (19)). The results of this summation are shown in Figs. 3a and 3c for the time intervals $[0, 2 \times 10^4$ yr] and $[0, 10^6$ yr], respectively. The individual expected dose curves in Fig. 3c are not as smooth as the individual expected dose curves in Fig. 3a. This difference results because the complexity of the calculations for the $[0, 10^6$ yr] time interval required the use of a sampling-based procedure to approximate expected dose from seismic ground motion events (see [1], Sect. J8.4); in contrast, the quadrature procedures used to determine expected dose for the individual scenarios for the $[0, 2 \times 10^4$ yr] time interval resulted in the smoother expected dose curves in Fig. 3a.

Ultimately, the NRC decided that their expected (i.e., "arithmetic mean") dose requirements of 15 mrem/yr and 100 mrem/yr for the time intervals $[0, 10^4$ yr] and $[10^4, 10^6$ yr], respectively, applied to the mean (i.e., solid line) doses in Figs. 3a and 3c (i.e., to results of the form defined in Eq. (14)). However, it is the spread of the individual dose curves in Figs. 3a and 3c that provide the NRC requested uncertainty information and thus a basis for a "reasonable expectation" that the requirements are being met despite the presence of substantial epistemic uncertainty. Further, sensitivity analyses of the form illustrated in Figs. 3b and 3d also contribute to "reasonable expectation" by enhancing understanding of the overall analysis and thus confidence in the numerical implementation of the analysis. Definitions for selected variables appearing Figs. 3b and 3d are given after Eq. (20).

Additional detailed information on uncertainty and sensitivity analysis in the 2008 YM PA is available in Apps. J and K of Ref. [1].

5   Concluding Message

Everyone cannot be expected to agree on the details of an analysis of a complex system, but everyone should be able to know what those details are. Without such knowledge, it is not possible to have informed and meaningful discussions involving the views of all parties interested in the analysis.

A necessary starting point for the design, computational implementation and, ultimately, communication of an analysis for a complex system is a clear conceptual structure. As described in this presentation, this structure for many, if not most, analyses, can be based on three basic entities: a probability space $(\mathcal{A}, \mathbb{A}, p_A)$ characterizing aleatory uncertainty, a probability space $(\mathcal{E}, \mathbb{E}, p_E)$ characterizing epistemic uncertainty, and a model that predicts system behavior (i.e., a function $f(t|\mathbf{a}, \mathbf{e})$, or more typically a vector function $\mathbf{f}(t|\mathbf{a}, \mathbf{e})$, that defines system behavior at time t conditional on elements $\mathbf{a}$ and $\mathbf{e}$ of the sample spaces $\mathcal{A}$ and $\mathcal{E}$ for aleatory and epistemic uncertainty). As illustrated with results from the 2008 YM PA, this conceptual view of the analysis of a complex system provides (i) a way to formally describe that analysis and then (ii) a clear path from formal description to computational implementation to written documentation.

Acknowledgements. Work performed at Sandia National Laboratories (SNL), which is a multiprogram laboratory operated by Sandia Corporation, a Lockheed Martin Company, for the U.S. Department of Energy's (DOE's) National Nuclear Security Administration under Contract No. DE-AC04-94AL85000. The United States Government retains and the publisher, by accepting this article for publication, acknowledges that the United States Government retains a non-exclusive, paid-up, irrevocable, world-wide license to publish or reproduce the published form of this article, or allow others to do so, for United States Government purposes. The views expressed in this article are those of the authors and do not necessarily reflect the views or policies of the DOE or SNL.

References

1. SNL (Sandia National Laboratories). Total System Performance Assessment Model/Analysis for the License Application. MDL-WIS-PA-000005 Rev. 00, AD 01. U.S. Department of Energy Office of Civilian Radioactive Waste Management, Las Vegas, NV (2008)
2. U.S. DOE (U.S. Department of Energy). Yucca Mountain Repository Safety Analysis Report. DOE/RW-0573. U.S. Department of Energy, Las Vegas, NV (2008)
3. U.S. DOE (U.S. Department of Energy). Yucca Mountain Repository License Application. DOE/RW-0573, Rev. 0. U.S. Department of Energy, Las Vegas, NV (2008)
4. Helton, J.C., Sallaberry, C.J.: Conceptual Basis for the Definition and Calculation of Expected Dose in Performance Assessments for the Proposed High-Level Radioactive Waste Repository at Yucca Mountain, Nevada. Reliability Engineering and System Safety 94(3), 677–698 (2009)
5. Helton, J.C., Sallaberry, C.J.: Computational Implementation of Sampling-Based Approaches to the Calculation of Expected Dose in Performance Assessments for the Proposed High-Level Radioactive Waste Repository at Yucca Mountain, Nevada. Reliability Engineering and System Safety 94(3), 699–721 (2009)
6. U.S. NRC (U.S. Nuclear Regulatory Commission). 10 CFR Part 63: Implementation of a Dose Standard after 10,000 Years. Federal Register 74(48), 10811–10830 (2009)
7. U.S. NRC (U.S. Nuclear Regulatory Commission). 10 CFR Parts 2, 19, 20, etc.: Disposal of High-Level Radioactive Wastes in a Proposed Geologic Repository at Yucca Mountain, Nevada; Final Rule. Federal Register 66(213), 55732–55816 (2001)
8. Helton, J.C., Burmaster, D.E.: Guest Editorial: Treatment of Aleatory and Epistemic Uncertainty in Performance Assessments for Complex Systems. Reliability Engineering and System Safety 54(2-3), 91–94 (1996)
9. Paté-Cornell, M.E.: Uncertainties in Risk Analysis: Six Levels of Treatment. Reliability Engineering and System Safety 54(2-3), 95–111 (1996)
10. Helton, J.C., Breeding, R.J.: Calculation of Reactor Accident Safety Goals. Reliability Engineering and System Safety 39(2), 129–158 (1993)
11. Helton, J.C., Anderson, D.R., Marietta, M.G., Rechard, R.P.: Performance Assessment for the Waste Isolation Pilot Plant: From Regulation to Calculation for 40 CFR 191.13. Operations Research 45(2), 157–177 (1997)
12. Helton, J.C., Anderson, D.R., Jow, H.-N., Marietta, M.G., Basabilvazo, G.: Conceptual Structure of the 1996 Performance Assessment for the Waste Isolation Pilot Plant. Reliability Engineering and System Safety 69(1-3), 151–165 (2000)
13. NAS/NRC (National Academy of Science/National Research Council). Evaluation of Quantification of Margins and Uncertainties for Assessing and Certifying the Reliability of the Nuclear Stockpile. National Academy Press, Washington, DC (2008)
14. Helton, J.C., Pilch, M.: Guest Editorial: Quantification of Margins and Uncertainties. Reliability Engineering and System Safety 96(9), 959–964 (2011)
15. Helton, J.C.: Quantification of Margins and Uncertainties: Conceptual and Computational Basis. Reliability Engineering and System Safety 96(9), 976–1013 (2011)
16. Hora, S.C., Iman, R.L.: Expert Opinion in Risk Analysis: The NUREG-1150 Methodology. Nuclear Science and Engineering 102(4), 323–331 (1989)
17. McKay, M.D., Beckman, R.J., Conover, W.J.: A Comparison of Three Methods for Selecting Values of Input Variables in the Analysis of Output from a Computer Code. Technometrics 21(2), 239–245 (1979)

18. Helton, J.C., Davis, F.J.: Latin Hypercube Sampling and the Propagation of Uncertainty in Analyses of Complex Systems. Reliability Engineering and System Safety 81(1), 23–69 (2003)
19. Helton, J.C., Johnson, J.D., Sallaberry, C.J., Storlie, C.B.: Survey of Sampling-Based Methods for Uncertainty and Sensitivity Analysis. Reliability Engineering and System Safety 91(10-11), 1175–1209 (2006)
20. Storlie, C.B., Helton, J.C.: Multiple Predictor Smoothing Methods for Sensitivity Analysis: Description of Techniques. Reliability Engineering and System Safety 93(1), 28–54 (2008)
21. Storlie, C.B., Swiler, L.P., Helton, J.C., Sallaberry, C.J.: Implementation and Evaluation of Nonparametric Regression Procedures for Sensitivity Analysis of Computationally Demanding Models. Reliability Engineering & System Safety 94(11), 1735–1763 (2009)

Discussion

Speaker: Jon Helton

Scott Ferson: Jon, the definition on your slide #16 says that epistemic uncertainty is a lack of knowledge about a fixed value. But it seems to me that the fixedness should not be part of the definition. Of course, some quantities are fixed by definition, but others are fixed only under assumptions we might make. Suppose some quantity is unknown to us. It may be a fixed value, or it may be changing its value through time, but we don't know which because of our great overall uncertainty about the quantity. Isn't that epistemic uncertainty too, or would you like to call it something else?

Jon Helton: The indicated slide states that epistemic uncertainty "arises from a lack of knowledge about the appropriate value to use for a quantity that is assumed to have a fixed value in the context of a specific analysis". The salient idea is that the assumptions and associated computational structure for a specific analysis are developed under whatever knowledge, time and resource constraints may exist. Once this development is complete at whatever level of sophistication is possible and before any calculations can be carried out, it is necessary to numerically define all the quantities required in the computational implementation of the analysis. In most analyses, many of these quantities will not be precisely known and are thus epistemically uncertain. This epistemic uncertainty is function of both available knowledge and the exact nature of each uncertain quantity in the analysis under consideration.

In the preceding, I have deliberately referred to uncertain "quantities" rather than using some expression that suggests that epistemic uncertain only involves analysis inputs that are real-valued parameters. Specifically, the uncertain quantities that are being referred to could be (i) real-valued parameters, (ii) functions of time or space, (iii) distributions, (iv) alternative models or model structures, or (v) some other possibility.

The "fixedness" that Scott refers to is unavoidably part of the assumptions of an analysis and thus central to the concept of epistemic uncertainty in the context of a specific analysis. Scott presents a modeling situation where a quantity may, or may not, be a function of time. This is an example of epistemic uncertainty where the uncertain quantity is actually a model (i.e., the appropriate model to use for the temporal behavior of a variable). In turn, there are several possible strategies that an analysis team might take to deal with this situation, including (i) define a distribution of possible models for temporal behavior, (ii) define a distribution of functions for temporal behavior, or (iii) use average temporal behavior as an analysis input and the treat this average as being epistemically uncertain. As indicated earlier, whatever is done will depend on existing knowledge, time and resource constraints, and in turn, will affect the epistemic uncertainty in individual quantities used as inputs to the analysis.

Bayes Linear Analysis for Complex Physical Systems Modeled by Computer Simulators

Michael Goldstein*

Department of Mathematical Sciences,
Durham University
Durham, United Kingdom

Abstract. Most large and complex physical systems are studied by mathematical models, implemented as high dimensional computer simulators. While all such cases differ in physical description, each analysis of a physical system based on a computer simulator involves the same underlying sources of uncertainty. These sources are defined and described below. In addition, there is a growing field of study which aims to quantify and synthesize all of the uncertainties involved in relating models to physical systems, within the framework of Bayesian statistics, and to use the resultant uncertainty specification to address problems of forecasting and decision making based on the application of these methods. We present an overview of the current status and future challenges in this emerging methodology, illustrating with examples drawn from current areas of application including: asset management for oil reservoirs, galaxy modeling, and rapid climate change.

Keywords: Bayes linear analysis, computer experiment, emulation, history matching, forecasting, reification.

1   Uncertainty in Complex Systems Represented by Computer Simulators

Most large and complex physical systems are studied by mathematical models, implemented as high dimensional computer simulators (like climate models). To use complex simulators to make statements about physical systems (like climate), we need to quantify the uncertainty involved in moving from the model to the system. The issues that we must address are methodological (how can we estimate what climate is likely to be?), computational (how can we ensure that our methods are tractable?) and foundational (why should our methods work and what do our answers mean?).

---

* Thanks to Engineering and Physical Sciences Research Council for funding through the Managing Uncertainty in Complex models project, and Natural Environment Research Council for funding through the RAPID programme. Thanks to Jonathan Cumming, Ian Vernon, Danny Williamson for the collaborations described in this paper.

A. Dienstfrey and R.F. Boisvert (Eds.): WoCoUQ 2011, IFIP AICT 377, pp. 78–94, 2012.

Applications range across all areas of science and technology. In this article, I will offer illustrations based on the following three applications, chosen as I have some personal experience with each, and they illustrate the wide range of areas of application for the methodology that we shall describe.

Oil Reservoirs. An oil reservoir simulator takes as inputs a physical description of the properties of a reservoir (permeabilities, porosities, faults, etc) and produces, as output, various well characteristics(pressure profiles, oil and gas production rates, etc.). The simulator is used to help manage assets associated with the reservoir. The aim is commercial: to develop efficient production schedules, determine whether and where to sink new wells, and so forth.

Galaxy Formation. The study of the development of the Universe is supported by using Galaxy formation simulators. These simulators take as input various parameters controlling physical processes which are thought to control the formation of galaxies and the simulators perform simulations of the development of the universe from the point of origin to the present producing as output various large scale quantities which can be compared to current cosmological measurements. The aim is scientific - to gain information about the physical processes underlying the Universe.

Climate Change. Large scale climate simulators are constructed to assess likely effects of human intervention upon future climate behaviour. Aims are both scientific - much is unknown about the large scale interactions which determine climate - and also very practical, as such simulators provide evidence for the importance of changing human behaviour before possibly irreversible changes are set into motion.

While each such model differs in all details of the scientific basis and mathematical implementation, there are various sources of uncertainty which are common across all such applications.

(i) **parametric uncertainty** (each model requires a, typically high dimensional, parametric specification, whose value is not known),

(ii) **condition uncertainty** (uncertainty as to boundary conditions, initial conditions, and forcing functions),

(iii) **functional uncertainty** (model evaluations often take a long time, so the function is unknown for almost all choices of inputs),

(iv) **stochastic uncertainty** (either the model is stochastic, giving different outcomes each time it is evaluated under the same choice of input parameters, introducing uncertainly directly, or aspects of the modelling which should involve such stochastic uncertainty have been reduced to a deterministic form, introducing uncertainty indirectly),

(v) **solution uncertainty** (the system equations can only be solved to some necessary level of approximation),

(vi) **structural uncertainty** (even taking into account all of the above sources of uncertainty, the model only approximates the physical system and this discrepancy introduces further uncertainty about system behaviour),

(vii) measurement uncertainty (as the model is calibrated against system data all of which is measured with error),

(viii) multi-model uncertainty (usually we have not one but many models related to the physical system),

(ix) decision uncertainty (to use the model to influence real world outcomes, we need to relate things in the world that we can influence to inputs to the simulator and through outputs to actual impacts. These links are uncertain).

Different physical models vary in many aspects, but the formal structures for analyzing all of these components of the uncertainty for the physical system, as derived from the study of computer simulators for the system, are very similar, which is why there is a common underlying methodology for such problems. In this article, we give an informal introduction to some important features of this methodology. First, we introduce some general features of the uncertainty structure, describe the Bayesian approach for addressing these uncertainties, and explain why we prefer, in certain cases, a Bayes linear approach to the uncertainty analysis. Then, we outline the rationale for history matching as a way to constrain the input space, and describe a simple forecasting methodology for future system outcomes. We illustrate the development with a brief description of examples arising from each of the simulation problems described above. Finally, we consider the reason why we should view a computer simulator as being informative for a physical system.

2   General Uncertainty Structure

Each simulator for a physical system can be conceived as a function $F(x)$, where x is an input vector, representing unknown properties of the physical system, and $F(x)$ is the corresponding output vector representing aspects of system behaviour, y.

Interest in the analysis concerns general qualitative insights as to the behaviour of the system plus some of the following: (i) the "appropriate" (in some sense) choice, x^*, for the system properties x; (ii) the use that we can make of historical observations z, observed with error on a subset y_h of y, both to test and to constrain the model; (iii) how informative $F(x^*)$ is for actual system behaviour, y, particularly for forecasting future system outcomes, y_p;(iv) the optimal assignment of any decision inputs, d, in the model.

For example, in a climate analysis, y_h might correspond to historical climate outcomes over space and time, y to past, current and future climate, and the "decisions" might correspond to different policy relevant choices such as carbon emission scenarios.

How can we solve such problems? If observations, z, are made without error and the model is a perfect reproduction of the system, then, in principle, we can write $z = F_h(x^*)$, invert f_h to find x^* and learn about all future components of $y = F(x^*)$. If x contains some control parameters, then these are set to optimize properties of future outcomes contained in y.

However, in practice, the inversion of slow, high dimensional and complex functions is a very hard problem. Further, the observations z are typically made with error, and the model always differs from the physical system, so we must separate the uncertainty representation into two relations, one expressing the data uncertainty and one expressing the structural uncertainty: for example, the simplest such representation is of the form

$$z = y_h \oplus e, \; y = F(x^*) \oplus \epsilon \tag{1}$$

where e, ϵ have some appropriate probabilistic specification, possibly involving parameters which require estimation, and the notation $U \oplus V$ denotes the addition of $U + V$, in the case where U is probabilistically independent of V, for a full probabilistic specification or U, V are uncorrelated, if we only make a second order specification of means, variances and covariances. We therefore need to make a statistical inversion of the data through the function and make statistical predictions as to future system behaviour. This is a much harder problem than the deterministic inversion, and we still haven't accounted for condition uncertainty, multi-model uncertainty, and so forth.

In practice it is extremely rare to find a serious quantification of the total uncertainty about a complex system arising from the all of the uncertainties in the model analysis that we have identified. Therefore, for almost all applications, no-one really knows the reliability of the model based analysis, so that there can be no sound basis for identifying appropriate real world decisions based on such analyses. The space between models and reality arises partly because modellers and scientists don't think about total uncertainty in a sufficiently systematic way and nor do most statisticians. Policy makers don't know how to frame the right questions for the modellers to identify the gap between their analyses and the likely outcomes in the real world and there are few funding mechanisms to address such issues. And, of course , such a full uncertainty analysis is difficult and time consuming.

3   Bayesian Uncertainty Analysis for Complex Models

In the subjectivist Bayesian view, the meaning of any probability statement is the uncertainty judgement of a specified individual, expressed on the scale of probability (by consideration of some operational elicitation scheme, for example by consideration of betting preferences); for a somewhat subjective introduction to the subjectivist position, see[6]. This interpretation has an agreed testable meaning, sufficiently precise to act as the basis of a discussion about the meaning of the analysis. In this interpretation, any probability statement is the judgement of a named individual, so we should speak not of the probability of rapid climate change, but instead of Anne's probability or Bob's probability of rapid climate change and so forth.

There is an important practical issue of perception, as most people expect something more authoritative and objective than a probability which is one person's judgement. However, the disappointing truth is that, in almost all cases, stated probabilities emerging from a complex analysis are not even the judgements of any individual. Nor do they have any other clear and well defined meaning.

So, it is not unreasonable that the objective of our analysis should be probabilities which are asserted by at least one person (more would be good!). The Bayesian formalism provides a way, at least in principle, to realize this aim. In the simplest form, the Bayesian approach requires the specification of the following ingredients:

- a prior probability distribution for best inputs x^*
- a probabilistic uncertainty description for the computer function F
- a probabilistic discrepancy measure relating $F(x^*)$ to the system y
- a likelihood function relating historical data z to y

This full probabilistic description provides a formal framework to synthesis expert elicitation, historical data and a careful choice of simulator runs. We may then use our collection of computer evaluations and historical observations to analyze the physical process in order to determine appropriate values for simulator inputs (calibration; history matching), to assess the future behaviour of the system (forecasting), and to optimize the performance of the system.

There is much current interest in this problem. Good starting points for entering the Bayesian literature in this area are [10], [12]. A great general resource, offering references, papers, discussion and a methodological toolkit, is the Managing Uncertainty in Complex Models (MUCM) web-site, http://www.mucm.ac.uk/. (MUCM is a consortium between the Universities of Aston, Durham, LSE, Sheffield, Southampton, developing general methodology for this general area with Basic Technology funding.)

This approach is very successful for problems of intermediate size and complexity. For very large scale problems, however, such a full Bayes analysis is very difficult because (i) it is hard to give a meaningful full prior probability specification over high dimensional spaces; (ii) the computations for learning from data (observations and computer runs), particularly for identifying informative ensembles of choices of parameter values at which to evaluate the simulator, may be technically difficult; (iii) the likelihood surface is extremely complicated, and any full Bayes calculation may therefore be extremely non-robust.

4   Bayes Linear Approach

The idea of the Bayesian approach, namely capturing our expert prior judgements in stochastic form and modifying them by appropriate rules given observations, is conceptually appropriate (and there is no obvious alternative). Bayes linear analysis is a practical alternative to the fully specified Bayesian approach, being based on a prior specification only of the means, variances and covariances of all quantities of interest, where we make expectation, rather than probability, the primitive for the theory, following de Finetti, [5]. For a full account of the Bayes linear approach, see [9].de Finetti chooses expectation over probability as, if expectation is primitive, then we can choose to make as many or as few expectation statements as we choose (including our choice of probabilities, which are simply expectations for the corresponding indicator functions), whereas, if probability is primitive, then we must make all of the probability statements before we can make any of the expectation statements. When there are many quantities that we must specify uncertainty judgements for, it is very helpful to have the option of restricting our attention to that sub-collection of specifications which we are most interested in analyzing carefully.

Corresponding to Bayes theorem, which is the basic updating tool for full Bayes analysis, is the operation of belief adjustment. $\mathsf{E}_z[y], \mathsf{Var}_z[y]$ are the expectation and variance for the vector y adjusted by the vector z, evaluated as

$$\mathsf{E}_z[y] = \mathrm{E}(y) + \mathrm{Cov}(y, z)\mathrm{Var}(z)^{-1}(z - \mathrm{E}(z)),$$
$$\mathsf{Var}_z[y] = \mathrm{Var}(y) - \mathrm{Cov}(y, z)\mathrm{Var}(z)^{-1}\mathrm{Cov}(z, y).$$

If $\mathrm{Var}(z)$ is not invertible, then we use an appropriate generalized inverse.

Bayes linear adjustment may be viewed as an approximation to a full Bayes analysis or the appropriate analysis given a partial specification based on expectation as primitive. The foundation for the approach is an explicit treatment of temporal uncertainty, and the underpinning mathematical structure is the inner product space, as opposed to the probability space, which is simply a special case. The adjusted expectation of y given z is the linear combination of the elements of z, plus the unit constant, which minimizes the expected squared distance to y. Observe that de Finetti's primitive definition for conditional expectation (see [5]) corresponds to this definition in the special case in which the vector $z = (z_1, \ldots, z_r)$ represents the elements of a partition (so that one and only one of the elements of z will equal 1, and all other elements will equal 0). In this special case, adjusted expectation is equivalent to conditional expectation, so that the definition of conditioning may be viewed as a special case of that for belief adjustment, in which the vector z is restricted to a partition vector. There are other special cases in which adjusted expectation and conditional expectation coincide, the most important being that of the multivariate Gaussian distribution.

Full Bayes analysis can be more informative than the Bayes linear counterpart, if done extremely carefully, both in terms of the prior specification and the analysis. Bayes linear analysis is partial but easier, faster, and often more robust particularly for history matching and forecasting. The examples discussed below were all carried out within the Bayes linear approach. However, the ideas and approaches are complementary and there are natural full Bayes counterparts for each of the analyses that we describe.

5   Function Emulation

Uncertainty analysis, for high dimensional problems, is even more challenging if the function $F(x)$ is expensive, in time and computational resources, to evaluate for any choice of x. For example, large climate models can take many weeks to evaluate on extremely powerful computers.

In such cases, $F(x)$ must be treated as uncertain for all input choices except the small subset for which an actual evaluation has been made. Therefore, we must construct a description of the uncertainty about the value of $F(x)$ for each possible choice of x. Such a representation is often termed an emulator of the function - the emulator both suggests an approximation to the function and also contains an assessment of the likely magnitude of the error of the approximation. We use the emulator either to provide a full joint probabilistic description of all of the function values (full Bayes) or to assess expectations variances and covariances for pairs of function values (Bayes linear).

There are many ways to construct emulators for computer models. A good introduction to this area is [11]. A common choice of form for the emulator is as follows. We express the emulator for component F_i of F as

$$f_i(x) = \sum_j \beta_{ij} g_{ij}(x) \oplus u_i(x)$$

where $B = \{\beta_{ij}\}$ are unknown scalars, g_{ij} are known deterministic functions of x, $u_i(x)$ is a weakly second order stationary stochastic process. There are many choices of correlation function for this process; a common choice is

$$\text{Corr}(u_i(x), u_i(x')) = \exp\left(-\left(\frac{\|x - x'\|}{\theta_i}\right)^2\right)$$

In this representation, $Bg(x)$ expresses global variation, i.e. aspects of the overall behaviour of the function that we can discover from a design which is well dispersed in parameter space, while $u(x)$ expresses local variation, i.e. those aspects of the behaviour of the function which can only be assessed by making function evaluations in the neighbourhood of x.

We fit the emulators, given a collection model evaluations, using our favourite statistical tools, such as generalized least squares, maximum likelihood, full Bayes or Bayes linear, aided wherever possible by detailed expert judgement. We need careful (multi-output) experimental design to choose informative model evaluations, and detailed diagnostics to check emulator validity.

If the simulator is really slow to evaluate, then a practical way to develop the emulator is to model jointly the simulator with a fast approximate version, F'. So, for example, based on many fast simulator evaluations, we build emulator

$$f_i'(x) = \sum_j \beta_{ij}' g_{ij}(x) \oplus u_i'(x)$$

We use this form as the prior specification for the emulator $f_i(x)$. Then a relatively small number of evaluations of $F_i(x)$, combined with relations such as

$$\beta_{ij} = \alpha_i \beta_{ij}' + \gamma_{ij}$$

enables us to adjust the prior emulator to an appropriate posterior emulator for $F_i(x)$. This approach exploits the heuristic that we need many more function evaluations to identify the qualitative form of the model (i.e. choose appropriate forms $g_{ij}(x)$, etc) than to assess the quantitative form of all of the terms in the model - particularly if we fit meaningful regression components to account for a large component of global variation.

6   History Matching

Model calibration aims to identify "true" input parameters x^*. However full Bayes calibration analysis may be technically difficult and non-robust. Further, we may not believe in a unique true input value for the model and, indeed, we may be unsure whether there are any good choices of input parameters (due to model deficiencies).

A conceptually simple alternative, or precursor, to calibration is "history matching", i.e. finding the collection of all input choices x for which we judge the match of the model to the data to be acceptable, using some 'implausibility measure' $I(x)$ based on a natural probabilistic metric, accounting for emulator uncertainty, condition uncertain, structural discrepancy, observational error and so forth.

We construct the implausibility measure as follows. Using the emulator we can obtain, for each set of inputs x, the mean and variance, $\mathrm{E}(F_h(x))$ and $\mathrm{Var}(F_h(x))$. If $x = x^*$, then , setting $F^* = F(x^*)$, we have

$$z_i = y_i \oplus e_i, \; y_i = F_i^* \oplus \epsilon_i$$

so that

$$\mathrm{Var}(z_i - \mathrm{E}(F_i(x))) = \mathrm{Var}(F_i(x)) + \mathrm{Var}(\epsilon_i) + \mathrm{Var}(e_i)$$

We can therefore calculate, for each output $F_i(x)$, the 'implausibility' if we consider the value x to be the best choice x^*, which is the standardized distance between z_i and $\mathrm{E}(F_i(x))$, given by

$$I_{(i)}(x) = |z_i - \mathrm{E}(F_i(x))|^2 / [\mathrm{Var}(F_i(x)) + \mathrm{Var}(\epsilon_i) + \mathrm{Var}(e_i)]$$

Large values of $I_{(i)}(x)$ suggest that it is 'implausible' that $x = x^*$.

The implausibility calculation can be performed univariately, or by multivariate calculation over sub-vectors for which we are prepared to make a full joint covariance specification for the emulator errors and for the structural discrepancy. With such a full joint specification, the implausibility criterion is a form of Mahalanobis distance between the system observations and the function outputs. The implausibilities are then combined, such as by using $I_M(x) = \max_i I_{(i)}(x)$, and can then be used to identify regions of x with large $I_M(x)$ as implausible, i.e., unlikely to be good choices for x^*.

Using this analysis, we can then refocus our efforts on the 'non-implausible' regions of the input space, by making more simulator runs and refitting our emulator over such sub-regions and repeating the analysis. This process is a form of iterative global search aimed at finding all choices of x^* which would give good fits to historical data.

We may find no choices at all which give good fits and that is a clear sign of problems with our physical simulator or with our data. Further, even if our ultimate goal is Bayesian model calibration, it is good practice to history match first, to check the model and (massively) reduce the search space for the Bayesian algorithm.

7 Forecasting

There are two basic sources of uncertainty that we must quantify in order to predict future system outcomes, y_p. Firstly, we are unsure as to the system prediction, $F_p(x^*)$, for y_p, as we are uncertain about both F and x^*, and secondly we are uncertain about the model discrepancy, ϵ_p, between $F_p(x^*)$ and y_p. The simplest Bayes linear forecasting system for taking account of these uncertainties is as follows; for details see [2].

The mean and variance of $F(x)$ are obtained from the mean function and variance function of the emulator f for F. Using these values, we compute the mean and variance of F^* by first conditioning on x^* and then integrating out x^*, typically over the parameter region identified by history matching. Given $E(F^*)$, $Var(F^*)$, and specification of the variances for model discrepancy, ϵ, and sampling error, e, it is straightforward to compute the joint mean and variance of the collection (y, z) (as $y = F^* \oplus \epsilon$, $z = y_h \oplus e$).

We can therefore evaluate the mean and variance for y_p adjusted by z using the Bayes linear adjustment formulae. This analysis is fast and tractable even for large systems. Further, because of the simple structure of the calculations, it is tractable to identify collections of simulator evaluations which are appropriate for minimizing adjusted forecast variance. Typically, this will be the second stage choice of simulator evaluations, as the first stage will be a design appropriate to identify the form of emulator, estimate coefficient matrices and refocus, once or several times.

This analysis exploits the global features of the emulator to construct the joint covariance structure and is effective when the local component of emulator variation is small. When the local component is large, then a more detailed analysis is required, either by full Bayes specification or using the approach of Bayes linear calibrated forecasting; for details, see [7].

8 Example: Emulating a Climate Simulator

(This uncertainty analysis is work with Danny Williamson, with NERC funding; details in [14].)

One of the aims of the NERC funded RAPID programme is to assess the risk of shutdown of the AMOC (Atlantic Meridionnal Overturning Circulation), which transports heat from the tropics to Northern Europe, and how this risk depends on the future emissions scenario for CO2. The RAPID sub-project aims to address aspects of this question by use of large ensembles of the UK Met Office climate model HadCM3, run through climate prediction.net. At an early stage of the project, as a preliminary demonstration of concept for the Met Office, we were asked to develop an emulator for HadCM3, based on 24 runs of the simulator, with a variety of parameter choices and future CO2 scenarios. We had access to some runs of FAMOUS (a lower resolution model), which consisted of 6 scenarios for future CO2 forcing, and between 40 and 80 runs of FAMOUS under each scenario, with different parameter choices. There was very little time to do the analysis.

The design that we chose was to match the inputs for 8 of the HadCM3 runs with corresponding inputs to a FAMOUS run (to help us to compare the models) and to construct a 16 run Latin hypercube over different parameter choices and CO2 scenarios (to extend the model across CO2 space). In this experiment only 3 parameters were varied (an entrainment coefficient in the model atmosphere, a vertical mixing parameter in the ocean, and the solar constant).

Our output of interest was a 170 year time series of AMOC values. The series is noisy and and the location and direction of spikes in the series was not important. Interest concerned aspects such as the value and location of the smoothed minimum of the series and the amount that AMOC responds to CO2 forcing and recovers if CO2 forcing is reduced.

To emulate the whole time series, we first smoothed by fitting splines $f^s(x,t) = \Sigma_j c_j(x) B_j(t)$ where $B_j(t)$ are basis functions over t and $c_j(x)$ are chosen to give the 'best' smooth fit to the time series. We emulate f^s by emulating each coefficient $c_j(x)$ in $f^s(x,t) = \Sigma_j c_j(x) B_j(t)$ (separately for each CO2 scenario). We test our approach by building emulators leaving out each observed run in turn, and checking whether the run falls within the stated uncertainty limits.

We now have an emulator for the smoothed version of FAMOUS, for each of the 6 CO2 scenarios. We extend the FAMOUS emulator across all choices of CO2 scenario using fast geometric arguments, exploiting the speed of working in inner product spaces. For example, we have a different covariance matrix for local variation at each of 6 CO2 scenarios. We extend this specification to all possible CO2 scenarios by identifying each covariance matrix as an element of an appropriate inner product space, and adjusting beliefs over covariance matrix space by projection.

We develop relationships between the elements of the emulator for FAMOUS and the corresponding emulator for HadCM3, using the paired runs, and expert judgements. This gives an informed prior for the HadCM3 emulator. We use the remaining runs of HadCM3 for Bayes linear adjustment of the emulator for HadCM3, and carry out further leave one out diagnostic checks and variance tuning. Our Met Office collaborators were happy with the resulting model emulations as a basis for further analysis given access to the larger ensemble.

9   Example: Oil Reservoir Simulators

(This uncertainty analysis is work with Jonathan Cumming, carried out with Basic Technology funding as part of the MUCM project; details of the application are in [4], and of the multi-level inference and design calculations are in [3].)

An oil reservoir is an underground region of porous rock which contains oil and/or gas. The hydrocarbons are trapped above by a layer of impermeable rock and below by a body of water, thus creating the reservoir. The oil and gas are pumped out of the reservoir and fluids are pumped into the reservoir (to boost production). The simulator models the flows and distributions of contents of the reservoir over time.

Each cell in the reservoir has a collection of associated input parameters, such as permeability and porosity. There are also other parameters, such as fault transmissibility, aquifer features and saturation properties. Since there are a huge number of cells in the reservoir, it is common to use scalar multipliers over subregions, to modify values.

The model outputs comprise the behaviour of the various wells and injectors in the reservoir Output, typically, is a time series on the following variables for each well; bottom-hole and tubing head pressure, production/injection rates and totals, for each of oil, water and gas, and fluid ratios for water cut and gas-oil ratio.

The term history matching, within the oil industry, refers to the identification of choices of input parameters for the simulator for which the simulator output is in close correspondence to the observed reservoir history. Our Bayes linear approach to reservoir history matching, based on the methodology described in [1], has been successfully implemented in software widely in use in the oil industry.

An example that we have provided to illustrate the methodology is given in [4]. This model, of a reservoir located in the North Sea, is based on grid size $38 \times 87 \times 25$, with 43 production and 13 injection wells, and simulates 10 years of production, taking up to three hours per simulation. The inputs, in the illustration, are field multipliers for porosity (ϕ), permeabilities (k_x, k_z), critical saturation (crw), and aquifer properties (A_p, A_h). The outputs that we use for history matching are oil production rates for a 3-year period, for the 10 production wells active in that period, described by four month averages over the time series.

The computer model is expensive to evaluate, so we use a 'coarse' model, F^c, based on coarsening vertical gridding by factor of 10, to capture qualitative features of F. F^c is substantially faster, allowing 1000 runs of F^c in a Latin Hypercube over the input parameters.

Because of the high level of correlation between the different outputs, we use the principal variables approach to screen the wells. This method identifies, sequentially, the output, or group of outputs, which accounts for most of the variation in the remaining outputs. Applied to the coarse model evaluations, we retain outputs from 4 of the wells. These capture 87% of the total variation in all outputs.

We consider the coarse and the full model emulators to have the form

$$f_i^c(x) = \boldsymbol{g}_i(x_{[i]})^T \boldsymbol{\beta}_i^c + w_i^c(x), \, f_i(x) = \boldsymbol{g}_i(x_{[i]})^T \boldsymbol{\beta}_i + w_i^c(x)\beta_{w_i} + w_i^a(x)$$

where $x_{[i]}$ is a subset of 'active inputs', i.e. the inputs which account for most of the variation in F. We fit emulators to each output individually, using stepwise regression and generalized least squares for the coarse model runs, to get emulator $f_i^c(x)$ for F_i^c. We found that the choice of three active inputs was adequate for expressing global variation in each output, for example achieving R^2 values in excess of 0.96 for all outputs but one. The porosity and critical saturation turned out to be active for all of the outputs, while each other output was active in a subset of the outputs. The two emulators are linked via equations relating corresponding pairs of coefficients as outlined in section (5). Careful choice of a small design of 20 evaluations for the full simulator, based on informative configuration over the active input collections, followed by Bayes linear adjustment, leads to the resulting emulator for F.

We now specify the observation and discrepancy variances and carry out the implausibility calculations for history matching. We find that working to a three standard deviation implausibility threshold eliminates about 90%of the input space, and corresponds to imposing a constraint on the upper value of ϕ. Since reducing the space, many of the old model runs are no longer relevant, so we supplement our emulation with further evaluations obeying the parameter constraint, namely an extra 100 coarse runs and 20 full simulator runs, and further adjust the emulator, using the old emulator structure as a starting point.

We now consider the final four time points in the three year period that we have emulated, and use the observed historical values to forecast the corresponding output values for an additional time point, one year beyond the end of this period. We have historical observations for the values to be forecast, which act as a quality check on the forecasts. We use the approach of section (7) effectively combining each model forecast with a correction for the estimated model discrepancy. In each case, the resulting forecast interval is within the measurement error of the actual historical measurement.

10 Example: Galaxy Formation Simulation

(This uncertainty analysis is work with Ian Vernon, carried out with with Basic Technology funding as part of the MUCM project; details in [13])

The Cosmologists at the Institute of Computational Cosmology at Durham University are interested in modelling galaxy formation in the presence of Dark Matter. First, a Dark Matter simulation is performed over a volume of $(1.63$ billion light years$)^3$. This takes 3 months on a supercomputer. Then, the simulator Galform takes the results of this simulation and models the evolution and attributes of approximately 1 million galaxies. Galform requires the specification of 17 unknown inputs in order to run. It takes approximately 1 day to complete 1 run (using a single processor).

The Galform model produces many outputs, some of which can be compared to observed data from the real Universe. Initially, we analyze luminosity functions giving the number of galaxies per unit volume, for each luminosity. These are Bj Luminosity, corresponding to density of young (blue) galaxies and K Luminosity, corresponding to density of old (red) galaxies. We choose 11 outputs that are representative of the Luminosity functions and emulate the functions $f_i(x)$.

We assess condition uncertainty, structural uncertainty, measurement uncertainty, and so forth. For example, we must account for the uncertainty resulting from the unknown configuration of dark matter in our universe. We can form judgements as to the magnitude of this uncertainty by making repeat simulations of Galform with the same input parameters and different choices of dark matter configuration.

We carry out the iterative history matching procedure, through four waves. For each wave, we evaluate the simulator many times, restricting parameter choices to those which have not yet been ruled out by earlier waves, emulate the simulator within the reduced space and carry out the implausibility calculations to reduce space further. A summary of the procedure, the number of active variables at each stage and the space removed at each stage is as follows.

	No. Model Runs	No. Active Vars	Space Remaining
Wave 1	1000	5	14.9 %
Wave 2	1414	8	5.9 %
Wave 3	1620	8	1.6 %
Wave 4	2011	10	0.12 %

In wave five, we evaluate many good fits to data, and we stop. Some of these choices give simultaneous matches to data sets that the Cosmologists have been unable to match before.

11 Linking Models to Reality

Each of the above examples, in common with most of the field of computer experiments, takes it as almost self-evident that the computer model is informative for the physical system. However, in most cases, the reason that the evaluations of the simulator are informative for the physical system is that the evaluations are informative about the general relationships between system properties, x, and system behaviour y. Therefore, our inference from model to reality should proceed in two parts.

We emulate the relationship between system properties and system behaviour. We call this relationship, F^*, the "reified model" (from reify: to treat an abstract concept as if it were real). We can then decompose the difference between our model and the physical system into two parts. The first is the difference between our simulator and the reified form, and the second is the difference between the reified form at the physically appropriate choice of x and the actual system behaviour y. We call this the "Reifying principle", namely that the simulator F is informative for y, because F is informative for F^* and $F^*(x^*)$ is informative for y. Similarly, a collection of simulators $F_1, F_2, \ldots$ is jointly informative for y, as the simulators are jointly informative for F^*.

We link F and F^* using emulators. Suppose that our emulator for F is

$$f(x) = Bg(x) \oplus u(x)$$

Our simplest emulator for F^* might be

$$f^*(x, w) = B^* g(x) \oplus u^*(x) \oplus u^*(x, w)$$

where we might model our judgements as $B^* = CB + \Gamma$ and correlate $u(x)$ and $u^*(x)$, while treating $u^*(x, w)$, with additional parameters, w, as uncorrelated with the remaining terms in the emulator. Structured reification improves on this with systematic modelling for all aspects of model deficiency whose effects we can consider explicitly. For an illustrated treatment of reification, see [8].

All of the Bayes linear history matching and forecasting methodology that we have described is unchanged by this extra layer of modelling. All that has changed is our description of the joint covariance structure which underlies each of the subsequent calculations.

12   Concluding Comments

To assess our uncertainty about complex systems, it is enormously helpful to have an overall (Bayesian) framework to unify all of the sources of uncertainty. Within this framework, all of the scientific, technical, computational, statistical and foundational issues can be addressed in principle. Such analysis poses serious challenges, but they are no harder than all of the other modelling, computational and observational challenges involved with studying complex systems.

In particular, Bayes and Bayes linear multivariate, multi-level, multi-model emulation, careful structural discrepancy modelling and iterative history matching gives a great first pass treatment for most large modelling problems.

References

1. Craig, P.S., Goldstein, M., Seheult, A.H., Smith, J.A.: Pressure matching for hydrocarbon reservoirs: a case study in the use of Bayes linear strategies for large computer experiments (with discussion). In: Gastonis, C., et al. (eds.) Case Studies in Bayesian Statistics, vol. III, pp. 37–93. Springer (1997)
2. Craig, P.S., Goldstein, M., Rougier, J.C., Seheult, A.H.: Bayesian Forecasting for Complex Systems Using Computer Simulations. Journal of the American Statistical Association 96, 717–729 (2001)
3. Cumming, J., Goldstein, M.: Small Sample Bayesian Designs for Complex High-Dimensional Models Based on Information Gained Using Fast Approximations. Technometrics 51, 377–388 (2009)
4. Cumming, J., Goldstein, M.: Bayes Linear Uncertainty Analysis for Oil Reservoirs Based on Multiscale Computer Experiments. In: O'Hagan, West, A.M. (eds.) The Oxford Handbook of Applied Bayesian Analysis, pp. 241–270. Oxford University Press (2009)

5. de Finetti, B.: Theory of Probability, vol. 1 & 2. Wiley (1974, 1975)
6. Goldstein, M.: Subjective Bayesian analysis: principles and practice. Bayesian Analysis 1, 403–420 (2006)
7. Goldstein, M., Rougier, J.C.: Bayes linear calibrated prediction for complex systems. Journal of the American Statistical Society 101, 1132–1143 (2006)
8. Goldstein, M., Rougier, J.C.: Reified Bayesian modelling and inference for physical systems (with discussion). Journal of Statistical Planning and Inference 139, 1221–1239 (2008)
9. Goldstein, M., Wooff, D.A.: Bayes Linear Statistics: Theory and Methods. Wiley (2007)
10. Kennedy, M.C., O'Hagan, A.: Bayesian calibration of computer models (with discussion). Journal of the Royal Statistical Society, B 63, 425–464 (2001)
11. O'Hagan, A.: Bayesian analysis of computer code outputs: a tutorial. Reliability Engineering and System Safety 91 (2006)
12. Santner, T., Williams, B., Notz, W.: The Design and Analysis of Computer Experiments. Springer, New York (2003)
13. Vernon, I., Goldstein, M., Bower, R.: Galaxy Formation: a Bayesian Uncertainty Analysis (with discussion). Bayesian Analysis 5, 619–670 (2010)
14. Williamson, D., Goldstein, M.: Fast linked analyses for scenario based hierarchies. Journal of the Royal Statistical Society: Series C (to appear, 2012)

Discussion

Speaker: Michael Goldstein

Kyle Hickmann: Could you speak a little bit more on in what sense the emulator converges to the simulator as more points are observed?

Michael Goldstein: The emulator is exactly equal to the simulator at each observation point, and uncertainty about the simulator increases for input choices far from any of the observed values. We reduce uncertainty in any region of parameter space by making function evaluations in that region, and the more evaluations that we make, the further will the uncertainty be reduced. How large a sample we must make to achieve a good measure of convergence across the whole input space depends on the dimension of the input and output spaces and the degree of regularity of the function over the range of the input space. For example, very small regions of input space, in which the function behaves quite differently from behaviour everywhere else, can be extremely difficult to identify and emulate appropriately.

Antonio Possolo: We have learned from Lindley that linear polling is one way of merging the conclusions multiple Bayesian analyses will have produced. What is the state of the art?

Michael Goldstein: The appropriate way to merge multiple Bayesian analyses depends on your judgements about the level of, and the relationship between, the information and expertise contained within each analysis. There is no automatic way to do this. The reification formalism described in this article is one way of structuring the joint analysis when dealing with Bayesian analyses based around computer simulators.

Antonio Possolo: An analysis that starts from expectations, variances, and covariances, is bound to produces results that are expectations, variances, and covariances. What additional assumptions would you regard as defensible to be able to quantify the conclusions probabilistically?

Michael Goldstein: Probabilities are themselves expectations, for the indicator functions corresponding to the events. If the analysis is described at a sufficient level of detail to identify some of these expectations, then we have a direct probabilistic inference. Alternately, we can use qualitative probabilistic judgements to make a low assumption bridge between the Bayes linear and the full probabilistic analysis. For example, when carrying out a history matching analysis as described in this article, it is useful to know that, for any continuous, unimodal probability density function, 95% of the probability will be contained within three standard deviations of the mean (the so-called 3 sigma rule).

Jeffrey Fong: Regarding Bayesian linear analysis, in your two equations, one expectation and the second variance, do they allow a user to derive a host of relationship (as in classical theory of error propagation) that are used to get expectation and variance of a sum, product, quotient, etc... of a complicated algebraic form?

Michael Goldstein: Bayes linear inferences obey all of the rules derived from the linearity of expectation. Therefore, it is necessary to ensure that the appropriate polynomial or other functional forms of the quantities of interest are introduced as elements of the adjusting vector and of the vector of terms to be adjusted. The Bayes linear Statistics volume ([9]) contains examples and discussions of this.

Verified Computation with Probabilities

Scott Ferson and Jack Siegrist

Applied Biomathematics

Abstract. Because machine calculations are prone to errors that can sometimes accumulate disastrously, computer scientists use special strategies called verified computation to ensure output is reliable. Such strategies are needed for computing with probability distributions. In probabilistic calculations, analysts have routinely assumed (i) probabilities and probability distributions are precisely specified, (ii) most or all variables are independent or otherwise have well-known dependence, and (iii) model structure is known perfectly. These assumptions are usually made for mathematical convenience, rather than with empirical justification, even in sophisticated applications. Probability bounds analysis computes bounds guaranteed to enclose probabilities and probability distributions even when these assumptions are relaxed or removed. In many cases, results are best-possible bounds, i.e., tightening them requires additional empirical information. This paper presents an overview of probability bounds analysis as a computationally practical implementation of the theory of imprecise probabilities that represents verified computation of probabilities and distributions.

Keywords: probability bounds analysis, probability box, p-box, verified computation, imprecise probabilities, interval analysis, probabilistic arithmetic.

1 Introduction

Many high-profile disasters are attributable to numerical errors in computer calculations. The self-destruction of the Ariane 5 rocket on its maiden test flight was caused by integer overflow (ESA 1996). The crash of the Mars Climate Orbiter during orbital insertion resulted from a units incompatibility (Isbell et al. 1999). The Sleipner A offshore drilling platform sank because of an inaccurate finite element approximation (Selby et al. 1997). The Aegis cruiser USS *Yorktown* was dead in the water for several hours because of a propagated divide-by-zero error (Slabodkin 1998). The Flash Crash in which the Dow Jones Industrial Average plunged 9% almost instantaneously was due to runaway computerized trading mediated by the interacting algorithms used by high-frequency traders (CFTC/SEC 2010). These errors can be worse than costly or embarrassing. The failure of a Patriot missile to intercept the SCUD missile that killed 28 people and injured 100 more was supposedly due to accumulated round-off error (GAO 1992). Miscalculations arising from a race-condition error in the medical

A. Dienstfrey and R.F. Boisvert (Eds.): WoCoUQ 2011, IFIP AICT 377, pp. 95–122, 2012.
© IFIP International Federation for Information Processing 2012

software controlling the Therac-25 used for radiation therapy caused multiple fatal radiation overdoses to patients (Baase 2008, 425).

The corny adage "To err is human, but to really foul things up requires a computer" is not merely bitterness of the underemployed. Because computers are so fast, errors can propagate and accumulate very quickly, and because they often lack a machine analog of human contextual common sense, dramatic errors can go unnoticed until damage is unavoidable. Subtle, even minor features can, in unlucky situations, interact to create disastrously bad numerical results. Ironically, the appearance of precision in computer results can often induce a human error in which users place undue trust in the computer's output.

To overcome these problems, computer scientists have developed methods for 'verified computing' by which users will always get reliably accurate results, or at least will be made aware of the problem when their results are not reliable. One basic task in verified computing is to find an enclosure that surely contains the exact result of a calculation. This problem is often addressed using the methods of interval analysis, which is a mathematically rigorous form of arithmetic that can be implemented in software even though computers can represent numbers with only finite precision (Kulisch et al. 1993; Hammer et al. 1997; Popova 2009; Tucker 2011). In fact, these interval calculations can have rigor corresponding to that of a mathematical proof, in spite of the fact that they are done automatically by machine. The approach guarantees that rounding error is limited, integer overflow is prevented, and division by zero as well as similar impossible operations are handled appropriately to ensure the integrity of the affected calculation. Of course, this means that real-valued answers cannot generally be represented precisely in finite machine number schemes. Instead, the answers are represented by enclosures consisting of two bounding machine-representable values. If this enclosure interval is narrow, we know the answer reliably and accurately. If the interval is wide, we have a transparent warning that the associated uncertainty is large, which implies that a more careful reanalysis may be useful.

Interval analysis is often offered as the primary—and one might think the only—method for verified computation, but verified computing requires a panoply of methods. Consistent application of mathematical rigor in the design of the algorithm, in the arithmetic operations it uses, and in the execution of the program allow an analyst to guarantee that a problem has a solution somewhere in the computed enclosing interval (or that no solution exists). To enable such consistency, methods must be developed for the wide variety of numerical and other operations that computers do for us. For instance, basic mathematical operations on floating-point numbers are replaced by interval analysis on intervals guaranteed to enclose scalar real values. Likewise, methods for vector and matrix operations have been developed that extend and generalize interval analysis with multidimensional arrays of interval ranges.

Similar methods of verified computing are needed for representing and calculating with probabilities and probability distributions on finite-precision machines. Unfortunately, the properties of probability distributions and the features of the laws of probability complicate the effort considerably. For instance, even representing a univariate continuous distribution is an infinite-dimensional problem, because it is a continuous function whose values at an infinity of points must be captured in the finite storage accessible by the computer. The critical role of assumptions about the stochastic dependence among variables is particularly complicating, and this is an issue even for total probabilities that can be represented by single scalar values. For example, if two events have probabilities 0.2 and 0.3 respectively, the AND operator commonly used in fault trees would only yield the product 0.06 when the probabilities are independent. Without specifying the dependence between the events, the result of the operator is undefined (although it can be bounded). The role of dependence assumptions is much more complicated still for distributions of random variables (Ferson et al. 2004; Nelsen 1999).

Nevertheless, we must undertake the effort, whatever its complexity. There is a pronounced need for verified computing methods to use with probabilities and probability distributions because they are becoming more and more pervasively used in engineering calculations including uncertainty analyses, risk assessments, sensitivity studies, and modeling of quantities with intrinsic aleatory uncertainties. They are being used across a host of fields as diverse as financial planning (Hertz 1964; Boyle 1977), human health risk analyses (McKone and Ryan 1989), ecological risk assessments (Suter 1993), materials and weapons safety calculations (Elliott 2005; Cooper 1994), extinction risk analysis for endangered species (Burgman et al. 1993; Ferson and Burgman 2000), and probabilistic risk assessment for nuclear power (Hickman et al. 1983) and other engineered systems (Vick 2002).

Engineers routinely face three crucial issues when they develop probabilistic models. The first is that their model uncertainty, i.e., their doubt about the proper mathematical form the model should have, is almost never articulated, much less accounted for in any comprehensive way. Modelers may recognize and acknowledge the limitations induced by this problem, yet they rarely conduct the sensitivity studies needed to fully assess the consequences of the uncertainty on model results. The second crucial problem is that there is often little or no quantitative information about possible correlations among the input variables, and in many cases the nature of the intervariable dependencies may not have been empirically studied at all. The typical response of analysts, even if they are aware of their uncertainty, is to nevertheless assume independence among variables, even though this assumption may be neither realistic nor conservative. In fact, using an incorrect assumption about dependence can strongly distort the output distributions, especially in their tails (Ferson et al. 2004; contra Smith et al. 1992).

The third crucial problem faced by engineers developing probabilistic models is that it is often impossible to fully justify a precise probability distribution to be used as input in the model, and sometimes the family of the distribution is only a guess. There is a huge literature on the subject of estimating probability distributions from empirical data, and there are several methods available for use including the method of matching moments, maximum likelihood estimation, the maximum entropy criterion and Bayesian methods to compute posterior predictive distributions. But these standard approaches are of limited practical reliability when few relevant data exist. Even when confidence limits are computed, little use can be made of them without an elaborate sensitivity study that is cumbersome to organize, computationally intense, and difficult to interpret. With limited empirical information, all of these methods for selecting input distributions require assumptions that cannot be justified by appeal to evidence and therefore may be false. These unsubstantiated assumptions can make a difference in the results. As Bukowski et al. (1995) showed, the choice about distribution shape can have a sizeable effect on the output distributions, especially at the tails.

It is generally assumed that the only solution to incomplete information is additional empirical effort to measure correlations, develop input distributions, and validate the model. As a practical matter, since such empirical information is typically incomplete—and indeed often quite sparse—analysts are forced to make assumptions without empirical justifications, leading to diminished credibility for the assessment and any subsequent decisions. There are, however, computational methods that allow analysts to sidestep a lack of information about the correlation and dependency structure among variables to obtain partial or complete solutions in many practical cases without having to make unjustified and possibly false assumptions. Likewise, when empirical information about the input distributions is limited, far more appropriate representations of uncertainty can be developed than are currently obtainable using techniques such as the maximum entropy criterion. These new methods allow us to compute bounds on estimates of probabilities and probability distributions that are guaranteed to be correct even when one or more of the assumptions is relaxed or removed. In many cases, the results obtained are the best possible bounds, which means that tightening them would require additional empirical information. This paper reviews probability bounds analysis (PBA, Ferson et al. 2003), as a computationally practical calculus of the theory of imprecise probabilities (IP, Walley 1991), that combines ideas from both interval analysis and probability theory to sidestep the limitations of each. Probability bounds analysis is logically and morally equivalent to a sensitivity analysis. Objecting to PBA implies an objection to sensitivity analysis. PBA uses exactly the same mathematical approach used in sensitivity analysis, but its computational methods are applicable to broader questions and are vastly more efficient.

2   Kinds of Uncertainty

In the past, uncertainty analysis considered the source of uncertainty to be its salient aspect, so modelers talked, for example, about their parametric uncertainty or their model-form uncertainty. A more modern view is that the nature of the uncertainty, rather than its source, is a more important characteristic. We can distinguish between two main forms of uncertainty: variability and incertitude. *Variability* refers to the stochastic fluctuations in a quantity through time, variation across space, manufacturing differences among components, genetic or phenotypic differences among individuals, or similar heterogeneity within some ensemble or population. Engineers often refer to variability as aleatory uncertainty, harkening to *alea*, the Latin word for dice. This is considered to be a form of uncertainty because the value of the quantity can change each time one looks, and one cannot predict precisely what the next value will be (although the distribution of values may be known). *Incertitude*, on the other hand, refers to the lack of full knowledge about a quantity that arises from imperfect measurement, limited sampling effort, or incomplete scientific understanding about the underlying processes that govern a quantity. Many engineers refer to incertitude as epistemic uncertainty. We might simply and non-euphemistically call it 'ignorance', except for the embarrassment or confusion that word might evoke should professionals need to mention it in front of their bosses or the laity.

These two forms of uncertainty have important differences. Incertitude can in principle be reduced by empirical effort; investing more in measurement should yield better precision. Variability, in contrast, can perhaps be better characterized, but cannot generally be reduced by empirical effort. Incertitude depends on the observer and the observations made. Variability does not depend on an observer at all. It exists whether or not anyone witnesses it, like the sound waves emanating from the proverbial tree falling unseen in the forest. Although variability and incertitude can sometimes be like ice and snow in that their distinction can be difficult to discern through complicating details, and sometimes one can change into the other depending on the scale and perspective of the analyst, the macroscopic differences between these two forms of uncertainty are usually obvious and often significant in practical settings.

There is a crucial difference between a quantity actually varying and our simply not being sure about its magnitude, and this difference affects how we should do calculations. Consider, for example, the following elementary question: Suppose we are told that a quantity A is some value or values between 2 and 4, and that B is a quantity inside the range between 3 and 5. What can be said about their sum $A + B$? When this exemplar question was posed on the Riskanal electronic mailing list, half the respondents suggested the proper answer can be computed by modeling A as a uniform distribution between 2 and 4, and modeling B with another uniform distribution between 3 and 5, and convolving these two uniforms together with Monte Carlo simulation to obtain the triangular distribution ranging between 5 and 9 with a mode at 7 shown as a probability density function in Fig. 1. This is the traditional answer from probabilists for such a question. Indeed, it is the answer that Laplace (1820) himself would have suggested. This answer says that the value 7 is the most likely magnitude of the sum, and also that the extreme values of 5 and 9 have vanishing probabilities. There is more than two-thirds probability that the sum falls in the middle interval [6.1, 7.9].

But what exactly justifies this concentration of probability mass in the central range? There is nothing in the statement of the elementary question that suggests that 2 is not a perfectly possible value of A, and likewise nothing to suggest that

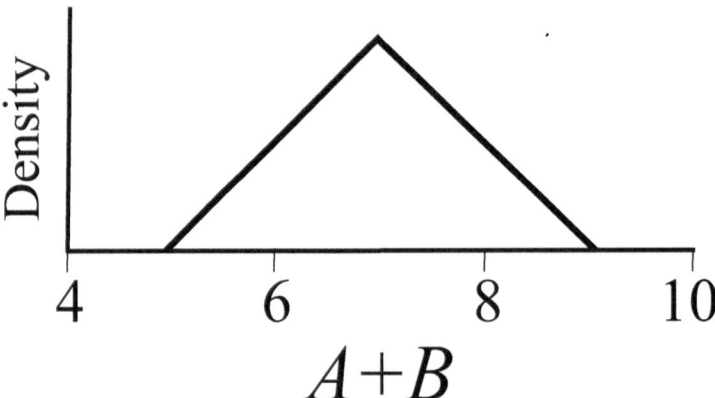

Fig. 1. Triangular distribution which is the traditional probabilist answer to the elementary question "What is the sum $A+B$ where A and B are in the respective intervals [2,4] and [3,5]?"

B might not simply just be 3. If so, then the sum is the scalar value 5, and the triangular distribution seems hard to explain. Given what is expressly known about the inputs, there is no reason to deprecate any of the possible values of the sum, or to distinguish one value as more probable than any other. But that may be a far cry from saying that all the values are equally probable. The other half of the respondents to the Riskanal poll said that the proper answer to the elementary question can be computed simply by adding together the intervals [2,4] + [3,5] using interval arithmetic (Moore 1966) to yield the interval [5,9]. Notice that this answer offers no concentration of mass in the central range, and suggests that the sum might simply be 5, and likewise might simply be 9, and there is nothing to suggest that these values, although extreme, are in any way unlikely.

The interval answer is a much looser statement than is any probability distribution. For instance, modeling the sum with a uniform probability distribution would say that all possible values within the range [5,9] are equally probable. Taking such a model seriously would suggest that one could profitably make bets about future values of the sum based on the probability. For instance, a probabilist would presumably be disposed to bet favorably, and big, on a gamble that the sum is larger than 5.01. A more sanguine view is that one had better not place any such bets, other than those that can be actually justified by the given knowledge. All that can be justified is that the probability distribution of the sum has its support within the range [5,9], but this admits a whole host of possible distributions. The interval answer can be identified with the entire class of such distributions.

Our view is that only one of these two answers to the elementary question is correct. We think the right answer is clearly the interval and not the triangular distribution, at least in practical contexts such as risk analysis and most uncertainty assessments. The triangular distribution traditionally given by probabilists is wrong because it implies or appears to imply more is known than is actually justifiable. It is the incertitude in the elementary problem that must be propagated by interval analysis or other bounding methods. Although these assertions are commonly met with nodding agreement from engineers and biologists, they sometimes evoke agitated criticism from probabilists. So let us hasten to point out some important tempering caveats. We are surely not saying we should only use intervals in risk or uncertainty analysis. We are not even saying that all uncertainty is incertitude. In fact, we would not be surprised that most of the uncertainty in some setting is not incertitude, and we agree that sometimes incertitude is entirely negligible, in which case probability theory is perfectly sufficient for modeling uncertainties and risks.

What we are saying, however, is that some analysts face non-negligible incertitude and handling this incertitude with standard probability theory requires assumptions that may not be tenable, including unbiasedness, uniformity or equiprobability, and independence. Because it will often be useful in practical situations to know what difference incertitude might make, it is important to have methods that can make probabilistic calculations without requiring the traditional assumptions. It turns out that this is possible with the theory of imprecise probabilities (Walley 1991) and a practical calculus for making computations with imprecisely specified probability distributions such as probability bounds analysis (Ferson 2002) which combines probability theory with interval analysis.

3   P-Boxes and Probability Bounds Analysis

A probability box, or p-box, is a characterization of an uncertain number which may have variability (aleatory uncertainty) or incertitude (epistemic uncertainty), or both. A p-box is specified by left and right bounds on the cumulative probability distribution function of a quantity and, optionally, additional information about the quantity's mean, variance and distributional shape (family, unimodality, symmetry, etc.). A p-box represents a class of probability distributions consistent with these constraints. Fig. 2 depicts an example for an uncertain number X consisting of a left (upper) bound and a right (lower) bound on the probability distribution for X. The bounds are coincident for values of X below 2 and above 29. The bounds may have almost any shapes, including step functions, so long as they are monotonically increasing and do not cross each other. A p-box simultaneously expresses incertitude (epistemic uncertainty), which is represented by the breadth between the left and right edges of the p-box, and variability (aleatory uncertainty), which is characterized by the overall slant of the p-box. This p-box suggests that the probability that X is below 10 is less than 25%. It might be as low as zero. We cannot say more than this because of the epistemic uncertainty about X's distribution function. The 95th percentile is somewhere between 18.5 and 26. We don't know where in that range it is because of the associated incertitude.

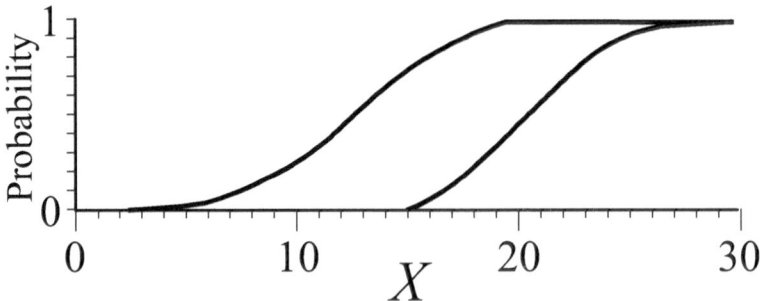

Fig. 2. A p-box specified by left and right bounding cumulative distribution functions and representing a class of probability distributions whose cumulative distribution functions can be drawn within the bounds

There are many ways that p-boxes can be constructed from the available information about uncertain numbers (Ferson et al. 2003). Fig. 3 illustrates six of these ways. The top, left graph depicts a distributional p-box for which the shape or family of the distribution is known (e.g., normal, uniform, beta, Weibull, etc.) but the parameters are known only to within intervals. For example, an analyst may know from mechanistic or physical considerations that the distribution is normal, but not be able to precisely identify the two parameters needed to specify it exactly. If the parameters can be bounded, then a distributional p-box can easily be constructed from enveloping all the possible distributions.

The top, right graph in Fig. 3 depicts what might be considered the opposite situation where the analyst is confident about some parameters describing the uncertain number, but is unsure about what shape or family of distributions it might be from. Such a situation arises frequently when distributions are developed from information obtained from scientific publications, where summary statistics are often reported without further details or the original data. Even though the available information might seem meager, what is known often suffices to define a nontrivial p-box that can be used in calculation. In some cases, classical results such as the Markov or Chebyshev inequalities can be used to derive formulas for p-boxes from a few parameters. In different situations, different sets of parameters may be known. Ferson et al. (2003) gave formulas for p-boxes for the following common situations:

{min, mean}	{min, max, mean}
{min, max, median}	{min, max, mean=median}
{min, max, mode}	{min, max, median=mode}
{mean, variance}	{min, mean, variance}
{min, max, mean, variance}	{min, max, mean, variance, mode}

These define what might be called distribution-free p-boxes because they make no assumption whatever about the family or shape of the uncertain distribution and yet enclose all distributions which match the given parameters.

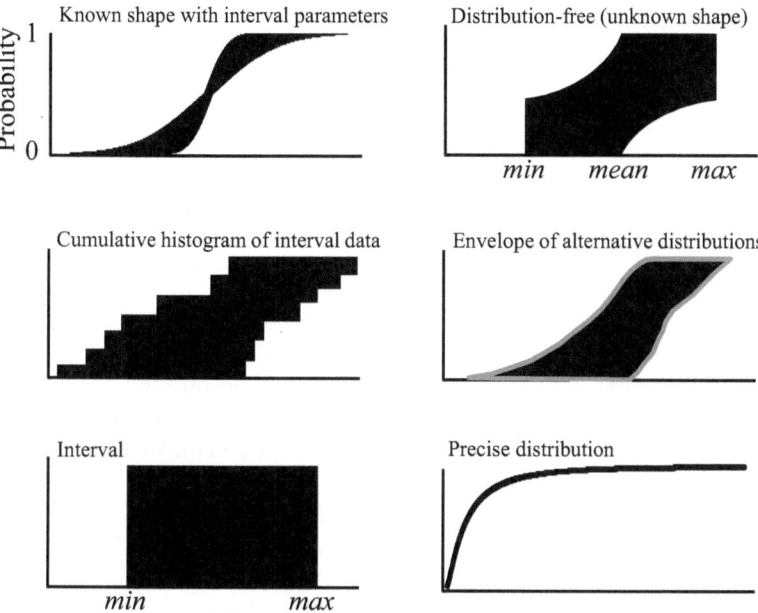

Fig. 3. A few ways p-boxes arise

The p-boxes are somewhat wider when the parameters are only known to within intervals. When qualitative information is available, such as that the distribution is symmetric or unimodal, the p-boxes can often be tightened substantially.

The middle, right graph in Fig. 3 depicts a situation in which one of two distributions is the correct one, but the analyst cannot discern which. By enveloping them into a single p-box, the analyst can represent the uncertainty in a single structure that does not require a cumbersome sensitivity analysis to propagate. This facility can become very important when there are multiple possible distributions and several variables have such uncertainty because exploring them in a sensitivity study requires a combinatorially complex effort. Collapsing the uncertainty into a single p-box per variable can simplify the problem considerably.

The middle, left graph in Fig. 3 shows a p-box from a situation in which there is no sampling uncertainty because the entire population has been measured, but there is substantial mensurational uncertainty that comes from our inability to measure individual values precisely. A similar p-box with both sampling and mensurational uncertainty can be formed by enclosing the empirical histogram of interval data with Kolmogorov–Smirnov confidence bands. These bands are distribution-free and merely assume independence of the sample data. Alternatively, there are potentially tighter confidence bands that can be used which do make assumptions about the shape of the distribution.

P-boxes include the special cases of intervals and precise probability distributions too. For example, in some situations an analyst may have no information about a distribution except its potential range, that is, knowledge that its values must be larger than min and smaller than max. In this case, Laplace (1820) used the Principle of Insufficient Reason (sometimes called the Principle of Indifference) to select a uniform distribution for the variable but, as argued above, an interval is a fuller characterization of this uncertainty than any particular probability distribution could be. An interval, illustrated in the bottom, left graph of Fig. 3, is a special case of a p-box whose left and right bounds are step functions at min and max respectively. Finally, it is also possible that the distribution for some variable actually is well specified. The bottom, right graph illustrates this case where the left and right bounds of the p-box are coincident.

This idea of bounding probability has a very long tradition throughout the history of probability theory. Indeed, George Boole (1854; Hailperin 1986) used the notion of interval bounds on probability. The classical inequality attributed to Chebyshev (1874) described bounds on a distribution when only the mean and variance of the variable are known, and the related inequality attributed to Markov (1886) found bounds on a positive variable when only the mean is known. Keynes (1921) argued that probabilities of some propositions cannot be ordered because they overlap due to uncertainty. Fréchet (1935; 1951) discovered how to bound calculations with total probabilities without assuming independence or making other dependence assumptions. Bounding probabilities has continued to the present day (e.g., Berger 1985; Walley 1991). Kyburg (1999) reviewed the history of interval probabilities and traced the development of the critical ideas over the last century.

Several authors have described strategies for computing with bounds on distribution functions (e.g., Makarov 1981; Yager 1986; Frank et al. 1987; Williamson and Downs 1990; Berleant 1993; 1996; 1998; Ferson 2002; Ferson et al. 2003; inter alia). Williamson and Downs (1990) described explicit algorithms to compute sums, products, differences and quotients. Since their effort, algorithms for essentially all the standard mathematical operations have been derived and implemented (Ferson 2002). These methods, collectively called probability bounds analysis (PBA), have been used to propagate p-boxes through mathematical expressions of widely varying complexity, ranging from simple arithmetic formulas common in risk analyses and logical expressions summarizing fault or event trees to finite-element computations (Zhang et al. 2010; 2012) and evaluations of nonlinear ordinary differential equations (Enszer et al. 2011). The calculations made with these methods can be shown to be *rigorous*, i.e., they are guaranteed to enclose the true outcome distribution whenever the input p-boxes enclose their respective distributions. In many cases, the calculations can also be shown to be pointwise *best-possible* in the sense that they could not be any narrower without excluding distributions that might arise as results given the inputs.

These calculations account for, and preserve the integrity of, both the incertitude and variability expressed by a p-box. Combining a p-box with a scalar, interval, probability distribution or other p-box in any arithmetic or logical calculation generally yields another p-box. Combining an interval with a probability distribution also generally yields a p-box, as the incertitude of the interval combines with the variability of the probability distribution. P-boxes also arise when two precise probability distributions are combined whenever their intervariable dependency is unknown or only partially known. Such combinations produce precise probability distributions only when the dependence function or copula (Nelsen 1999) is completely specified.

The approach of Williamson and Downs (1990) includes a way to rigorously represent continuous probability distributions by using outward-directed rounding on finitely many interval discretizations of the bounds on cumulative distribution functions. Bounding in the cumulative domain, rather than the density domain, allows representation error to be completely contained and propagated using the same algorithms that handle mathematical combinations. When operations on the interval discretizations are handled with interval analysis, the method constitutes verified computation for probability distributions.

The methods of probability bounds analysis are available in several software implementations, including multiple free demonstration programs (e.g., Berleant and Zhang 2004), a full-featured stand-alone commercial program (Ferson 2002), an advanced add-in for Microsoft Excel developed for NASA (Ferson et al. 2011), and a package in development for the statistical computing language R (R Development Core Team 2010).

4   Correlations and Dependencies

Independence can be a dangerous assumption for analysts to make. Stochastic dependence is far more pervasive—and important—than many analysts seem to recognize. For instance, placing backup generators side by side makes their failure probabilities dependent and reduces the redundancy they were intended to provide because they become susceptible to the common-cause failure from flooding, as was realized too late in New Orleans and Fukushima. Even in relatively sophisticated analyses of uncertainty, the most common assumption about the dependence among variables is independence, although there may be no actual empirical evidence or serious theoretical justification to support this assumption. In truth, despite warnings about falsely assuming statistical independence, some analysts routinely ignore correlations for the sake of computational convenience. And conscientious analysts who would like to include them in analyses are often stymied by the difficulty of measuring correlations when data are sparse. As a consequence, correlations are commonly omitted from analyses and the default assumption of independence is used even when there is no evidence whatsoever in support of this assumption.

Although central tendencies may be generally insensitive to correlations of small to moderate strength (Smith et al. 1992), the *tails* of distributions can be extremely sensitive to even small or moderate correlations. Of course, decision makers are often especially concerned with these tails. They represent the risks of extreme events, which might be the probability of some mechanical stress exceeding the engineered strength intended to resist it, or the probability of exposing people to large doses of a carcinogen, or perhaps the risk of extinction for an endangered species. It is these extreme adverse events in the distribution tails that are often the whole focus of the analysis, so it may be very important that the tail probabilities not be underestimated. Unfortunately, the common practice of assuming independence among all input variables can lead directly to such underestimations.

Moreover, many analysts seem to be unaware that consideration of correlation is only the tip of the iceberg. The issue of dependence is much broader than correlation because it includes all nonlinear relationships. This is the reason, of course, that lack of correlation does not guarantee independence. What we might call linear dependence, which can be fully characterized by a single correlation coefficient, is only a small subspace of the forms of statistical dependence. Consequently, it is impossible to use a sensitivity study to characterize the effect of uncertainty about dependence on an uncertainty projection. Varying correlations, even all the way from +1 to –1, over a single family of dependence functions does not come close to capturing the diversity of possible interactions the variables may have. This is similar to supposing that one has characterized the variability of *all possible functions* through a point simply by representing *all linear functions* through a point. (And there is no analog of Taylor's theorem for dependence, so the mistake is severe on all scales.) We note that no popular software packages for probabilistic calculations support more than a single family of dependence functions, if they support intervariable dependence at all.

Makarov (1981) and Frank et al. (1987) showed, however, that it is possible to compute bounds on results of probabilistic calculations no matter what correlations or statistical dependencies may exist among the variables. The algorithms of Williamson and Downs (1990) include this no-assumptions case. The top, left-hand graph of Fig. 4 shows an example calculation. In this example, X and Y are random variables each drawn from uniform distributions between 1 and 25. Any distribution of the sums $X + Y$ that could result from adding these uniformly distributed random values together must lie entirely inside the p-box shaped like a parallelogram ranging between 2 and 50. This does not mean of course that any distribution within the bounds could be the sum of these two distributions, but the bounds are pointwise best-possible, which means the black region could not be any smaller without excluding some distributions that could arise as sums of these two uniform variables. This elementary calculation shows that the probability that $X + Y$ is smaller than 10 could be as high as one third, or as low as zero. This is a very different characterization of the distribution tail than what comes from a conventional Monte Carlo analysis that falsely assumes independence which, in this case, would suggest the chance the sum is less than 10 is about 5%. The PBA result makes no false assumptions about independence because it makes no assumptions at all about dependence, which potentially makes it very useful in applications such as risk analysis where it is critical not to underestimate tail risks.

The simple parallelogram shape of the example result is a consequence of the uniformity of the marginal distributions and the simplicity of the addition function that combines them. The algorithms can be applied equally well to virtually any finite distributions, including theoretically infinite distributions such as the

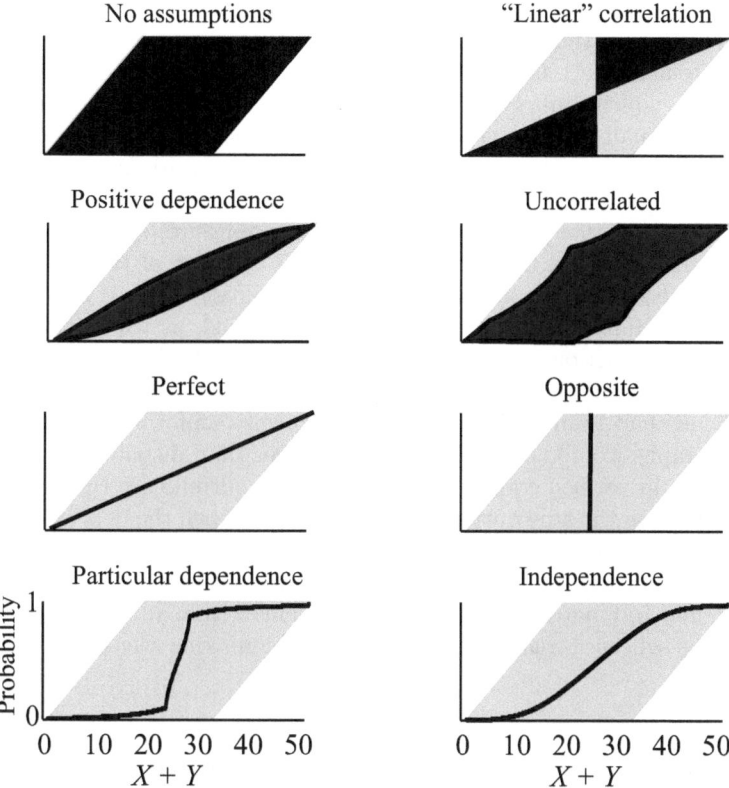

Fig. 4. The effect of various assumptions about the dependence between X and Y on the convolution $X + Y$, where X and Y are both uniformly distributed over the range [1, 25]. For comparison, the no-assumptions p-box is shown in gray.

normal that are truncated to some practical range, and extend beyond addition to all the basic mathematical operations. Surprisingly, the algorithms for the no-assumptions cases are computationally less expensive than Monte Carlo simulation methods that assume independence. More importantly, Monte Carlo methods cannot generally compute these no-assumptions bounds, no matter how many replications are used. Strategies that vary the correlation coefficient in a Monte Carlo sensitivity study will be limited to outputs like the cone shown in the top, right graph of Fig. 4, which grossly understates the uncertainty including possible tail risks.

Partial knowledge about the dependence can tighten the output p-box substantially (Ferson et al. 2004). For example, the left-hand graph in the second row of Fig. 4 shows the bounds on the distribution that can be inferred from the qualitative knowledge that the dependence function has positive sign. It is also possible to compute bounds on the result implied by a given correlation coefficient, which might have been reported among summary statistics published without original data. Berleant and Goodman-Strauss (1998) showed how the bounds in this case can be computed using mathematical programming. The right-hand graph of the second row of Fig. 4 shows an example with correlation zero. Remarkably, the resulting p-box almost fills up the no-assumptions parallelogram, which confirms the intuition that knowing only that variables are uncorrelated actually tells us very little about their possible dependence. In fact, the bounds are widest when the correlation is close to zero. When the correlation is extremal, corresponding to the third row of graphs in the figure, and the dependence is either perfect (comonotonic) or opposite (countermonotonic), then the resulting p-box becomes very tight. In the case of the example with precise addends, the result is also a precise distribution. This is true whenever the dependence function (copula) is precisely specified, as when it is given by some particular function such as that yielding the result shown in the bottom, left graph, or when it is taken to be independence as in the bottom, right graph. If the addends had been p-boxes rather than precise uniform distributions, the lower four graphs would also depict p-boxes rather than distributions.

As required in verified computation, all of these outputs are rigorous so they are sure to enclose the true output distribution even when the dependence is not precisely known, and their representations are pointwise best-possible so they could not be any tighter without excluding some possible distributions. It is possible to mix and match dependence assumptions within an analysis, assuming independence where it really is appropriate and making weaker assumptions where it is not.

5   A Calculus for Uncertainty Beyond Probability

It is possible to conceive of probability bounds analysis as an entirely traditional application of sensitivity analysis for conventional probability theory. Indeed, the same can be said of the whole of the theory of imprecise probability. Although this conception should therefore be acceptable even to strict Bayesians, PBA and IP are sometimes met with distrust by probabilists. The reason is that they recognize in it an essential heterodoxy: that there exists a kind of uncertainty that should not be characterized by a unique probability measure, which is a notion they perceive as wrong, or subversive.

There are many misconceptions about what justifies what we might call the precisionist school of probability. Some people have argued that probability theory is the only logically possible calculus of uncertainty (e.g., Lindley 1982). Lindley (2006, 71) asserted, "Whatever way uncertainty is approached, probability is the *only* sound way to think about it." Neapolitan (1992) pointed out, however, that the arguments about the inevitability of probability theory say nothing to deny the utility of interval bounding. Some in the precisionist school mention a 'proof' due to Cox (1946), but Cox's theorem has been proven to be incorrect (Halpern 1999), and perhaps more importantly it has been shown to be irrelevant to the question of whether probability is the only possible calculus for uncertainty because it uses as assumptions the very questions that are at stake in the debate (Colyvan 2004).

Some critics of p-boxes and imprecise probabilities in general argue that precisely specified probability distributions are in any case *sufficient* to characterize uncertainty of all kinds. These critics argue that it is therefore meaningless to talk about 'uncertainty about probability' and that traditional probability is a complete theory. Under this criticism, users of p-boxes have simply not made the requisite effort to identify the appropriate precisely specified distribution functions. This argument may have been reasonable when probability distributions were only used to characterize uncertain scalar values, that is, real numbers. In modern uncertainty analyses, however, the objects of study are often themselves probability distributions deployed in risk assessments already involving aleatory uncertainty. In such cases, a richer characterization of uncertainty seems useful.

Other people think that requiring inferences to be consistent and coherent is logically equivalent to requiring the use of precise probabilities. This is not the case, as was suggested in the expansive review of the axiomatics of decision theory under subjective probability by Fishburn (1986). Walley (1991) showed how the theory of imprecise probabilities enjoys the same properties of consistency and coherence championed by probabilists as the definition of rationality, and how proper inferences in the IP context also avoid sure losses from Dutch books just as coherent Bayesian decisions do.

Luce (1992) noted that a major finding in the theory of subjective expected utilities is that humans do not make decisions according to the theory of subjective expected utilities. One of the reasons for this is that a traditional probabilist must always be able to discern which of two events is more probable, unless they are equally probable, and which of two options is preferable, unless she is indifferent to the choice between them. The axiom that this is always possible is called "completeness" by Fishburn (1986), or the "ordering postulate" by Seidenfeld (1988). A relaxed, more general theory of uncertainty recognizes that some events or options may be *incomparable* and does not demand an agent be able to distinguish between any two, even beyond indifference. The notion that one might not be able to compare every two probabilities dates back at least to Keynes (1921), and it is another of his great ideas that has remarkable salience today.

Getting rid of the completeness axiom—which some might argue is far from self-evident anyway—induces the theory of imprecise probabilities. In this case, the definition of the probability of an event can be operationalized as the interval between the highest buying price and the lowest selling price for a gamble that pays one dollar if the event occurs (but nothing otherwise). This generalizes de Finetti's notion of a fair price to an interval whenever the probability is uncertain. Traditional Bayesians in principle *must agree to either buy or sell any gamble at the same fair price*. Under a relaxed theory, an agent may elect to neither buy nor sell a gamble if the price is not sufficiently favorable. Of course, if one knows all the probabilities (and utilities) perfectly, then IP reduces to Bayes.

Seidenfeld (1988) holds that a relaxed theory based on imprecise probabilities can provide a unified treatment of group decisions, where Bayesians admit their theory does not apply. This is important because most engineering decisions cannot be described as personal decisions, but rather must be in line with collective verdicts by teams of collaborators. Under IP, their decisions can be rational and coherent so long as indecision is admitted occasionally.

A parallel to the status of imprecise probability today can be found in the non-Euclidean revolution in geometry nearly two hundred years ago (Bardi 2009). The basic question at the time was this: Given a line in a plane, how many parallel lines in the plane can be drawn through a point in the plane not on the line? This was the subject of Euclid's fifth axiom, which had prescribed the answer 'one'. For over twenty centuries in mathematics since Euclid, this was the only answer to the question. Although the answer 'one' was itself never in doubt, the fifth axiom had long been controversial because it did not seem to be self-evident in the same way the other axioms were.

Many geometers attempted to prove the fifth axiom as a consequence of the other Euclidean axioms, and, in the early 1800s, tried a proof by *reductio ad absurdum* in which they denied the fifth axiom and looked for logical inconsistencies. But they found no logical inconsistencies from denying the fifth axiom, and, almost accidentally, they developed non-Euclidean geometries in which the answer to the question about the number of parallels could also be either 'zero' or 'many' rather than 'one'.

The advent of non-Euclidean geometry created a tumult in mathematics. Because it rests on what seemed to many to be an obvious fallacy, it was controversial and maddening to some mathematicians. Some proponents of the new geometry were ignored or faced condescension, and others, including even the eminent Gauss, hid what they had realized for fear of ridicule (Bardi 2009). Nevertheless, the new non-Euclidean geometries were in time accepted as legitimate and eventually heralded as greatly enriching the discipline of geometry by vastly expanding its scope and broadening its applications. Famously, Einstein used non-Euclidean geometry in his general theory of relativity.

We believe that probability theory may be in the early throes of a revolution similar in some ways to the non-Euclidean revolution in geometry. The debate today between the traditional precisionists and the imprecise probability community hinges on the adoption or rejection of a single axiom. Omitting that axiom relaxes the earlier, unnecessarily strict theory and opens it up to a richer perspective on a wider array of applications. The new theory, although an obvious generalization of the older theory, has met some strong and sometimes dismissive criticism. The advantages of the newer theory arise because of its greater flexibility, and its usefulness will likely be cemented by applications to significant problems beyond the reach of the older theory, even though the older theory persists as an important and still extremely widely used special case.

6   Conclusions

Given the increasing and critical use of Monte Carlo simulation and other probabilistic calculations, it is important to be able to assess the reliability of their numerical outputs as a function of the error in finite computer representation of probabilities and distributions, as well as the express uncertainty about the input distributions, their interdependencies, and model assumptions. Until recently, the only way to assess this reliability has been to develop an elaborate sensitivity study, which, because of its combinatorial complexity, is rarely even attempted. Consequently, in practice residual uncertainties about the selection of input probability distributions and the nature of interdependencies among the variables in an analysis are usually neglected, and the sufficiency of the numerical methods used in the calculations is not even addressed.

Practical computational methods in probability bounds analysis now exist to assess the potential impact on probabilistic calculations from (i) computer representation error of distributions, (ii) incomplete knowledge about the parameters and shapes of input distributions, (iii) imperfect understanding of the correlation and dependency structure among variables, and (iv) several kinds of model-form uncertainty. Probability bounds analysis is essentially an efficient and comprehensive form of sensitivity analysis that is compatible with both frequentist and Bayesian perspectives. Its methods allow many analysts to make routine and relatively inexpensive bounding estimates for calculations involving probabilities or probability distributions. In many cases, the bounds will be exact in the sense that they are the pointwise best-possible bounds. In all cases, the bounds will enclose the probability distributions and therefore provide a conservative expression of the reliability of the results.

The Panglossian view that every probabilistic calculation can be solved with precisely specified inputs determined from available information does not seem plausible in many cases, especially in new environments or novel situations with highly constrained scientific knowledge and limited engineering experience. When there is insufficient empirical information available to justify precise probabilistic calculations, the methods of probability bounds analysis naturally yield bounds on output distributions. Different input variables can be specified either as particular distributions or as wide or narrow bounds on distributions as appropriate to represent the state of empirical information, whatever it may be, that is available about each variable. Likewise, dependence functions can be precisely modeled with a specific copula, qualitatively modeled with available information, or discharged entirely by making no assumptions about the dependence. The degree of specificity about the marginal distributions and dependencies does not affect how they are combined in subsequent arithmetic operations, and the result appropriately represents its own level of uncertainty, rather than a false precision gained by unjustified assumptions.

Sound scientific and engineering analyses ought to be based on objective, documented, and verifiable information. Whenever assumptions are used to make up the difference between what is empirically known and what is needed to obtain a working answer, those assumptions should be subjected to quantitative assessment with methods such as probability bounds analysis. Without such assessment, the calculations are unsound, and the resulting answers are merely wishful thinking.

Acknowledgements. We gratefully acknowledge helpful discussions with Lev Ginzburg, Andrew Dienstfry, Bill Oberkampf, Tony O'Hagan, Michael Goldstein, Jason O'Rawe, Helen Regan, and Michael Balch. This paper was presented at the IFIP Working Conference on Uncertainty Quantification and Scientific Computing Boulder, Colorado, 2 August 2011. Great thanks are due to the organizers, Ronald Boisvert and Andrew Dienstfry of the National Institute of Standards and Technology. Support was provided by the National Library of Medicine, a component of the National Institutes of Health (NIH), under Award Number RC3LM010794, and the National Institute of Standards and Technology. The views and opinions expressed herein are solely those of the authors and not those of the National Library of Medicine, National Institutes of Health, National Institute of Standards and Technology, or other sponsors.

References

1. Baase, S.: A Gift of Fire. Pearson Prentice Hall (2008)
2. Bardi, J.S.: The Fifth Postulate: How Unraveling a Two-thousand-year-old Mystery Unraveled the Universe. John Wiley & Sons, Hoboken (2009)
3. Beer, M.: Fuzzy probability theory. In: Meyers, R. (ed.) Encyclopedia of Complexity and System Science, vol. 6, pp. 4047–4059. Springer, New York (2009)
4. Ben-Haim, Y.: Information Gap Decision Theory: Decisions under Severe Uncertainty. Academic Press, San Diego (2001)
5. Berger, J.O.: Statistical Decision Theory and Bayesian Analysis. Springer, New York (1985)
6. Berleant, D.: Automatically verified reasoning with both intervals and probability density functions. Interval Computations (2), 48–70 (1993)
7. Berleant, D.: Automatically verified arithmetic on probability distributions and intervals. In: Kearfott, B., Kreinovich, V. (eds.) Applications of Interval Computations, pp. 227–244. Kluwer Academic Publishers, Dordrecht (1996)
8. Berleant, D., Goodman-Strauss, C.: Bounding the results of arithmetic operations on random variables of unknown dependency using intervals. Reliable Computing 4, 147–165 (1998)
9. Berleant, D., Zhang, J.: Representation and problem solving with distribution envelope determination (DEnv). Reliability Engineering and System Safety 85, 153–168 (2004)
10. Boole, G.: An Investigation of the Laws of Thought, On Which Are Founded the Mathematical Theories of Logic and Probability. Walton and Maberly, London (1854)
11. Boyle, P.P.: Options: a Monte Carlo approach. Journal of Financial Economics 4, 323–338 (1977)
12. Bukowski, J., Korn, L., Wartenberg, D.: Correlated inputs in quantitative risk assessment: the effects of distributional shape. Risk Analysis 15, 215–219 (1995)
13. Burgman, M., Ferson, S., Akçakaya, H.R.: Risk Assessment in Conservation Biology. Chapman & Hall, London (1993)
14. Burmaster, D.E., Harris, R.H.: The magnitude of compounding conservatisms in superfund risk assessments. Risk Analysis 13, 131–134 (1993)

15. CFTC/SEC (U.S. Commodity Futures Trading Commission/U.S. Securities and Exchange Commission): Findings Regarding the Market Events of May 6, 2010: Report of the Staffs of the CFTC and SEC to the Joint Advisory Committee on Emerging Regulatory Issues (2010),
    http://www.sec.gov/news/studies/2010/marketevents-report.pdf
16. Chebyshev [Tchebichef], P.: Sur les valeurs limites des intégrales. Journal de Mathématiques Pures et Appliquées. Ser. 2 19, 157–160 (1874)
17. Colyvan, M.: The philosophical significance of Cox's theorem. International Journal of Approximate Reasoning 37, 71–85 (2004)
18. Cooper, J.A.: Fuzzy algebra uncertainty analysis for abnormal environment safety assessment. Journal of Intelligent and Fuzzy Systems 2, 337–345 (1994)
19. Cox, R.T.: Probability, frequency and reasonable expectation. American Journal of Physics 14, 1–13 (1946)
20. de Finetti, B.: Theory of Probability. Wiley, New York (1974)
21. Elliott, G.: US nuclear weapon safety and control. MIT Program in Science, Technology, and Society (2005),
    http://web.mit.edu/gelliott/Public/sts.072/paper.pdf
22. Enszer, J.A., Lin, Y., Ferson, S., Corliss, G.F., Stadtherr, M.A.: Probability bounds analysis for nonlinear dynamic process models. AIChE Journal 57, 404–422 (2011)
23. EPA (U.S. Environmental Protections Agency (Region I)): Human Health Risk Assessment (2005a), http://www.epa.gov/region1/ge/thesite/restofriver-reports.html#HHRA
24. EPA (U.S. Environmental Protections Agency (Region I)): Ecological Risk Assessment (2005b), http://www.epa.gov/region1/ge/thesite/restofriver-reports.html#ERA
25. EPA (U.S. Environmental Protections Agency (Region 6 Superfund Program)): Calcasieu Estuary Remedial Investigation (2006),
    http://www.epa.gov/region6/6sf/louisiana/calcasieu/la_calcasieu_calcri.html
26. ESA (European Space Agency): Ariane 5 Flight 501 Failure. Report by the Inquiry Board, Paris (1996), http://esamultimedia.esa.int/docs/esa-x-1819eng.pdf
27. Ferson, S.: RAMAS Risk Calc 4.0 Software: Risk Assessment with Uncertain Numbers. Lewis Publishers, Boca Raton (2002)
28. Ferson, S., Burgman, M. (eds.): Quantitative Methods for Conservation Biology. Springer, New York (2000)
29. Ferson, S., Tucker, W.T.: Probability boxes as info-gap models. In: Proceedings of the North American Fuzzy Information Processing Society. IEEE, New York City (2008)
30. Ferson, S., Kreinovich, V., Ginzburg, L., Sentz, K., Myers, D.S.: Constructing probability boxes and Dempster–Shafer structures. Sandia National Laboratories, SAND2002-4015, Albuquerque, New Mexico (2003),
    http://www.ramas.com/unabridged.zip
31. Ferson, S., Nelsen, R., Hajagos, J., Berleant, D., Zhang, J., Tucker, W.T., Ginzburg, L., Oberkampf, W.L.: Dependence in Probabilistic Modeling, Dempster–Shafer Theory, and Probability Bounds Analysis. Sandia National Laboratories, SAND2004-3072, Albuquerque, New Mexico (2004), www.ramas.com/depend.pdf
32. Ferson, S., Mickley, J., McGill, W.: Uncertainty arithmetic on Excel spreadsheets: add-in for intervals, probability distributions and probability boxes. In: Vulnerability, Uncertainty, and Risk: Analysis, Modeling, and Management, Proceedings of the ICVRAM 2011 and ISUMA 2011 Conferences. ASCE, Reston (2011)

33. Fishburn, P.C.: The axioms of subjective probability. Statistical Science 1, 335–358 (1986)
34. Frank, M.J., Nelsen, R.B., Schweizer, B.: Best-possible bounds for the distribution of a sum—a problem of Kolmogorov. Probability Theory and Related Fields 74, 199–211 (1987)
35. Fréchet, M.: Généralisations du théorème des probabilités totales. Fundamenta Mathematica 25, 379–387 (1935)
36. Fréchet, M.: Sur les tableaux de corrélation dont les marges sont données. Annales de l'Université de Lyon. Section A: Sciences Mathématiques et Astronomie 9, 53–77 (1951)
37. GAO (General Accounting Office): Patriot Missile Defense: Software Problem Led to System Failure at Dhahran, Saudi Arabia. GAO/IMTEC-92-26 (1992), http://www.fas.org/spp/starwars/gao/im92026.htm
38. Grove, A.J., Halpern, J.Y.: Updating sets of probabilities. In: Proceedings of the Fourteenth Conference on Uncertainty in AI, pp. 173–182 (1998), http://arxiv.org/abs/0906.4332
39. Hailperin, T.: Boole's Logic and Probability. North-Holland, Amsterdam (1986)
40. Halpern, J.Y.: Cox's theorem revisited. Journal of AI Research 11, 429–435 (1999)
41. Hammer, R., Hocks, M., Kulisch, U., Ratz, D.: C++ Toolbox for Verified Computing I: Basic Numerical Problems. Springer, Heidelberg (1997)
42. Hertz, D.B.: Risk analysis in capital investment. Harvard Business Review 42, 95–106 (1964)
43. Hickman, J.W., et al.: PRA Procedures Guide: A Guide to the Performance of Probabilistic Risk Assessments for Nuclear Power Plants, vol. 2, NUREG/CR-2300-V1 and -V2. National Technical Information Service, Washington (1983)
44. Isbell, D., Hardin, M., Underwood, J.: Mars Climate Orbiter team finds likely cause of loss (1999), http://mars.jpl.nasa.gov/msp98/news/mco990930.html
45. Jeffreys, H.: Scientific Inference. Cambridge University Press, Cambridge (1931)
46. Kaufmann, A., Gupta, M.M.: Introduction to Fuzzy Arithmetic: Theory and Applications. Van Nostrand Reinhold Company, New York (1991)
47. Keynes, J.M.: A Treatise on Probability. Macmillan, New York (1921), http://www.archive.org/details/treatiseonprobab007528mbp
48. Kolmogorov, A.N.: Grundbegriffe der Wahrscheinlichkeitsrechnung. Julius Springer, Berlin (1933); Translated into English as Kolmogorov, A.N.: Foundations of the Theory of Probability. Chelsea Publishing, New York (1950)
49. Kriegler, E., Held, H.: Utilizing belief functions for the estimation of future climate change. International Journal of Approximate Reasoning 39, 185–209 (2005)
50. Kulisch, U., Hammer, R., Ratz, D., Hocks, M.: Numerical Toolbox for Verified Computing I: Basic Numerical Problems: Theory, Algorithms, and Pascal-XSC Programs. Computational Mathematics, vol. 21. Springer, Heidelberg (1993)
51. Kyburg, Jr., H.E.: Interval valued probabilities. SIPTA Documention on Imprecise Probability (1999), http://www.sipta.org/documentation/interval_prob/kyburg.pdf
52. Marquis de Laplace, P.S.: Théorie analytique de probabilités, (edition troisième). Courcier, Paris (1820); The introduction (Essai philosophique sur les probabilités) is available in an English translation in A Philosophical Essay on Probabilities. Dover Publications, New York (1951)
53. Lavine, M.: Sensitivity in Bayesian statistics: the prior and the likelihood. Journal of the American Statistical Association 86, 396–399 (1991)
54. Lindley, D.V.: Scoring rules and the inevitability of probability. International Statistical Review 50, 1–26 (1982)

55. Lindley, D.V.: Understanding Uncertainty. John Wiley & Sons, Hoboken (2006)
56. Luce, R.D.: Where does subjective expected utility fail descriptively? Journal of Risk and Uncertainty 5, 5–27 (1992)
57. Makarov, G.: Estimates for the distribution function of a sum of two random variables when the marginal distributions are fixed. Theory of Probability and its Applications 26, 803–806 (1981)
58. Markov [Markoff], A.: Sur une question de maximum et de minimum proposée par M. Tchebycheff. Acta Mathematica 9, 57–70 (1886)
59. McKone, T.E., Ryan, P.B.: Human exposures to chemicals through food chains: an uncertainty analysis. Environmental Science and Technology 23, 1154–1163 (1989)
60. Möller, B., Beer, M.: Fuzzy Randomness—Uncertainty in Civil Engineering and Computational Mechanics. Springer, Berlin (2004)
61. Moore, R.E.: Interval analysis. Prentice Hall, Englewood Cliffs (1966)
62. Neapolitan, R.E.: A survey of uncertain and approximate inference. In: Zadeh, L., Kacprzyk, J. (eds.) Fuzzy Logic for the Management of Uncertainty, pp. 55–82. John Wiley & Sons, New York (1992)
63. Nelsen, R.B.: An Introduction to Copulas. Lecture Notes in Statistics, vol. 139. Springer, New York (1999)
64. Nong, A., Krishnan, K.: Estimation of interindividual pharmacokinetic variability factor for inhaled volatile organic chemicals using a probability-bounds approach. Regulatory Toxicology and Pharmacology 48, 93–101 (2007)
65. Oberguggenberger, M., King, J., Schmelzer, B.: Imprecise probability methods for sensitivity analysis in engineering. In: Proceedings of the 5th International Symposium on Imprecise Probability: Theories and Applications, Prague, Czech Republic (2007), http://www.sipta.org/isipta07/proceedings/papers/s032.pdf
66. Popova, E.D.: *Mathematica* Connectivity to Interval Libraries filib++ and C-XSC. In: Cuyt, A., Krämer, W., Luther, W., Markstein, P. (eds.) Numerical Validation in Current Hardware Architectures. LNCS, vol. 5492, pp. 117–132. Springer, Heidelberg (2009)
67. R Development Core Team: R: A Language and Environment for Statistical Computing. R Foundation for Statistical Computing, Vienna, Austria (2010), http://www.R-project.org
68. Seidenfeld, T.: Decision theory without "independence" or without "ordering": what is the difference? Economics and Philosophy 4, 267–290 (1988)
69. Seidenfeld, T., Wasserman, L.: Dilation for sets of probabilities. The Annals of Statistics 21, 1139–1154 (1993)
70. Selby, R.G., Vecchio, F.J., Collins, M.P.: The failure of an offshore platform. Concrete International 19, 28–35 (1997)
71. Slabodkin, G.: Smart ship inquiry a go. Government Computer News (August 31, 1998), http://gcn.com/articles/1998/08/31/smart-ship-inquiry-a-go.aspx
72. Smith, A.E., Ryan, P.B., Evans, J.S.: The effect of neglecting correlations when propagating uncertainty and estimating the population of risk. Risk Analysis 12, 467–474 (1992)
73. Suter II, G.W.: Ecological Risk Assessment. Lewis Publishers, Boca Raton (1993)
74. Tucker, W.: Validated Numerics: A Short Introduction to Rigorous Computations. Princeton University Press, Princeton (2011)
75. Tuyl, F., Gerlachy, R., Mengersen, K.: Posterior predictive arguments in favor of the Bayes-Laplace prior as the consensus prior for binomial and multinomial parameters. Bayesian Analysis 4, 151–158 (2009)
76. Vick, S.G.: Degrees of Belief: Subjective Probability and Engineering Judgment. American Society of Civil Engineers, Reston (2002)

77. Walley, P.: Statistical Reasoning with Imprecise Probabilities. Chapman and Hall, London (1991)
78. Walley, P.: Inferences from multinomial data: learning about a bag of marbles. Journal of the Royal Statistical Society, Series B 58, 3–57 (1996)
79. Walley, P., Gurrin, L., Barton, P.: Analysis of clinical data using imprecise prior probabilities. The Statistician 45, 457–485 (1996)
80. Williamson, R.C., Downs, T.: Probabilistic arithmetic I: numerical methods for calculating convolutions and dependency bounds. International Journal of Approximate Reasoning 4, 89–158 (1990)
81. Yager, R.R.: Arithmetic and other operations on Dempster–Shafer structures. International Journal of Man-Machine Studies 25, 357–366 (1986)
82. Zhang, H., Mullen, R.L., Muhanna, R.L.: Finite element structural analysis using imprecise probabilities based on p-box representation. In: Beer, M., Muhanna, R.L., Mullen, R.L. (eds.) Proceedings of the 4th International Workshop on Reliable Engineering Computing: Robust Design—Coping with Hazards, Risk and Uncertainty, Singapore, March 3-5. Research Publishing (2010), http://rpsonline.com.sg/rpsweb/9789810851187.html, http://www.eng.nus.edu.sg/civil/REC2010/documents/papers/013.pdf
83. Zhang, H., Mullen, R., Muhanna, R.: Safety structural analysis with probability-boxes. International Journal of Reliability and Safety 6, 110–129 (2012)

Discussion

Speaker: Scott Ferson

Jon Helton: What is the current status of computational capability to propagate p-boxes through complex models, e.g., long running and/or containing repeated variables?

Scott Ferson: I would say it is significantly better than the computational capability currently available for propagating Dempster–Shafer (DS) structures, and in some instances, better than that for Monte Carlo simulations. There are several reasons for this. Firstly, as a bounding approach it can effectively propagate some kinds of uncertainties that cannot be comprehensively addressed by sampling approaches even with infinitely many samples. For instance, if an analyst does not know the distribution family for some input, she can use a distribution-free p-box that bounds all possible distribution families consistent with the other information available about that variable. Zhang et al. (2012) noted that the computational burden for p-box propagation when no assumptions are made about intervariable dependencies is actually smaller than even that for simple Monte Carlo simulation, because it does not require a full convolution of all possible combinations of values for the various input variables.

There are also a variety of tricks and shortcuts available for p-box calculations, including and extending the conjugacy rules familiar to Bayesians. If a problem is computationally challenging when p-boxes are used to characterize the inputs, one or more of the p-boxes can be coarsened in a way that preserves conservativism yet radically lessens the computational burden. This coarsening is also possible with DS structures, but it will alter the internal features of the output uncertainty structure, the elucidation of which is the whole point of using DS structures in the first place.

The computational capability to propagate p-boxes can be judged by its practical applications. Probability bounds analysis has been used in cases with many dozens of inputs, although I have seen no practical applications yet with many hundreds of inputs. It has been used in uncertainty analyses of substantial complexity in a wide variety of contexts ranging from Superfund human health and ecological risk assessments (EPA 2005a; 2005b; 2006) to finite element models in engineering (Oberguggenberger et al. 2007; Zhang et al. 2010; 2012), and on scales from lab bench chemistry and pharmacokinetics (Enszer et al. 2011; Nong and Krishnan 2007) to the planet's climate in global circulation models (Kriegler and Held 2005).

Certainly there are challenges in computing with p-boxes, especially concerning the appearance of repeated uncertain quantities, and developing strategies to meet these challenges is a current area of research. There are no particular computational challenges associated with "long-running" calculations per se, and p-box operations can be applied iteratively in a straightforward way. However, in cases where sequential iterations involve repeated uncertain quantities such as in calculating solutions to differential equations, difficulties can arise. Enszer et al.

(2011) described special methods to solve nonlinear ordinary differential equations with parameters or initial conditions expressed as p-boxes. This is another area where what we can do computationally with p-boxes exceeds what can be done with Monte Carlo simulation which quickly runs into instability in such problems.

William Kahan: Your acceptance of diverse bounds upon probabilities (wherever those bounds may come from) reminds me of the "degrees of belonging" of fuzzy sets and Lotfi Zadeh's fuzzy logic; but you do not mention fuzzy sets at all. Why not?

Scott Ferson: The theory I'm talking about here is purely probabilistic and conforms with the Kolmogorov (1933) axioms. Of course the bounds can come from many places just because there are many sources of information and data and many disparate reasons for uncertainty, but the quantities we work with are bounds on probabilities which are interpreted in only one way, as Kolmogorov probabilities.

I have considered fuzzy sets and possibility theory in the past, and many colleagues still use these ideas extensively (Möller and Beer 2004; Beer 2009). I don't use them now myself mostly because they evoke such a very negative reaction among probabilists. I do not think the visceral reaction from probabilists is legitimate at all, but this is not my battle. My objection to fuzzy numbers and their arithmetic as proposed by Kaufmann and Gupta (1991) is that there is still no way to ensure that the result will be meaningful from level-wise combining fuzzy numbers that came from different formulations with distinct possibility scales. This is the same objection I have for possibility theory and, by the way, for info-gap decision theory (Ben-Haim 2001).

Pasky Pascual: How does one use p-boxes to formulate priors within a Bayesian framework?

Scott Ferson: The most common way is to use p-boxes to characterize an analyst's uncertainty about the appropriate prior to employ. Consider, for example, the problem of estimating a binomial probability which is perhaps the most elementary and fundamental problem in all of risk or uncertainty analysis. Amazingly, Bayesians cannot agree on the prior to use for this problem, even in the basic case when they all agree no relevant prior information is available (Tuyl et al. 2009; Berger 1985, page 89). The search for a so-called "uninformative prior" has produced several candidates, including Haldane's improper prior beta(0,0), Jeffreys' reference prior beta($\frac{1}{2},\frac{1}{2}$), the uniform distribution favored by Laplace which can be modeled as beta(1,1), and other distributions such as Zellner's binomial prior (which is not from the beta family). Unless the sample size is pretty large, which is rare in many practical situations, these different priors yield noticeably different results.

Peter Walley (1991; Walley et al. 1996) has suggested using an imprecise beta model (IBM) which is effectively a p-box of all beta distributions that could be good priors. In the degenerate case, when the sample size is zero, the IBM yields a vacuous posterior that says the probability could be anywhere in the interval [0,1]. Isn't that a reasonable result for an analysis that *uses no data at all?* When the sample size is very large, the posterior is a tight p-box that tends to the observed frequency, as most all Bayesian analyses do. In the practical intermediate cases of small sample sizes, the posterior from the IBM is a p-box containing a range of beta distributions whose breadth reflects the uncertainty about the prior that a traditional Bayesian analysis ignores. Contrary to what Tony O'Hagan suggests in his comment below, this breadth is not too wide to be useful, but yields answers whose imprecision is roughly what one might expect to see across a community of competent Bayesians.

The imprecise beta model generalizes in the multivariate case to an imprecise Dirichlet model (IDM, Walley 1996). The IBM and IDM are examples of Bayesian sensitivity analysis (Lavine 1991) or robust Bayes analysis (Berger 1985), the idea of which originated with Jeffreys (1931) and de Finetti (1974). Walley (1991) has demonstrated that robust Bayes analysis is part of a more general theory based on imprecise probabilities of very broad scope and flexibility, for which there is a firm theoretical foundation based on respecting consistency and coherence requirements but which avoids making unwarranted assumptions to obtain quantitative answers. Probability bounds analysis is a computationally convenient method within this general theory

Michael Goldstein: While I agree that notions of imprecision have a valuable role in considering uncertainty, I am not convinced that the approach advocated by the speaker can be viewed as a complete theory. In particular, it is quite possible for the analyst to face a situation where there is available data which, for every possible outcome, will increase uncertainty about some key quantity in such a way that the information has negative value to the analyst. Effectively, this turns the analyst into a money-pump, as the analyst appears to need to keep paying to avoid receiving the data. This is counter-intuitive to me, and makes me uncomfortable about the notion that the theory is sufficient to deal with all uncertainty models.

Scott Ferson: Michael is talking about a phenomenon described by Seidenfeld and Wasserman (1993) known as dilation. It occurs when new evidence leads different Bayesian investigators into greater disagreement than they had prior to their getting the new evidence. Such evidence is not merely surprising in the sense that it contradicts one's prior conceptions; it expands everyone's uncertainty. It is counterintuitive because it does not depend on what the new information is actually saying. Michael's criticism is perhaps the pot calling the kettle black, because dilation occurs among Bayesians too. They simply don't recognize it because Bayesians don't have to agree with each other (or with the world for that matter).

It's hard to explain dilation with a simple example, but let me try. Suppose Lucius Malfoy tosses a fair coin twice, but the second 'toss' depends on the outcome of the first toss. It could be that Malfoy just lets the coin ride, and the second outcome is exactly the same as the first outcome. Or he could just flip the coin over so that the second outcome is the opposite of the first. You don't know what he will do. The outcome of the first toss is either heads H1 or tails T1. Because the first toss is fair (and no spells are cast midair), you judge the probability $P(H1) = 0.5$. Whether Malfoy lets the coin ride or flips it, you judge the probability the second 'toss' ends up heads to be the same,

P(H2) = 0.5. What happens when you see the outcome of Malfoy's first toss? Suppose it was a head. What is your probability now that the second 'toss' will also be a head? It turns out that once you condition on the first observation, the probability of the second toss being a head dilates. It is now either zero or one, but you don't know which. It doesn't depend on chance now; it depends on Malfoy's choice, about which you have no knowledge (unless perhaps you too dabble in the dark arts). Dilation occurs because the observation H1 has caused the earlier precise unconditional probability P(H2) = 0.5 to devolve into the vacuous interval P(H2 | H1) = [0,1].

This issue may be a pretty esoteric theoretical concern. I have yet to see examples of dilation in practice that would create any problems for analysts or decision makers. Although dilation seems highly counterintuitive to some people, others consider it a natural consequence of the interactions of partial knowledge (Walley 1991, 298f). My attitude is far from Michael's worry that the theory of imprecise probabilities might somehow be incomplete because it recognizes this phenomenon. Instead, I think it is rather evidence of its being a *richer* theory.

One way to avoid dilation is not to use conditionalization as the updating rule for new information. Interestingly, it is possible to do this with imprecise probabilities. Grove and Halpern (1998) point out that the standard justifications for conditionalization may no longer apply when we consider sets of probabilities. And it may turn out that conditionalization may not be the most natural way to update sets of probabilities in the first place. Instead, a constraint-based updating rule may sometimes be more sensible. We note that dilation does not occur in interval analysis (Seidenfeld and Wasserman 1993), which is a kind of constraint analysis.

Anthony O'Hagan: First I would like to thank Scott for an entertaining presentation. Unfortunately, much of what he says is in my opinion wrong–entertainingly wrong, but nonetheless *wrong*. In these comments I would just like to pick out what seem to me to be the two most important errors.

First he repeatedly confuses epistemic uncertainty with what he calls incertitude. Epistemic uncertainty relates to a quantity that has a unique (albeit unknown) value but which is not random in the usual sense of that word. In particular, we cannot observe a series of repetitions or 'trials' and so its uncertainty cannot be described by the conventional relative frequency form of probability. Bayesian statistics quantifies epistemic uncertainty with subjective probabilities. Frequentist statistics cannot do this because it only acknowledges relative frequency probability, so its "quantifications" of epistemic uncertainty are oblique, using such convoluted devices as confidence intervals.

What Scott calls incertitude is not completely clear, but I think I can define it as those things that he would represent using intervals of probabilities or p-boxes. His primary idea of a p-box is to express incertitude about a probability distribution. In his examples, those distributions are often conventional frequency probability distributions (for quantities with aleatory uncertainty or randomness), but he also discuss putting p-boxes on Bayesian prior and posterior distributions, which relate to epistemic uncertainty.

To my mind, his incertitude is an attempt to do something that is perfectly sensible, namely to quantify in some way the fact that when we specify probability distributions (whether they be aleatory sampling distributions or epistemic prior distributions, for instance) we can never do so precisely. All judgments are imprecise, and probability distributions are nearly always specified partly by convenience. So what he calls incertitude seems to me to be addressing imprecision in judgments. And that's an important issue.

My second point, however, is that I have reservations about whether p-boxes and interval arithmetic are the way to handle incertitude. His approach is a kind of half-way house between formal treatment of imprecision with a second-order (or hierarchical) probability quantification on the one hand, and on the other a purely informal 'sensitivity analysis' in which we simply explore a few possible alternative distributions. As such, it is certainly more comprehensive than sensitivity analysis and may indeed have a role to play. However, it requires one to specify bounds, and these bounds are almost always arbitrary. If set very wide so as to be quite sure of encompassing whatever the 'true' distributions might be, then the resulting derived bounds on quantities of interest and decisions are likely to be hopelessly wide. If set narrower so as to encompass just the more likely 'true' distributions, then he can no longer claim that the p-boxes are exhaustive.

Despite these reservations, I welcome the basic idea of p-boxes and interval arithmetic. I just wish that Scott would not oversell it. These are *not* tools for quantifying epistemic uncertainty (although they may have a role in addressing the imprecision in subjective distributions that *do* quantify epistemic uncertainty). I might also add that they have nothing to do with utility theory or prospect theory.

Scott Ferson: Tony seems to want to dismiss the presentation as wrong, wrong, *wrong*, yet he agrees that my concern with incertitude is "perfectly sensible" and that it is an "important issue". Moreover he concedes that p-boxes "may indeed have a role to play" and "welcome[s] the idea of p-boxes". So let us try to clarify the sources and details of our differences.

The first of two disagreements that Tony highlights is my use of the phrase 'epistemic uncertainty'. I'm not at all sure why Tony insists that a quantity about which we are epistemically uncertain must be a unique, fixed quantity. There is nothing in the definition of the phrase that *requires* the quantity to be fixed underneath all our uncertainty about it. A quantity might be a fixed value, but it also might not be, and indeed our epistemic uncertainty about a quantity might often include whether it is in fact fixed or varying. Whether we know it is fixed or not, we could still have epistemic uncertainty about it.

It is true that Bayesians use probability distributions to model epistemic uncertainty, but it is simply not true that these objects represent the only possible way to model epistemic uncertainty. Clearly, intervals and p-boxes are general and flexible tools to quantify and propagate epistemic uncertainty, the latter specifically designed to do so, even when the nature of the underlying quantity is itself unknown. PBA can also handle Tony's case where the unknown quantity does have some unique fixed (but unknown) value. A p-box conveniently represents this case as extra information that the quantity's variance is zero. This is a special case compared to the general situation in which the possible range of the variance includes zero. Knowing a p-box's variance is exactly zero may have implications for the left and right bounds of the p-box, and it will usually have implications for mathematical results that depend on the p-box. You might also know the quantity must vary, in which case the p-box's variance might exclude zero as a possible value, even though you may not know its distribution precisely. Knowing a minimum range for the quantity or knowing a lower bound on its variance can improve the p-box and calculations that depend on it.

I do not know the origin or purpose of Tony's restrictive definition of 'epistemic uncertainty'. Our view is more expansive, and perhaps more useful. In addition to our not requiring the underlying value to be fixed, our usage of the phrase need not refer to a *quantity* at all, but includes uncertainty about the mathematical form of the model, which can also be captured in probability bounds analysis.

The second of Tony's two complaints is about whether intervals and p-boxes are a practical way to handle incertitude (epistemic uncertainty). He asserts that the bounds of p-boxes are "almost always arbitrary" and suggests that setting them very wide to be sure to encompass the true distribution will make the results vacuous, and narrowing them will lose the claim that p-boxes are comprehensive. In fact, however, there are many ways to construct p-boxes, and many of these ways constitute constraint analyses that are best-possible and in no way arbitrary. They don't even depend on parameters that might be varied arbitrarily. The subsequent calculations are also essentially constraint analyses that include no arbitrariness.

There are other ways to construct p-boxes that do involve decisions by the analyst that might have to be made arbitrarily. For instance, picking the confidence level in a p-box defined as a confidence band. Most analysts consider these decisions to be part of the modeling task and therefore to be the responsibility of the analyst, as they are in many exercises involving modeling or analysis. There are various strategies to avoid arbitrariness including appealing to conventions such as Fisher's 0.05 level, or considering would-be arbitrary parameters to be part of the analysis by nesting p-boxes at different levels (Ferson and Tucker 2008) or enveloping all tenable levels. Whether the uncertainty overwhelms the analyst's ability to make decisions depends on the details of the application and the available empirical information. In practical cases, analysts have generally found that useful inferences and decisions can be obtained.

One further fundamental point should be emphasized in reaction to Tony's complaint. Uncertainty analysis shouldn't be a game. Analysts are invited to be honest in expressing what they know and what they don't know. If it turns out that so little is known about a system that the p-boxes characterizing it are wide to the point that the results are vacuous, then it seems to me that a proper uncertainty analysis should reveal this fact. It is, after all, the very point of an uncertainty analysis. The alternative—which is to squeeze *unwarranted* conclusions or decisions out of a tenuous model that is not actually supported by evidence—is of course possible, but does not seem desirable, especially in an engineering context.

Defects, Scientific Computation
and the Scientific Method

Les Hatton

Computing and Information Systems, Kingston University
London, United Kingdom
lesh@oakcomp.co.uk

Abstract. Computation has rapidly grown in the last 50 years so that
in many scientific areas it is the dominant partner in the practice of
science. Unfortunately, unlike the experimental sciences, it does not ad-
here well to the principles of the scientific method as espoused by for
example, the philosopher Karl Popper. Such principles are built around
the notions of deniability and reproducibility. Although much research
effort has been spent on measuring the density of software defects, much
less has been spent on the more difficult problem of measuring their effect on the output of a program. This paper explores these issues with
numerous examples suggesting how this situation might be improved to
match the demands of modern science. Finally it develops a theoretical
model based on Shannon information which suggests that software systems have strong implementation independent behaviour and presents
supporting evidence.

Keywords: Scientific method, reproducibility, unquantifiable computation.

1 Introduction

The thesis of this paper is that many scientific computations are tainted by the
presence of unquantifiable software defects. To understand how this has come to
pass, it is important to realise two things:-

- Computer science is historically not a particularly critical discipline. In experimental terms, it appears to be considerably less mature than the natural
sciences as for example was demonstrated by [35], [36] when assessing the
degree to which experiment played a part in typical computer science publications.
- The majority of the empirical research carried out into software defects has
concerned itself with quantifying the density of such defects rather than the
much more difficult problem of quantifying the *effects* those defects have on
the output of scientific computations. For a thorough review, see [3]. The
end product of this research suggests that typical residual defect densities in
released software seem to be between 1 and 10 per thousand lines of code.

A. Dienstfrey and R.F. Boisvert (Eds.): WoCoUQ 2011, IFIP AICT 377, pp. 123–138, 2012.

Some very good systems may be as good as 0.1 per thousand lines of code, [18], although it is not always clear if like is being compared with like, (for example, there are numerous ways of measuring lines of code - source with or without comment, or executable lines - and it is rarely clear which one is in use).

1.1   A Small Diversion on Lines of Code

I mentioned above that the use of the phrase "line of code" is problematic. It occurs in a number of guises. The simplest way of counting them is to use the number of newlines giving a value known as *SLOC* (Source Line of Code). This is normally shown in text editors and can be counted very simply indeed.

The presence of comments and language pre-processors complicates this leading to alternative measures such as *PPSLOC*, (Pre-Processed Source Line of Code) and *XLOC*, (Executable Lines of Code), neither of which are readily available when code is compiled and require either special tools or hand-coded tools to measure. As a result, most lines of code measured are SLOC. It is possible to understand the relationships between them by correlating them for a given population of code. As a simple example, Fig. 1 illustrates SLOC v. XLOC and also bytes for a typical C application. Repeating on larger populations reveals similar relationships allowing us to move between SLOC, PPSLOC, XLOC and bytes with relative ease normalising defect densities as appropriate.

However, as I will show later in a token-based development using Hartley-Shannon Information Theory as eloquently described in [4], lines of code is too crude a measure.

1.2   Software Testing and Deniability

Finally, it is also worth stating the central tenets of Popperian deniability here cast into a software context.

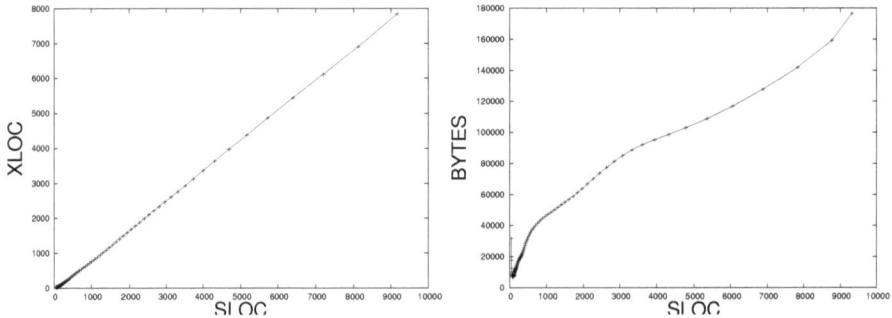

Fig. 1. The left hand diagram is a plot of SLOC count against XLOC count for a typical C application of around 140,000 SLOC in total. The right hand diagram shows the SLOC count against the object code size in bytes generated by compilation with the GNU C compiler. For this application, 1 XLOC = 0.8 SLOC very accurately and 1 SLOC = 25 +/- 3 bytes.

- Truth cannot be verified by software testing, it can only be falsified.
- Falsification requires quantification of computational modelling error.
- Deniability is at the heart of progress in scientific modelling. We are always seeking to deny the truth of a result and a continued failure to deny such truth simply adds weight to a result but not verification.

It will become clear that scientific source code plays a key part in this process.

2 Quantification of Defect

I have distinguished above between the relative success of quantifying defect density, and the much more difficult problem of quantifying their effects. I will now expand on this.

2.1 Defect Density and Static Program Properties

Even though calculating defect density has been more successful, teasing out any relationships with statically measurable software properties such as the numerous software metrics which have been described in the literature, [8], [33], [24] has been rather less successful.

Complexity. For a long time, a considerable amount of hope has been pinned on using statically measured structural properties of a program to predict the occurrence of defect after release, with probably the earliest and most well known being cyclomatic complexity, [22]. Whilst it has value because of its relationship with the number of test cases, [8], there remain difficulties and its originally suggested relationship as a predictor of defect seems illusory at best as can be seen in a study carried out by [13] on the NAG Fortran library. Figure 2 illustrates the lack of predictive power for these two metrics.

Programming Language. Programming language definitions historically reflect the continuing tension between performance and verifiability. Simultaneously, they embody elements of fashion in the form of a need to present the latest features and paradigms to the end-user, even when those features are perhaps not well understood in terms of their capability for injecting defects. A perfect example is the inclusion of object-orientated features into virtually all programming languages in the last twenty years.

The effect of these, coupled with long-term difficulties in removing features of dubious benefit from internationally-standardised languages because of the need to preserve backwards compatibility, has resulted in programming languages which have grown dramatically in size. Furthermore, they are often punctuated with significant numbers of features which have no defined behaviour and for which there is no requirement for compiler writers to diagnose. Examples include the 191 undefined features of ISO/IEC 9899:1999 (C99), (one of the few languages which actually bothers to list them as an appendix). In addition to these, languages contain features which often lead to erroneous behaviour as exemplified in C by [20] for example.

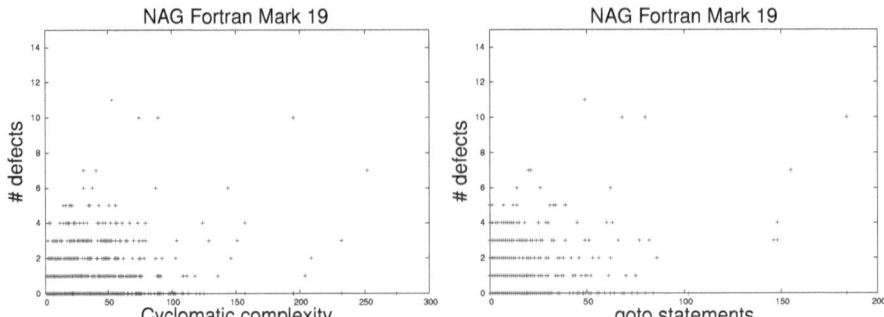

Fig. 2. The left hand diagram is a plot of historical defect against cyclomatic complexity for approximately 20 years history of the NAG Fortran library leading up to Mark 19 shown here. The right hand plot shows the same defects plotted against occurrences of the eponymous goto statement. Neither plot presents any significant statistical correlation of any dependability.

Although to my knowledge, there has been no published modern attempt to quantify the occurrence of these in released code, [9] demonstrated occurrence rates of around 8 per KSLOC in a study of several MSLOC several years ago, with a number of these packages still in use, whilst [29] demonstrated that these failed with some frequency by measuring an air-traffic control system over several years.

On top of these static fault modes, there are enduring problems with implementations of floating point arithmetic, [17], [16]. These are of fundamental importance to scientists as floating point arithmetic is at the very heart of scientific computation due to the enormous scale over which physical phenomena manifest themselves.

2.2   Quantification of the Effect of Defect

Whilst we have been fairly successful at understanding the density of such fault modes, (if not preventing them), little progress has been made in quantifying their effect on the computational results themselves, because the problem appears difficult. Several factors contribute to this.

Delayed Defect Discovery. A surprisingly large number of defects take an extraordinarily long time to appear for the first time. In a definitive study, Adams [1] demonstrated in an analysis of faults and failures in a number of IBM products, that around a third of all faults *took longer than 5,000 executable years to fail for the first time.* This immediately compromises the possible effectiveness of dynamic testing. Based on the kinds of product he analysed, Adams states:

"It may well be that as software engineering techniques improve, the population of DEs (Design Errors) will balance at a lower level; but absent development methods that generate truly error-free code,

the same sort of error rate distribution may well persist in future
large products"

This was written almost thirty years ago and we are certainly still "absent
methods that generate truly error-free code".

Unknown Answers. In many if not most areas of scientific computation, we
don't know what the answer is except perhaps in the broadest terms. This
is particularly a problem in remote sensing where corroborating physical
experiments on the target phenomena cannot actually be carried out at
all because they are simply inaccessible, either temporally (for example in
back-casting numerical climate models) or spatially, (seismological data).
This latter will be the topic of an experiment I will describe shortly. In such
cases, rough order of magnitude estimates may be all that is available and as
will be seen, this is insufficient to diagnose significant long-present defects.

Access to Source Code. It is only relatively recently, since the real advent of
open source, that source code has been widely available in any area. However,
in spite of the fact that there is very significant evidence of its pivotal part in
defect discovery, it is still not a requirement to parcel up the source code with
the algorithmic research, the data and the means to reproduce the results,
the very essence of the scientific method. Some research groups, for example,
[5] have led the way but progress is slow and even prestigious journals such
as Nature remain ambivalent, [6] stating:-

"Nature does not require authors to make code available, but we
do expect a description detailed enough to allow others to write their
own code to do similar analysis"

Software Testing. Software testing remains the Cinderella profession in Computing. It is not usually a significant part of the CS curriculum in universities,
[15] and it is unclear whether this deficiency is ever addressed successfully
in organisations.

N-Version. One methodology which at least casts some light on the magnitude
of errors in computation is known as N-version or back-to-back testing. In this
approach, the same program specifications are given to N different groups who
develop one version each independently, sometimes in different programming languages. These N versions are then given the same input data and any differences
in the outputs must be explained. There are two significant disadvantages.

- Cost. Since they must be independently developed, there are no economies
  of scale so the cost of development is effectively N times the cost of a single
  version.
- Independence. Important experiments such as those of [19] and [23] have
  demonstrated that there are dependent failures even in packages developed
  completely independently.

In spite of these deficiencies, N-version experiments have demonstrated their
value in flushing out very long-lived defects which had evaded any other technique. In [12], nine different seismic data processing packages which had evolved

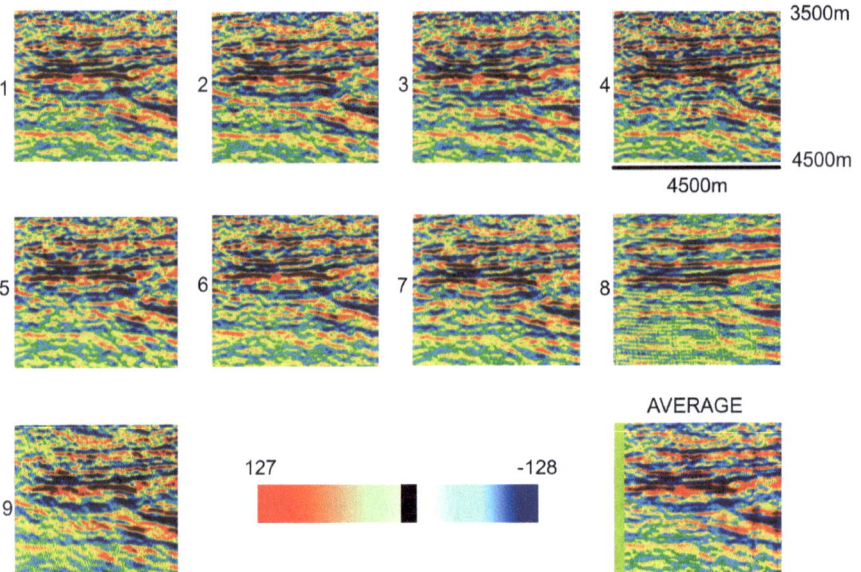

Fig. 3. A comparison of nine independently developed packages in the same programming language on the same input seismological data shown by [12]. The y-axis is depth of burial in the earth and the x-axis is distance along the surface of the earth. The colourscale shows relative echo intensities derived from acoustic sounding experiments after significant processing of raw data. The outputs vary in the second and sometimes first significant digit whereas three digits are desired to position a well reliably.

independently in a commercial environment to very well-specified standard algorithms were tested by giving them an identical set of 32 bit floating point input data. After an identical processing sequence, the individual results differed in the 2nd and sometimes 1st significant figure. The results can be seen in Fig. 3. In the figure, the y-axis is depth of burial in the earth and the x-axis is distance along the surface of the earth. The colourscale shows relative echo intensities derived from acoustic sounding experiments after significant processing of raw data. The outputs vary in the second and sometimes first significant digit whereas three significant digits of accuracy are deemed necessary to resolve the geological features (in this case an unconformity trap for a gas field in the North Sea) sufficiently accurately for reliable positioning of a well.

Amongst other things, the paper concluded

- The differences were due to previously undiscovered software faults, in some cases remaining hidden for many execution years.
- The initial 6 significant figures of agreement had shrunk to 1-2 by the time the data was passed to the scientist end-user for interpretation.
- The differences in the final datasets were non-random and therefore more likely to mislead.

– Each software fault which was identified and corrected caused the differences to reduce, so there was convergence although of course it is not possible to say what it was converging to as this is a remote sensing environment with the end product effectively inaccessible. (Drilling a gas well does not validate data as the act of drilling itself interferes with the lithology.)

Although conducted almost twenty years ago, the language used by all participants is still widely used in one form or another (Fortran), the software and test processes used by the participants are also still used and software engineers haven't changed. In other words, it seems likely that the lessons of this experiment are just as valid today.

Open Source. It is believed that open source has a beneficial ameliorating effect on defect, [26], [31], [27] and numerous other authors. This is simply an extension of the quoted effectiveness of code inspections, [7] and [14] amongst many. Although in some senses obvious, the mechanisms are not clear although it may be a simple analogue of N-version experiments where there is one version but N independent sets of eyes rather than N independent versions. This is coupled in the open source world with a form of Darwinian overturn whereby the same feature set may appear many times but the best ones are adopted by the community and further strengthened. As in nature, the unsuccessful ones simply disappear.

Whatever model we ascribe to this process, there seems little doubt of its effectiveness. I have included it under the topic of quantifying the effects of defect as it is also commonly associated with a very close relationship between development and testing as occurs in the Linux kernel[1].

3   A Theory of Defect

One of the things engineers often note about software systems is that the same things occur again and again, [2]. To take one particular example, it is very often observed that defects appear to cluster, [34], [2], independently of either programming language or application area. Following on from [11], I will investigate this using an information theoretic model to avoid the straitjacket of dependence on line of code measures. This does require the development of tools to extract the tokens so is rather more effort than extracting SLOC but that effort proves to be important.

All languages are specified by such tokens, which are extracted at the lexical analysis stage of a language compiler or interpreter. In this sense a token of a programming language takes one of two forms:-

Fixed Token. Fixed tokens of programming languages are those tokens specified by the language designer whose form cannot be altered - the programmer either uses them or not. Examples include language keywords such as

---

[1] http://www.ibm.com/developerworks/linux/library/l-stress/index.html,   accessed 18-Oct-2011

if, then, while; structural tokens such as [,] and operators such as +, -, *
and so on.

Variable Token. These are the user-specified tokens invented by the programmer in order to implement an algorithm. Examples include identifier names,
constants such as 3.14159265 and strings. Apart from some mild lexical constraints such as limiting the length of an identifier to 31 characters and its
starting character to be alphabetic, the programmer has complete freedom
to invent what he or she chooses.

From this token model, all algorithms in all programming languages are constructed.

3.1   An Information Theoretic Model

Suppose a software system is split up into M components, with the ith component
containing t_i tokens altogether from an alphabet consisting of a_i tokens. In
simple procedural languages such as Fortran, components would correspond to
a function or a subroutine. In an object-oriented language, they would be the
outer classes. No finer granularity will be used as the mathematical development
considers only one level.

Following the discussion above, the alphabet can be decomposed as

$$a_i = a_f + a_v(i) \tag{1}$$

where a_f is the alphabet of fixed tokens and $a_v(i)$ is the alphabet of variable
tokens and is clearly dependent on i, since programmers are free to create them
as and when desired.

The number of ways of arranging the tokens of this alphabet in the ith component is therefore $a_i^{t_i}$. Following Hartley, the quantity of information in the ith
component I_i will therefore be defined as

$$I_i = \log(a_i^{t_i}) = t_i \log a_i \tag{2}$$

We can then see that the total amount of information in a system I, can be
written as

$$I = \sum_{i=1}^{M} I_i = \sum_{i=1}^{M} t_i \left(\frac{I_i}{t_i}\right) \equiv \sum_{i=1}^{M} t_i I_i' \tag{3}$$

where I_i' is the information density in the ith component. We will see the reason
for this transformation shortly. We can also see that the total system size T is
given by

$$T = \sum_{i=1}^{M} t_i \tag{4}$$

Equations 3 and 4 will provide constraints in the analysis below.

We can envisage a software system as a fixed level of functionality within some
fixed size. Now functionality is intimately related to choice which as Cherry

points out [4], is itself intimately related to Hartley-Shannon information. *It therefore makes sense to find the most likely way in which tokens can be arranged in components subject to the twin constraints that total size and total amount of information are fixed.* This can be solved using basic principles from statistical mechanics as follows.

The total number of different ways of distributing tokens amongst the components is given by:

$$W = \frac{T!}{t_1! t_2! .. t_M!} \tag{5}$$

We will now suppose that the information density of the ith component is externally imposed by the nature of the algorithm and therefore in common with variational principles is kept constant during variation.

The most likely distribution of the t_i's is defined as the one maximizing 5 subject to the constraints in equations (4) and (3). Using the method of Lagrange multipliers this is equivalent to maximising the following

$$F \equiv T \log T - \sum_{i=1}^{M} t_i \log(t_i) + \lambda \left(T - \sum_{i=1}^{M} t_i \right) + \beta \left(I - \sum_{i=1}^{M} t_i I_i' \right) \tag{6}$$

where λ and β are the Lagrange multipliers, and the first term of Sterling's Formula is used to simplify the factorials under the assumption that $t_i \gg 1$. Setting $\delta F = 0$ leads to

$$0 = - \sum_{i=1}^{M} \delta t_i \left(\log(t_i) + \alpha + \beta I_i' \right) \tag{7}$$

where $\alpha = 1 + \lambda$. This must be true for all variations δt_i and so

$$\log(t_i) = -\alpha - \beta I_i' \tag{8}$$

Defining $p_i = \frac{t_i}{T}$ using (3), p_i can be interpreted as the probability that a component is found with a share of I equal to I_i'. Cancelling the common factor of $e^{-\alpha}$ in numerator and denominator p_i is given by

$$p_i \equiv \frac{t_i}{T} = \frac{e^{-\beta I_i'}}{\sum_{i=1}^{M} e^{-\beta I_i'}} \tag{9}$$

In other words, the probability of finding a component with a large amount of I_i' is correspondingly small. Given the assumed externally imposed nature of I_i', *p_i can then be taken to be the probability that a component of t_i tokens actually occurs.*

Using (3) and (9), we define

$$Q(\beta) = \sum_{i=1}^{M} e^{-\beta \frac{I_i}{t_i}} \tag{10}$$

and can finally write

$$p_i = \frac{(a_i)^{-\beta}}{Q(\beta)} \qquad (11)$$

Thus this information theoretic argument predicts a power-law distribution for the probability of token number as a function of alphabet length.

So far this is a similar development to that followed in [30] and [10] for example, although it generalises the argument by using tokens of programming languages, which are the natural currency of information theory.

Note that this overall process does not care about the tokens themselves - all individual microstates are equally likely. It simply says that if total size and choice in the Hartley-Shannon sense is conserved during the process of distributing the tokens, (and programming is all about choices), then power-law distribution of component size in tokens is overwhelmingly likely to emerge since it occupies the vast majority of the microstates. As will be seen in the data analysis, the specific contribution made by the fact that choice is being made from programming language tokens is represented by the behaviour implicit in (1). This contrasts nicely with monkeys pounding on keyboards as eloquently described by [25]. The ergodic nature of (11) simply accumulates all possible programmers pounding on keyboards. Although not shown here, it also works well with much smaller numbers, i.e. individual systems, a characteristic of classical statistical mechanics.

Finally, I will observe that every language has a *fixed token overhead* in order to implement even the simplest of algorithms. In other words, smaller components must use a higher proportion of fixed tokens than variable tokens. In contrast, larger components use a higher proportion of user-specified tokens because the finite fixed token alphabet quickly stabilises. This can easily be measured. In the very large amount of data reported shortly, the $a_v(i)/a_f$ ratio is typically around 0.2 for smaller components and at least 5 for large components.

It turns out that computing p_i is fundamentally noisy in the tail of power-law distributions and [28] recommends using the equivalent cumulative density function c_i instead. We can then anticipate the final shape of (11) as follows.

Combining (1) and (11) gives

$$c_i \sim (a_f + a_v(i))^{-\beta+1} \qquad (12)$$

For small components, as has been seen, it is reasonable to assume that the number of fixed tokens will tend to dominate the total number of tokens. In other words, $a_f \gg a_v(i)$. (12) can then be written

$$c_i \sim (a_f)^{-\beta+1}(1 + \frac{a_v(i)}{a_f})^{-\beta+1} \qquad (13)$$

In other words,

$$c_i \sim (a_f)^{-\beta+1} \qquad (14)$$

which implies that c_i will be tend to a constant for small components on a log-log plot. For large components, using the same arguments,

$$c_i \sim (a_v(i))^{-\beta+1} \qquad (15)$$

The generic shape of the predicted curve on a log-log plot is shown in Figure 4.

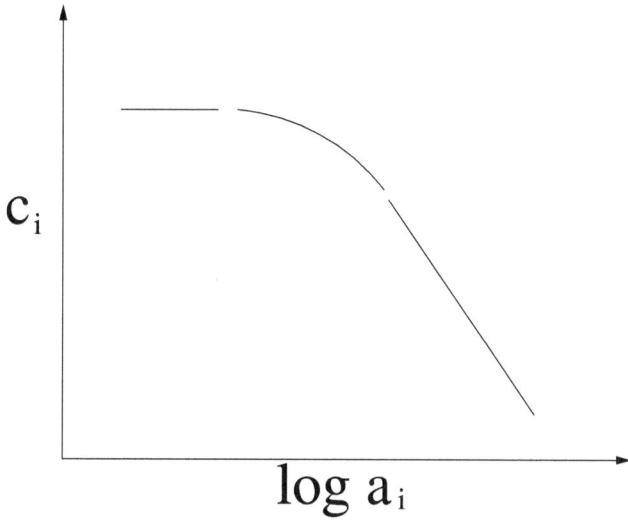

Fig. 4. The predicted cdf using the model described in this paper. The cdf is predicted to be approximately constant for small components and power-law for large ones with a merging zone between.

3.2 Results

To give a sufficiently broad analysis, many software systems comprising multiple languages, (Java, C, C++, Ada, Fortran, Tcl-Tk) were analysed. A generic token extractor was developed for each and calibrated against existing parsing engines in Fortran and C which I had developed in previous projects and which had been tested against the appropriate validation suites, (FCVS and FIPS160 respectively). 75 systems totalling 34 million lines of code (around half a billion tokens) were analysed and the results are shown in Figure 5.

Although the tail of the distribution shown in Figure 5 looks decidedly linear, this was confirmed using the linear modelling function (lm()) in the widely-used R statistical package, (http://www.r-project.org/) which reported a very high degree of linearity with a linear-fit correlation of 0.998 between token counts of 30 and 1500, a span of almost two decades. The same analysis reports a slope of -2.404 +/ 0.004, which is squarely in the range -2 $\rightarrow$ -3 reported for most natural phenomena by [28].

If we now use the simplest model of defect, that we make a mistake every N tokens on average, $d_i \sim t_i \sim a_i$ (using Zipf's law [32]), then

$$c_i \sim (a_i)^{-\beta+1} \sim (t_i)^{-\beta+1} \sim (d_i)^{-\beta+1} \tag{16}$$

So defects will also statistically be distributed as a power-law and should exhibit clustering. As discussed above, this has been widely observed, and also exploited, [21].

34 million lines of Ada,C,C++,Fortran,Java,Tcl

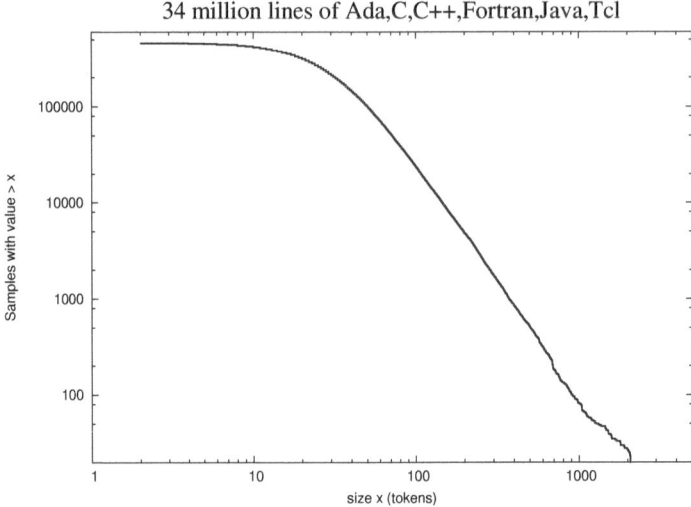

Fig. 5. The measured cdf for 75 systems combining 34 MSLOC into one super-system. This comprises around 15% Java, 15% C++, 15% Fortran, Ada and Tcl combined and around 55% C. This very roughly reflects the amount of each language freely available under open source.

4   Conclusions

This paper gives a guide to some of the problems of quantifying defect in scientific computation. It also demonstrates that software systems appear to have implementation independent properties in which power-laws strongly figure and suggests that defects might be fundamentally statistical in nature rather than predictive. The development gives theoretical support to the observation that defects cluster and this phenomenon can be exploited.

N-version experiments to measure difference are formidably expensive although can emphasise that we have a problem but perhaps the only real way forward is through open source and open data so that reproducibility can be consistently achieved as in other parts of science.

Perhaps I can best sum up this paper by the following aphorism:-

We make progress in science by peer review. To make progress in scientific computation we must extend this to code review.

References

1. Adams, E.: Optimising preventive service of software products. IBM Journal of Research and Development 1(28), 2–14 (1984)
2. Boehm, B., Basili, V.: Software defect reduction top 10 list. IEEE Computer 34(1), 135–137 (2001)

3. Boehm, B., Romback, H., Zelkowitz, M.: Foundations of empirical software engineering: the legacy of Victor R. Basili, 1st edn. Springer (2005) ISBN 3-540-24547-2
4. Cherry, C.: On Human Communication. John Wiley Science Editions (1963), library of Congress 56-9820
5. Donoho, D., Maleki, A., Rahman, I., Shahram, M., Stodden, V.: Reproducible research in computational harmonic analysis. Computing in Science and Engineering 8(18) (2009)
6. Editorial: Devil in the details. Nature 470, 305–306 (2011)
7. Fagan, M.: Design and code inspections to reduce errors in program development. IBM Systems Journal 2, 182–211 (1976)
8. Fenton, N., Pfleeger, S.: Software Metrics: A Rigorous and Practical Approach, 2nd edn. PWS (1997)
9. Hatton, L.: The T experiments: Errors in scientific software. IEEE Computational Science and Engineering 4(2), 27–38 (1997)
10. Hatton, L.: Power-law distributions of component sizes in general software systems. IEEE Transactions on Software Engineering (July/August 2009)
11. Hatton, L.: Power-laws, persistence and the distribution of information in software systems (January 2010), preprint available at
    http://www.leshatton.org/variations_2010.html
12. Hatton, L., Roberts, A.: How accurate is scientific software? IEEE Transactions on Software Engineering 20(10) (1994)
13. Hopkins, T., Hatton, L.: Defect correlations in a major numerical library. Submitted for publication (2008), preprint available at
    http://www.leshatton.org/NAG01_01-08.html
14. Humphrey, W.: A discipline of software engineering. Addison-Wesley (1995) ISBN 0-201-54610-8
15. Jones, E.: Software testing in the computer science curriculum – a holistic approach. In: Proceedings of the Australasian Conference on Computing Education, ACSE 2000. ACM, New York (2000)
16. Kahan, W.: Desperately needed remedies for the Undebuggability of Large Floating-Point Computations in Science and Engineering. In: IFIP/SIAM/NIST Working Conference on Uncertainty Quantification in Scientific Computing (2011)
17. Kahan, W., Darcy, J.: How java's floating point hurts everyone everywhere. Originally presented at ACM 1998 Workshop on Java for High–Performance Network Computing (July 2004)
18. Keller, T.: Achieving error-free man-rated software. In: Second International Software Testing, Analysis and Review Conference, Monterey, USA (1993)
19. Knight, J., Leveson, N.: An experimental evaluation of the assumption of independence in multi-version programming. IEEE Transactions on Software Engineering 12(1), 96–109 (1986)
20. Koenig, A.: C Traps and Pitfalls. Addison-Wesley (1989) ISBN 0-201-17928-8
21. Koru, A.G., Emam, K.E., Zhang, D., Liu, H., Mathew, D.: Theory of relative defect proneness. Empirical Softw. Engg. 13(5), 473–498 (2008)
22. McCabe, T.: A software complexity measure. IEEE Transactions on Software Engineering 2(4), 308–320 (1976)
23. van der Meulen, M., Revilla, M.A.: The effectiveness of software diversity in a large population of programs. IEEE Trans. Software Eng. 34(6), 753–764 (2008)
24. van der Meulen, M.J., Revilla, M.A.: Correlations between internal software metrics and software dependability in a large population of small c/c++ programs. In: International Symposium on Software Reliability Engineering, pp. 203–208 (2007)

25. Mitzenmacher, M.: A brief history of generative models for power-law and lognormal distributions. Internet Mathematics 1(2), 226–251 (2003)
26. Mockus, A., Fielding, R.T., Herbsleb, J.: A case study of open source software development: the apache server. In: ICSE 2000: Proceedings of the 22nd International Conference on Software Engineering, pp. 263–272. ACM, New York (2000)
27. Mockus, A., Fielding, R.T., Herbsleb, J.D.: Two case studies of open source software development: Apache and mozilla. ACM Trans. Softw. Eng. Methodol. 11(3), 309–346 (2002), http://dx.doi.org/10.1145/567793.567795
28. Newman, M.E.J.: Power laws, pareto distributions and zipf's law. Contemporary Physics 46, 323–351 (2006)
29. Pfleeger, S., Hatton, L.: Do formal methods really work? IEEE Computer 30(2), 33–43 (1997)
30. Rawlings, P., Reguera, D., Reiss, H.: Entropic basis of the pareto law. Physica A 343, 643–652 (2004)
31. Raymond, E.S.: The cathedral and the bazaar. O'Reilly (February 2001)
32. Shooman, M.: Software Engineering, 2nd edn. McGraw-Hill (1985)
33. Subramanyam, R., Krishnan, M.: Empirical analysis of CK metrics for object-oriented design complexity: Implications for software defects. IEEE Transactions on Software Engineering 29(4), 297–310 (2003)
34. Tian, J., Troster, J.: A comparison of measurement and defect characteristics of new and legacy software systems. Journal of Systems and Software 44(2), 135–146 (1998), http://www.sciencedirect.com/science/article/B6V0N-3VHWGDW-6/2/408ff89b3fca0948041d218f40ee0509
35. Tichy, W.F., Lukowicz, P., Prechelt, L., Heinz, E.A.: Experimental evaluation in computer science: a quantitative study. J. Syst. Softw. 28, 9–18 (1995), http://portal.acm.org/citation.cfm?id=209090.209093
36. Tichy, W.: Should computer scientists experiment more? IEEE Computer 31(5), 32–40 (1998)

Discussion

Speaker: Les Hatton

William Oberkampf: Experience with the ineffectiveness of unit and regression testing in the ASC program at Sandia National Labs is similar to your documented experience in software quality. We have found that the method of manufactured solutions has been extremely effective at detecting software bugs and numerical algorithm deficiencies in scientific software. Have you used this method to detect software bugs and numerical algorithm deficiencies?

Les Hatton: Not in recent times. I actually didn't know it under this name although I used something similar to this in studying the flow regime in the core of a tornado in my Ph.D thesis many years ago, (and it was successful in flushing out some numerical problems in the non-linear matched boundary value systems I was trying to solve !). I would imagine that it is not a well-known technique amongst scientists though.

Philip Starhill: Do you think that testing can have a positive impact on defect counts or other measures of software quality?

Les Hatton: It depends really on the quality of the testing. It is very variable in my experience, (from casual and ineffective all the way up to a determined and highly skilled assault on a program). However, I think there is little doubt that experienced testers can have an extraordinarily positive impact, particularly if they are involved as early as the design stage where such experience can be highly beneficial to the eventual testability. Too often, testing is an afterthought and I usually picture it as a crumple zone between developer creep and intransigent delivery deadlines.

William Kahan: Of the 19 languages you have used, you mentioned that 18 were not your own choice. Which was the one you would choose? And was its capture cross-section for error lower than the others'?

Les Hatton: An interesting question. I finished up with C but its capture cross-section for error is not one of its most advertised features - a stated philosophy of "trust the programmer" is in itself a little unnerving. However, I had built by then a considerable arsenal of tools to control some of its worst excesses and it remains at heart a simple, elegant and astonishingly versatile language of great longevity. With this tool support, I can write portably and with a gratifyingly low defect density but its taken a long time to get there. C is a great tribute to the skill and insight of its inventor, the late Dennis Ritchie, although like a number of languages, it has suffered somewhat in the hands of standards committees. Last but not least it is well-implemented with the redoubtable GNU C compiler.

William Kahan: Have you noticed that effective testing requires rather more cleverness than writing the program to be tested?

Les Hatton: Absolutely. For most of my career, I have been struck by this. For some reason however, it remains the Cinderella of Computing technologies and is still not considered a good career direction. We don't really teach it in universities and we rarely seem to carry it out well in spite of its importance. In my personal experience, good testers are much rarer than good programmers.

Mladen Vouk: How is the work you describe related to Halstead Theory and metrics: token counting, fault generation, . . . etc?

Les Hatton: Token extraction and counting is identical in Halstead's work and is fundamental in programming languages where it forms the initial lexical analysis stage of all language translation. Where I take a different slant from Halstead (and later, Shooman), is using variational methods to find the most likely distribution of tokens under the twin constraints of size and Hartley-Shannon information. I make no effort to fit defect curves. This approach leads very naturally to the observed implementation independent power-law behaviour and suggests under a simple model of defect that they are also distributed amongst components according to a power-law. This adds some theoretical support to the widely-observed and exploitable phenomenon of defect clustering.

Mladen Vouk: N-version has been extensively studied both as run-time fault tolerance tool and as a testing tool (called back-to-back testing, BBT). One of the "blindness" issues with BBT are common-cause and/or correlated faults and failures. Please comment on this in the context of your work.

Les Hatton: Indeed. There are a number of well-known studies (Knight and Leveson, 1986, van der Meulen and Revilla 2008), which demonstrate non-independent behaviour in BBT. However, in spite of this, there appears to be sufficient independent behaviour that such experiments are very effective at quantifying and flushing out defects which have evaded other techniques. Having said that, by far the biggest barrier to BBT is its cost which is basically N times the cost of a single version. For this reason, as I state in the paper, such methods only serve to highlight the problem. Open source, which is related to N-version in subtle ways, is a much more likely general purpose tool although BBT is used successfully today in some safety-critical systems such as railway signalling and communication systems.

Parametric and Uncertainty Computations with Tensor Product Representations

Hermann G. Matthies, Alexander Litvinenko, Oliver Pajonk, Bojana V. Rosić, and Elmar Zander

Institute of Scientific Computing,
Technische Universität, Braunschweig, Germany
`wire@tu-bs.de`

Abstract. Computational uncertainty quantification in a probabilistic setting is a special case of a parametric problem. Parameter dependent state vectors lead via association to a linear operator to analogues of covariance, its spectral decomposition, and the associated Karhunen-Loève expansion. From this one obtains a generalised tensor representation The parameter in question may be a tuple of numbers, a function, a stochastic process, or a random tensor field. The tensor factorisation may be cascaded, leading to tensors of higher degree. When carried on a discretised level, such factorisations in the form of low-rank approximations lead to very sparse representations of the high dimensional quantities involved. Updating of uncertainty for new information is an important part of uncertainty quantification. Formulated in terms or random variables instead of measures, the Bayesian update is a projection and allows the use of the tensor factorisations also in this case.

Keywords: uncertainty quantification, parametric problems, low-rank tensor approximation, Bayesian updating.

1 Introduction

Situations where one is concerned with uncertainty quantification often come in the following guise: we are investigating some physical system which is modelled by an evolution equation for its state:

$$\frac{\partial}{\partial t} u(t) = A(p; u(t)) + f(p; t), \tag{1}$$

where $u(t) \in \mathcal{V}$ describes the state of the system at time $t \in [0, T]$ lying in a Hilbert space $\mathcal{V}$ (for the sake of simplicity), A is an operator modelling the physics of the system, and f is some external influence (action / excitation / loading). The model depends on some parameter $p \in \mathcal{P}$; in the context of uncertainty quantification the actual value of p is uncertain. Often this uncertainty is modelled by giving the set $\mathcal{P}$ a probability measure. Evaluation and quantification of the uncertainty will often involve functionals of the state $\Psi(u(p; t))$, and the functional dependence of u on p becomes important. Similar situations arise in

A. Dienstfrey and R.F. Boisvert (Eds.): WoCoUQ 2011, IFIP AICT 377, pp. 139–150, 2012.

design, where p may be a design parameter still to be chosen, and one may seek a a design such that a functional $\Psi(u(p;t))$ is e.g. maximised.

The situation just sketched involves a number of objects which are functions of the parameter values. While evaluating $A(p)$ of $f(p)$ for a certain p may be straightforward, one may easily envisage situations where evaluating $u(p)$ or $\Psi(u(p))$ may be very costly as it may involve some very time consuming simulation or computation, like for example running a climate model.

As will be shown in the following Section 2, any such parametric object like $u(p)$, $A(p)$, or $f(p)$ may be seen as an element of a tensor product space. This in turn can be used to find very sparse approximations to those objects, and in turn much cheaper ways to evaluate the for other parameter values. In particular this may be used in the uncertainty quantification to large advantage, like computing means, covariances, exceedance probabilities, etc. For this the dependence of $A(p)$ and $f(p)$ on p has to be propagated to the solution or state vector $u(p)$. This is called the *forward* problem, the resolution of which will be sketched in Section 3, e.g. see [14,15] and the references therein.

The situation we would like to address finally is actually a bit more complicated: In a situation as just described, we observe a function of the state $Y(u(p), p)$, and from this observation we would like to identify the corresponding p. This is called the *inverse* problem, and as the mapping $p \mapsto Y$ is usually not invertible, it is usually *ill-posed*. We embed this in a larger class by modelling our knowledge about p with the help of probability theory, and in a Bayesian manner our task becomes to estimate conditional expectations, e.g. see [21] and the references therein. The problem is now *well-posed*, but at the price of 'only' obtaining probability distributions on the possible values of p. The resolution of the inverse or identification problem will be addressed in Section 4.

2   Parametric Problems

Let $r : \mathcal{P} \to \mathcal{V}$ be a parametric description of one of the objects alluded to in the introduction, where $\mathcal{P}$ is some set, and $\mathcal{V}$ for the sake of simplicity is assumed as a separable Hilbert space with inner product $\langle \cdot | \cdot \rangle_\mathcal{U}$ (the meaning of the index $\mathcal{U}$ will soon become clear). What we desire is a simple representation / approximation of that function, which avoids solving Eq. (1) every time one wants to know $r(p)$ for a new $p \in \mathcal{P}$, i.e. a *response surface* or surrogate model, sometimes also called an *emulator*, whereas the solver for (1) is termed a *simulator*.

One relatively well-known way, particularly in statistical estimation [9], turns the problem into one of approximation of a linear mapping: let $\mathcal{U} = \overline{\operatorname{span}}\, r(\mathcal{P}) = \overline{\operatorname{span}}\, \operatorname{im} r \subseteq \mathcal{V}$ be the smallest closed subspace of $\mathcal{V}$ which is spanned by all the vectors $\{r(p) |\, p \in \mathcal{P}\}$. Then to each such function $r : \mathcal{P} \to \mathcal{U}$ one may associate a linear map

$$R : \mathcal{U} \ni u \mapsto \langle r(\cdot) | u \rangle_\mathcal{U} \in \mathbb{R}^\mathcal{P}. \tag{2}$$

By construction, R is injective. This may be used to define an inner product on $\operatorname{im} R$ as

$$\forall \phi, \psi \in \operatorname{im} R : \quad \langle \phi | \psi \rangle_\mathcal{R} := \langle R^{-1}\phi | R^{-1}\psi \rangle_\mathcal{U}, \tag{3}$$

and let $\mathcal{R}$ be the completion of im R with that inner product. It is obvious that R is a unitary map between the Hilbert spaces $\mathcal{U}$ and $\mathcal{R}$.

Up to now, no structure on the set $\mathcal{P}$ has been assumed, whereas on $\mathcal{U}$ the inner product is assumed to measure what is important for the state $r(p) \in \mathcal{U}$. This is carried via the map R in (2) onto the space of scalar functions $\mathcal{R}$ on the set $\mathcal{P}$, and the inner product there measures essentially the same thing as the one on $\mathcal{U}$.

2.1 Reproducing Kernel Hilbert Space

This is a first representation, and $\mathcal{R}$ is called a reproducing kernel Hilbert space (RKHS) [7] with *reproducing kernel* $\varkappa \in \mathbb{R}^{\mathcal{P}} \times \mathbb{R}^{\mathcal{P}}$

$$\varkappa(p_1, p_2) := \langle r(p_1) | r(p_2) \rangle_{\mathcal{U}}. \tag{4}$$

It is straightforward to verify that it defines an obviously continuous (on $\mathcal{R}$) point-evaluation functional $\mathcal{R} \ni \phi \mapsto \phi(p) = \langle \varkappa(p, \cdot) | \phi \rangle_{\mathcal{R}} \in \mathbb{R}$, hence the name.

In other settings like classification or machine learning, e.g. with support vector machines, where $p \in \mathcal{P}$ has to be classified as belonging to certain subsets of $\mathcal{P}$, the space $\mathcal{V}$ and the map $r : \mathcal{P} \to \mathcal{V}$ may often be freely chosen. This is then referred to as the "kernel trick", and classification may be achieved by mapping these subsets with r into $\mathcal{U}$ and separating them with hyperplanes—a linear classifier.

In terms of representation, one may now choose a basis $\{\varphi_m\}_{m \in \mathbb{N}}$ in $\mathcal{R}$, which may be assumed to be a complete orthonormal system (CONS). With the CONS $\{y_m \mid y_m = R^{-1}\varphi_m\}_{m \in \mathbb{N}}$ in $\mathcal{U}$, the operator R, its inverse R^{-1}, and the parametric element $r(p)$ become

$$R = \sum_m \varphi_m \otimes y_m; \quad R^{-1} = \sum_m y_m \otimes \varphi_m; \quad r(p) = \sum_m y_m \varphi_m(p), \tag{5}$$

exhibiting the tensorial nature of the representation mapping. With such a basis one may define a unitary map from ℓ_2 to $\mathcal{R}$ and via R^{-1} further to $\mathcal{U}$:

$$\ell_2 \ni \boldsymbol{a} = (a_1, a_2, \ldots) \mapsto \sum_m a_m \varphi_m \mapsto \sum_m a_m y_m \in \mathcal{U}. \tag{6}$$

Note that this representation is linear in the new 'parameters' $(a1, a2, \ldots) \in \ell_2$. Model reductions may be achieved by choosing only subspaces of $\mathcal{R}$ or ℓ_2, or by approximating the map R^{-1}. This pattern of (5) or (6) repeats itself for all representations to follow.

2.2 Spectral Decomposition

As a way of measuring of what is important on the set $\mathcal{P}$, assume that there is another inner product $\langle \cdot | \cdot \rangle_{\mathcal{W}}$ for scalar functions $\phi \in \mathbb{R}^{\mathcal{P}}$, and denote the Hilbert space of functions with that inner product by $\mathcal{W}$. With this, one may define [9] a densly defined map C in $\mathcal{U}$ through the bilinear form

$$\forall u, v \in \mathcal{U}: \quad \langle Cu|v\rangle_{\mathcal{U}} := \langle Ru|Rv\rangle_{\mathcal{W}}. \tag{7}$$

The map $C = R^*R$ (the adjoint is taken w.r.t. the $\mathcal{W}$-inner product, by abuse of notation we shall still call the map R) may be called the 'correlation' operator. By construction it is injective, positive, and self-adjoint.

Often the inner product $\langle\cdot|\cdot\rangle_{\mathcal{W}}$ comes from a measure ϖ on $\mathcal{P}$, so that $\mathcal{W}$ may be taken as $L_2(\mathcal{P}, \varpi)$. One important class of problems is when ϖ is a probability measure on $\mathcal{P}$, i.e. $\varpi(\mathcal{P}) = 1$. Often the set has more structure, like being in a topological space, differentiable (Riemann) manifold, or a Lie group, which then may induce the choice of σ-algebra or measure. In all such cases one has $C = R^*R = \int_{\mathcal{P}} r(p) \otimes r(p) \, \varpi(dp)$. It is the factorisation of $C = R^*R$ which paves the way for further possibilities of representation. Most common is to use the spectral decomposition (e.g. [19,3]) of C:

$$Cu = \int_0^\infty \lambda \, dE_\lambda(u), \tag{8}$$

where E_λ is the corresponding projection valued spectral measure, with the spectrum $\sigma(C) \subseteq \mathbb{R}_+$. For the sake of simplicity assume that C has a pure point spectrum $\sigma_p(C) = \sigma(C)$—the important case where C has also a continuous spectrum requires too many technical tools such as Gel'fand triplets (rigged Hilbert spaces) [3] and generalised eigenvectors to be treated in this short note— such that (8) may be written with the CONS of unit-$\mathcal{U}$-norm eigenvectors v_m:

$$Cu = \sum_m \lambda_m \langle v_m|u\rangle_{\mathcal{U}} v_m = \sum_m \lambda_m (v_m \otimes v_m) u. \tag{9}$$

From this follows the singular value decomposition of R, with $\lambda_m^{1/2} s_m := Rv_m$:

$$R = \sum_m \lambda_m^{\frac{1}{2}} (s_m \otimes v_m); \quad R^* = \sum_m \lambda_m^{\frac{1}{2}} (v_m \otimes s_m); \quad r(p) = \sum_m \lambda_m^{\frac{1}{2}} s_m(p) v_m, \tag{10}$$

where the last relation is the so-called Karhunen-Loève or proper orthogonal decomposition (POD). Observe that r—as well as R^*—is linear in the s_m. Similarly to (6), we have the—linear in $\boldsymbol{a}$—representation:

$$\ell_2 \ni \boldsymbol{a} = (a_1, a_2, \ldots) \mapsto \sum_m a_m s_m \mapsto \sum_m \lambda_m^{1/2} a_m v_m \in \mathcal{U}. \tag{11}$$

An alternative formulation of the spectral decomposition (8) is [19] that C is unitarily equivalent with a multiplication operator:

$$C = VM_kV^* = (VM_k^{1/2})(VM_k^{1/2})^*, \tag{12}$$

where V is unitary between some $L_2(\mathcal{T})$ and $\mathcal{U}$, M_k is a multiplication operator on the measure space $\mathcal{T}$ with a positive function $k(s) > 0$, and $M_k^{1/2} = M_{\sqrt{k}}$.

The essential range of k is the spectrum of C. This gives in the now familiar manner a representation on $L_2(\mathcal{T})$ through the choice of a CONS $\{\varsigma_m\}$. Setting $u_m := VM_{\sqrt{k}}\varsigma_m$, one obtains

$$(VM_k^{1/2}) = (VM_{\sqrt{k}}) = \sum_m u_m \otimes \varsigma_m \qquad (13)$$

as tensorial representation.

2.3 Other Factorisations of C

Other factorisations $C = B^*B$—which are all unitarily equivalent—lead to analogous representations. Let $B : \mathcal{U} \to \mathcal{H}$ be an injective mapping into another Hilbert space $\mathcal{H}$. Pick a CONS $\{e_m\}$ in $\mathcal{H}$ and set $f_m := B^*e_m$, then

$$B^* = \sum_m f_m \otimes e_m, \qquad (14)$$

again a tensorial representation. All the representations considered so far are of this type. Similarly to (6), we have the—linear in a—representation:

$$\ell_2 \ni a = (a_1, a_2, \ldots) \mapsto \sum_m a_m e_m \mapsto \sum_m a_m f_m \in \mathcal{U}. \qquad (15)$$

For finite dimensional spaces, a favourite choice for such a decomposition of C is the Cholesky factorisation $C = LL^T$.

Another often used possibility to consider factorisations of the reproducing kernel, i.e. *kernel decompositions*. In an abstract way, one considers $\hat{C} := RR^*$ instead of $C = R^*R$, and $\hat{C} : \mathcal{W} \ni \phi \mapsto \langle \varkappa(p, \cdot), \phi \rangle_\mathcal{W} \in \mathcal{W}$. In case $\mathcal{W} = L_2(\mathcal{P}, \varpi)$ one has here the Fredholm equation $(\hat{C}\phi)(p_1) = \int_\mathcal{P} \varkappa(p_1, p_2)\phi(p_2)\,\varpi(\mathrm{d}p)$. As this is similar to the decompositions considered for C it will be omitted for the sake of brevity.

2.4 Examples and Interpretations

Some examples are now in order, so that one may see that the above description is in many cases an abstract statement of already very familiar constructions. For a general parameter space, the constructions provide a 'response surface', but in some cases this is known under a different name:

- If $\mathcal{V}$ is a space of centred random variables (RVs), r is a random field or stochastic process indexed by $\mathcal{P}$, and the reproducing kernel is the covariance function.
- If the measure ϖ on $\mathcal{P}$ is a probability measure ($\varpi(\mathcal{P}) = 1$), and r is a centred $\mathcal{V}$-valued RV, then C is the covariance operator.
- If $\mathcal{P} = \{1, 2, \ldots, n\}$ and $\mathcal{R} = \mathbb{R}^n$, then $\varkappa$ is the Gram matrix of the vectors $v_1, \ldots, v_n$.

- If $\mathcal{P} = [0, T]$ and $r(t), t \in [0, T]$, is the response of a dynamical system with state space $\mathcal{V}$, the R^* leads to the POD.
- If the two preceeding items are combined, this gives the method of snapshots for the POD.
- If $\mathcal{P} = \{\omega_s | \ \omega_s \in \Omega\}$ are samples from some probability space Ω, then one gets the POD method for samples.

For the sake of simplicity we had restricted ourselves in the spectral decomposition (8) in Subsection 2.2 to the case of a pure point spectrum (9). If in the first item $\mathcal{P} = \mathbb{R}^d$, and the covariance function / reproducing kernel satisfies $\varkappa(\boldsymbol{r}_1, \boldsymbol{r}_2) = c(\boldsymbol{r}_1 - \boldsymbol{r}_2)$ for $\boldsymbol{r}_1, \boldsymbol{r}_2 \in \mathbb{R}^d$, one calls the covariance *translation invariant*, and the random process for $d = 1$ *stationary* or the the random field *homogeneous*. In that case the eigenvalue equation with the operator $\hat{C} = RR^*$ is a convolution equation $(\hat{C}\phi)(\boldsymbol{r}_1) = \int_{\mathbb{R}^d} c(\boldsymbol{r}_1 - \boldsymbol{r}_2)\phi(\boldsymbol{r}_2) \, \mathrm{d}\boldsymbol{r}_2$, which is well-known to be diagonalised by the (real) Fourier transform. This is an example of the spectral decomposition in (12), the function k for the multiplication operator M_k is the Fourier transform of c, and the point spectrum is typically empty [14].

3   Approximation and Propagation

When it comes to computing, two kinds of approximations will usually be employed: one is that the parametric dependence of the entities in (1) needs to be simplified to make it computationally accessible—often this is also termed a representation; the other approximation derives from the fact that even for a fixed parameter $p \in \mathcal{P}$ the system modelled by (1) can not be computationally treated without further approximation, i.e. because it is often an equation in an infinite dimensional space, e..g. a partial differential equation.

3.1   Representation, Approximation, and Model Reduction

In Section 2 were a number of examples on how to construct representations of the type
$$S : \mathcal{S} \to \mathcal{U} \tag{16}$$
with a Hilbert space $\mathcal{S}$ which is used for the representation, such that $SS^* = C$, or equivalently $C = B^*B$ as in Subsection 2.3. At the core of all constructions was the mapping R in (2) and (10), which led to the operator $C = R^*R$ on $\mathcal{U}$, see (7). This mapping 'linearises' the problem, as one may choose new parameters on which the representation depends linearly, as was pointed out repeatedly, e.g. (11). Most representations are connected with the spectral decomposition in Subsection 2.2 or equivalently with the spectral kernel decomposition. All the representations shown could be written in a tensor product format. The possibilities alluded to for $\mathcal{S}$ were the RKHS $\mathcal{R} \subset \mathbb{R}^\mathcal{P}$, see Subsection 2.1 and (3), with the reproducing kernel (4) and representation R^{-1} in (5), or the Hilbert space $\mathcal{Q} \subset \mathbb{R}^\mathcal{P}$ connected with the correlation operator C in (7), or Hilbert spaces induced by the spectral decomposition like $L_2(\sigma(C))$ implicitly appearing in (8), or $L_2(\mathcal{T})$ in (12).

Other factorisations of C such as in Subsection 2.3, or of the reproducing kernel $\varkappa$ with integral transforms, lead to representations which are not necessarily connected to the spectral decomposition and may be more convenient in certain circumstances.

On the other hand, through the magnitude of spectral values the spectral decomposition gives guidance on the relative importance of different subspaces of $\mathcal{S}$, and approximations of the representation map which may be computationally more advantageous, like low-rank approximations.

All these representations may be carried onto ℓ_2 in the manner of (14) in Subsection 2.3. Model reductions may be achieved through choice of a subspace of $\mathcal{S}$, and / or by approximating the representation map as alluded to above. Thus the quantity $r(p)$ is in all cases approximated by a tensor expression $r \approx \sum_j r_j \otimes \tau_j$, and the number of terms in the sum is termed the *rank* of the tensor, and this kind of versatile sparse approximation is also called a low-rank approximation [6].

3.2   Discretisation and Propagation

For brevity we follow [14], where more references may be found, cf. also the recent monograph [11]. For the sake of simplicity, let us concentrate on the time-independent or stationary version of (1), namely $A(p; u) = f(p)$. Usually this is some partial differential equation and has to be discretised, approximated, or somehow projected onto some finite dimensional subspace $\mathcal{V}_N \subset \mathcal{V}$, with $\dim \mathcal{V}_N = N$. The entities of (1) which are projected or induced on the corresponding $\mathbb{R}^N$ will be denoted by boldface, such that the stationary, projected equation reads as

$$\boldsymbol{A}(p; \boldsymbol{u}) = \boldsymbol{f}(p). \tag{17}$$

To propagate the parametric dependence, choose a finite dimensional subspace of the Hilbert spaces mentioned in Subsection 3.1, say $\mathcal{S}_M \subset \mathcal{S}$ for the solution $\boldsymbol{u}(p)$ in (17). Via Galerkin projection or collocation, or other such techniques, the still parametric model (17) is thereby formulated on the tensor product $\mathcal{V}_N \otimes \mathcal{S}_M$, denoted as

$$\mathbf{A}(\mathbf{u}) = \mathbf{f}. \tag{18}$$

The solution of (18) is often computationally challenging, as $\dim \mathcal{V}_N \otimes \mathcal{S}_M = N \times M$ may be very large. One possibility for such high-dimensional problems are the low-rank approximations alluded to at the end of Subsection 3.1, by representing the entities in (17) such as $\boldsymbol{A}$, $\boldsymbol{u}$, and $\boldsymbol{f}$ in a low-rank format. Several numerical techniques [17,4,10,16] have been developed recently to obtain an approximation to the solution $\mathbf{u} \approx \sum_j \boldsymbol{u}_j \otimes \boldsymbol{z}_j$ to (18) in this format by only ever operating on the data-sparse low-rank representation, thus allowing for an efficient resolution of the high-dimensional problem.

Once this has been computed, any other functional such as $\Psi(u(p))$ mentioned in Section 1 may be computed with relative ease. In case there is a probability measure on $\mathcal{P}$ as given in the examples in Subsection 3.1, for example to quantify some uncertainty in the parameters, the functionals usually take the form of expectations, such that $\Psi(u) = \mathbb{E}(\psi(\mathbf{u}))$ becomes a mean, a variance, an exceedance probability, or other such quantity needed in an *uncertainty quantification*.

4 Identification and Inverse Problems

In the setting of (1) let us pose the following problem: Some components—let us denote these by q—of the parameters $p \in \mathcal{P}$ are not only uncertain, but we would like to infer what they are by making observations y_k at times $0 < t_1 < \cdots < t_k \cdots \in [0, T]$. But we can not observe the entity q directly—like in Plato's cave allegory we can only see a 'shadow' of it, formally given by

$$Y : \mathcal{Q} \times \mathcal{V} \ni (q, u(t_k)) \mapsto z_k = Y(q; u(t_k)) \in \mathcal{Y}; \tag{19}$$

at least this is our model of what we are measuring. Usually the observation will deviate from what we expect to observe even if we knew the right q as (1) is only a *model*—so there is some model error ϵ, and the measurement will be polluted by some measurement error ε. Hence we observe $y_k = z_k + \epsilon + \varepsilon$. From this one would like to know what q and $u(t_k)$ are.

4.1 Identification

The mapping in (19) is usually not invertible and hence the problem is called ill-posed. One way to address this is via regularisation, but here we follow a different track. Modelling out lack-of-knowledge about q and $u(t_k)$ in a Bayesian way [21] by replacing them with a $\mathcal{Q}$- resp. $\mathcal{V}$-valued random variable (RV), the problem becomes well-posed [20]. But of course one is looking now at the problem of finding a probability distribution that best fits the data; and one also obtains a probability distribution, not just *one* pair q and $u(t_k)$.

The mathematical setup then is as follows: we assume that Ω is a measure space with σ-algebra $\mathfrak{A}$ and with a probability measure $\mathbb{P}$, and that $q : \Omega \to \mathcal{Q}$ and $u : \Omega \to \mathcal{U}$ are random variables. For simplicity, we shall also require $\mathcal{Q}$ to be a Hilbert space where each vector is a possible realisation. This is in order to allow to measure the distance between different q's as the norm of their difference, and to allow the operations of linear algebra to be performed.

In case the q's are not without constraints, or not in a vector space, then they should be mapped to such quantities. For example, if q is a diffusion tensor field, then it has to be symmetric and positive definite. The symmetric tensors are of course a subspace, but the manifold of positive definite ones is not, nor is it closed. But they can be given the structure of a Lie group and a Riemannian manifold [1], and then distance is measured as a the length of a path along a geodesic. But the associated Lie algebra—the tangent space at the neutral element of group—is in one to one correspondence with the geodesics; hence one play everything back to a vector space. A simple case of this are positive scalars; through the logarithm they are transformed into a vector space without constraints. The computations to be described should be performed in such a vector space.

4.2   Bayesian Updating

Bayes's theorem is commonly accepted as a consistent way to incorporate new knowledge into a probabilistic description [21]. The textbook statement of the theorem is about conditional probabilities

$$\mathbb{P}(I_q|M_y) = \frac{\mathbb{P}(M_y|I_q)}{\mathbb{P}(M_y)}\mathbb{P}(I_q), \tag{20}$$

where I_q is some subset of possible q's, and M_y is the information provided by the measurement. As this becomes problematic when the set M_y has vanishing probability measure, Kolmogorov already defined conditional probabilities via conditional expectation [2]. But most computational approaches compute via the measures [13,20] Given the conditional expectation $\mathbb{E}(\cdot|M_y)$, the conditional probability is easily recovered as $\mathbb{P}(I_q|M_y) = \mathbb{E}\left(\chi_{I_q}|M_y\right)$, where χ_{I_q} is the characteristic function of the subset I_q.

The easiest point of departure for conditional expectation in our setting is to define it not just for one piece of measurement M_y, but for sub-σ-algebras of $\mathfrak{A}$. The connection with an event M_y is then that we take $\sigma(Y)$, the σ-algebra generated by Y. Observe that if $\mathfrak{S} \subseteq \mathfrak{A}$ is a sub-σ-algebra, then $L_2(\Omega, \mathfrak{S}, \mathbb{P})$ is a closed subspace of $L_2(\Omega, \mathfrak{A}, \mathbb{P})$.

For RVs with finite variance (elements of $L_2(\Omega, \mathfrak{A}, \mathbb{P})$) the conditional expectation $\mathbb{E}(\cdot|\mathfrak{S})$ is defined as the orthogonal projection onto onto $L_2(\Omega, \mathfrak{S}, \mathbb{P})$. It can then be extended as a contraction onto all $L_p(\Omega, \mathfrak{A}, \mathbb{P})$ with $p \geq 1$ [2]. In other words the Bayesian update may now be simply shown to be

$$\mathbb{E}(q|\sigma(Y)) = P_{\mathcal{Q}_n}(q) = \mathrm{argmin}_{\tilde{q}\in\mathcal{Q}_n}\|\tilde{q} - q\|_{L_2}^2, \tag{21}$$

where $\mathcal{Q}_n := L_2(\Omega, \sigma(Y), \mathbb{P})$ represents the new information, and $P_{\mathcal{Q}_n}$ is the orthogonal projector onto $\mathcal{Q}_n$. Already in [8] it was noted that the conditional expectation is the best estimate not only for the *loss function* 'distance squared', but for a much larger range of loss functions under certain distributional constraints. But for the above loss function this is independent of what distribution q might have.

Requiring the derivative of the loss function in (21) to vanish—equivalently remembering from elementary geometry that the line to the closest point is perpendicular to the approximating subspace—one arrives at the Galerkin orthogonality conditions

$$\forall \tilde{q} \in \mathcal{Q}_n : \quad \langle q - \mathbb{E}(q|\sigma(Y)), \tilde{q}\rangle_{L_2} = 0. \tag{22}$$

To continue, note that the Doob-Dynkin lemma [2] assures us that if a RV like $\mathbb{E}(q|\sigma(Y))$ is measurable w.r.t. $\sigma(Y)$, then $\mathbb{E}(q|\sigma(Y)) = \psi(Y)$ for some measurable $\psi \in L_0(\mathcal{Y}; \mathcal{P})$. Hence $L_2(\Omega, \sigma(Y), \mathbb{P}) = L_2(\Omega, \mathfrak{A}, \mathbb{P}) \cap \overline{\mathrm{span}}\{\phi(y) \mid y \in \mathcal{Y}, \phi \in L_0(\mathcal{Y}; \mathcal{Q})\}$, where $L_0(\mathcal{Y}; \mathcal{Q})$ is the vector space of measurable maps from $\mathcal{Y}$ to $\mathcal{Q}$. In particular one sees that $\mathbb{E}(q|\sigma(Y)) \in L_0(\mathcal{Y}; \mathcal{Q})$. In the light of (22) the task of computing $\psi(y) := \mathbb{E}(q|\sigma(Y))$ may be phrased as: find $\psi \in L_0(\mathcal{Y}; \mathcal{Q})$ such that

$$\forall \phi \in L_0(\mathcal{Y}; \mathcal{Q}), y \in \mathcal{Y} : \quad \langle q - \psi(z), \phi(y)\rangle_{L_2} = 0. \tag{23}$$

The value $q_a := \psi(y) = \mathbb{E}\,(q|\sigma(Y))$ is called the *analysis, assimilated,* or *posterior* value, incorporating the new information. In case one has some *prior* approximation, also called a *forecast* q_f, this results in an affine shift of the subspaces involved, and hence $q_a = q_f + \psi(y - z_k)$ with $z_k = Y(q_f; u_f(t_k))$, e.g. see [8].

We would like to emphasise that it is the vector space setting of $\mathcal{Q}$ and $\mathcal{Y}$ which has made this well-known formulation possible, and it will also allow for easy numerical computation. To work with measures as in (20) is cumbersome, as probability measures are on the intersection of the unit sphere and the positive cone in the space of signed finite measures. A bit easier would be to work with RVs which are in a metric space, the conditional expectation then minimises the metric distance squared; but the Hilbert space setting is certainly the simplest instance of this. As we work in a vector space, we make another approximation to simplify the computations by replacing $L_0(\mathcal{Y}; \mathcal{Q})$ above by $\mathscr{L}(\mathcal{Y}, \mathcal{Q})$, the space of linear continuous maps. The Galerkin orthogonality condition (23) is then translated to: find $K \in \mathscr{L}(\mathcal{Y}, \mathcal{Q})$ such that [8,12]

$$\forall H \in \mathscr{L}(\mathcal{Y}, \mathcal{Q}), y \in \mathcal{Y}: \quad \langle q - Kz, Hy \rangle_{L_2} = 0, \tag{24}$$

and we set $\mathbb{E}\,(q|\sigma(Y))_\ell := K$, a linear approximation to $\mathbb{E}\,(q|\sigma(Y))$. As the projection is now onto the smaller space $\mathcal{Q}_\ell := L_2(\Omega, \mathfrak{A}, \mathbb{P}) \cap \overline{\mathrm{span}}\{Hy \mid y \in \mathcal{Y}, H \in \mathscr{L}(\mathcal{Y}, \mathcal{Q})\} \subset L_2(\Omega, \sigma(Y), \mathbb{P})$, we are not using all the information available but the computation is simpler. In the case of prior information this is extended as before to [8,12]

$$q_a = q_f + K(y - z_k), \text{ with } K = C_{q,z}(C_z + C_\varepsilon)^{-1}. \tag{25}$$

This includes the errors $\epsilon + \varepsilon$ with covariance operator C_ε, and it is not difficult to show [8,12] that the optimal K is given by the well-known Kálmán gain in (25), where $C_z := \mathbb{E}\,(Y(q, u) \otimes Y(q, u))$ and $C_{q,z} = \mathbb{E}\,(q \otimes Y(q, u))$. In case $C_z + C_\varepsilon$ is not invertible or close to singularity, its inverse in (25) should be replaced by the Moore-Penrose pseudo-inverse. This update is in some ways very similar to the 'Bayes linear' approach [5].

4.3   Computing the Bayesian Update and an Example

As before in Subsection 3.2 for an actual computation the forward model is discretised like in (17) and (18). The space of possible measurements is also discretised by $\mathcal{Y}_I \subset \mathcal{Y}$, as is the space of entities to be identified $\mathcal{Q}_J \subset \mathcal{Q}$, giving a discrete forward model and measurement operator

$$\frac{\partial}{\partial t}\boldsymbol{u}(\omega; t) + \boldsymbol{A}(\boldsymbol{q}(\omega); \boldsymbol{u}(\omega; t)) = \boldsymbol{f}(\omega; t); \quad \boldsymbol{z_k} = \boldsymbol{Y}(\boldsymbol{q}(\omega); \boldsymbol{u}(\omega; t_k)). \tag{26}$$

The update (25) is in this way also discretised to

$$\boldsymbol{q_a}(\omega) = \boldsymbol{q_f}(\omega) + \boldsymbol{K}(\boldsymbol{y}(\omega) - \boldsymbol{z_k}(\omega)), \text{ with } \boldsymbol{K} = \boldsymbol{C_{q,z}}(\boldsymbol{C_z} + \boldsymbol{C_\varepsilon})^{-1}. \tag{27}$$

Completely analogous to how the dependence on $p \in \mathcal{P}$ was treated in Section 2 and Section 3, we represent the dependence on $\omega \in \Omega$ through a subspace

$\mathcal{S} \subset L_0(\Omega, \mathbb{R}) \subset \mathbb{R}^\Omega$, i.e. a *random variable*, and in the discrete form by a finite dimensional subspace $\mathcal{S}_M \subset \mathcal{S}$. A popular choice for $\mathcal{S}_M$ is Wiener's *polynomial chaos*, orthogonal multi-variate Hermite polynomials in standard Gaussian RVs [7,14]. Looking at tensor products of these finite dimensional spaces, with the results of Section 2 and Section 3 the model and measurement equation (26) change to an analogue of (18), and (27) becomes $\mathbf{q}_a = \mathbf{q}_f + \mathbf{K}(\mathbf{y} - \mathbf{z}_k)$, where $\mathbf{K} = \boldsymbol{K} \otimes \boldsymbol{I}$ with $\boldsymbol{K}$ from (27). Hence the update equation is naturally in a tensorised form, allowing to apply it directly to low-rank approximations as introduced in Section 3.

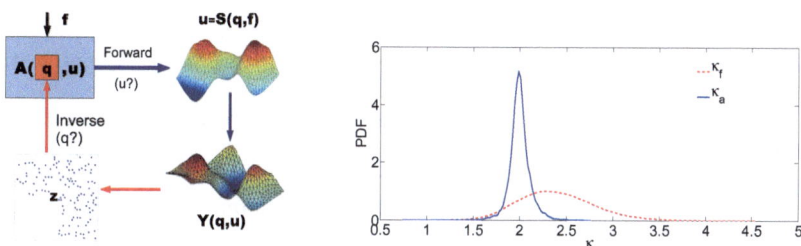

Fig. 1. Updating schema and prior-posterior comparison

The following example of a diffusion equation on an L-shaped domain shows how the method works, for details and additional references please see [18]. A schematic view of the process is shown on the left picture in Fig. 1. The diffusion coefficient κ is the quantity to be identified; this is a positive quantity as alluded to before, so we consider its logarithm $q = \log(\kappa)$. The forward model in the top left produces a forecast on the system behaviour, shown on the top right. From this a forecast for the measurement is deduced, shown on the bottom right, and then compared with measurements—shown on the bottom left—to produce the Bayesian update. In the depiction on the right in Fig. 1, one may see how the prior or forecast distribution for $\kappa_f = \exp(q_f)$ is updated to the posterior or assimilated distribution $\kappa_a = \exp(q_a)$. For details please refer to [18].

References

1. Arsigny, V., Fillard, P., Pennec, X., Ayache, N.: Geometric means in a novel vector space structure on symmetric positive-definite matrices. SIAM Journal on Matrix Analysis and Applications 29(1), 328–347 (2006)
2. Bobrowski, A.: Functional Analysis for Probability and Stochastic Processes. Cambridge University Press, Cambridge (2005)
3. Dautray, R., Lions, J.L.: Spectral Theory and Applications, Mathematical Analysis and Numerical Methods for Science and Technology, vol. 3. Springer, Berlin (1990)
4. Doostan, A., Iaccarino, G.: A least-squares approximation of partial differential equations with high-dimensional random inputs. Journal of Computational Physics 228, 4332–4345 (2009)

5. Goldstein, M., Wooff, D.: Bayes Linear Statistics—Theory and Methods. Wiley Series in Probability and Statistics. John Wiley & Sons, Chichester (2007)

6. Hackbusch, W., Khoromskij, B.N.: Tensor-product approximation to operators and functions in high dimensions. J. Complexity 23(4-6), 697–714 (2007)

7. Janson, S.: Gaussian Hilbert spaces. Cambridge Tracts in Mathematics, vol. 129. Cambridge University Press, Cambridge (1997)

8. Kálmán, R.E.: A new approach to linear filtering and prediction problems. Transactions of the ASME—J. of Basic Engineering (Series D) 82, 35–45 (1960)

9. Krée, P., Soize, C.: Mathematics of Random Phenomena. D. Reidel Publishing Co., Dordrecht (1986)

10. Krosche, M., Niekamp, R.: Low rank approximation in spectral stochastic finite element method with solution space adaption. Informatikbericht 2010-03, TU Braunschweig, Brunswick (2010), http://www.digibib.tu-bs.de/?docid=00036351

11. Le Maître, O.P., Knio, O.M.: Spectral Methods for Uncertainty Quantification. Scientific Computation. Springer, Berlin (2010)

12. Luenberger, D.G.: Optimization by Vector Space Methods. John Wiley & Sons, Chichester (1969)

13. Marzouk, Y.M., Najm, H.N., Rahn, L.A.: Stochastic spectral methods for efficient Bayesian solution of inverse problems. Journal of Computational Physics 224(2), 560–586 (2007)

14. Matthies, H.G.: Uncertainty quantification with stochastic finite elements. In: Stein, E., de Borst, R., Hughes, T.J.R. (eds.) Encyclopaedia of Computational Mechanics. John Wiley & Sons, Chichester (2007)

15. Matthies, H.G., Keese, A.: Galerkin methods for linear and nonlinear elliptic stochastic partial differential equations. Computer Methods in Applied Mechanics and Engineering 194(12-16), 1295–1331 (2005)

16. Matthies, H.G., Zander, E.: Sparse representations in stochastic dynamics. In: Papadrakakis, M., Stefanou, G., Papadopoulos, V. (eds.) Computational Methods in Stochastic Dynamics, pp. 247–265. Springer (2011)

17. Nouy, A., Le Maître, O.P.: Generalized spectral decomposition for stochastic nonlinear problems. Journal of Computational Physics 228(1), 202–235 (2009)

18. Rosić, B.V., Litvinenko, A., Pajonk, O., Matthies, H.G.: Direct Bayesian update of polynomial chaos representations. Informatikbericht 2011-02, TU Braunschweig, Brunswick (2011), http://www.digibib.tu-bs.de/?docid=00039000

19. Segal, I.E., Kunze, R.A.: Integrals and Operators. Springer, Berlin (1978)

20. Stuart, A.M.: Inverse problems: A Bayesian perspective. Acta Numerica 19, 451–559 (2010)

21. Tarantola, A.: Inverse Problem Theory and Methods for Model Parameter Estimation. SIAM, Philadelphia (2004)

Using Emulators to Estimate Uncertainty in Complex Models

Peter Challenor

National Oceanography Centre
Southampton, United Kingdom
p.challenor@noc.ac.uk

Abstract. The Managing Uncertainty in Complex Models project has been developing methods for estimating uncertainty in complex models using emulators. Emulators are statistical descriptions of our beliefs about the models (or simulators). They can also be thought of as interpolators of simulator outputs between previous runs. Because they are quick to run, emulators can be used to carry out calculations that would otherwise require large numbers of simulator runs, for example Monte Carlo uncertainty calculations. Both Gaussian and Bayes Linear emulators will be explained and examples given. One of the outputs of the MUCM project is the MUCM toolkit, an on-line recipe book for emulator based methods. Using the toolkit as our basis we will illustrate the breadth of applications that can be addressed by emulator methodology and detail some of the methodology. We will cover sensitivity and uncertainty analysis and describe in less detail other aspects such as how emulators can also be used to calibrate complex computer simulators and how they can be modified for use with stochastic simulators.

Keywords: emulator, Gaussian process, Bayes linear, sensitivity, uncertainty, calibration.

1 Introduction

The increase in computing power over the last few decades has led to an explosion in the use of complex computer codes in both science and engineering. Almost every area of science and engineering now uses complex numerical simulators to solve problems that could not have been tackled only a few years ago. Examples include engineering [2], climate science, [4], and oceanography [12]. At the present time most of these simulators are deterministic but there is an increasing use of stochastic simulators as well [24]. These computer codes comprise many thousands (or even millions) of lines of code and take long times to compute even on the fastest supercomputers available today. Questions we would like to ask of the simulators include: for a given uncertainty in the inputs to the simulator what is the uncertainty in the outputs; which inputs have the most effect on the outputs; are all the inputs important; how can we relate the simulator to reality? In this paper we look at how emulators can be used to

A. Dienstfrey and R.F. Boisvert (Eds.): WoCoUQ 2011, IFIP AICT 377, pp. 151–164, 2012.

address these problems. For a more detailed examination of the background and theory of what is discussed here see [5].

2   Uncertainty

Assuming for the moment that our simulators are deterministic rather than stochastic where does the uncertainty come from? We know that the predictions we make are not exact. (If you are tempted to think of your simulator as perfect, how much you would be prepared to bet on its result being the same as a physical experiment?) Some of the 'error' may be numerical, running on different computer architectures or in different precisions will give different answers. But for well written code these differences will be small. In general the uncertainty in our simulator outputs comes from two sources: from uncertainty in its inputs and from uncertainty in its structure. By inputs we mean all the external inputs to the simulator; these include the initial conditions, boundary conditions and parameters. Although these may have very different properties, for example initial conditions are often spatial fields while the parameters are normally collections of single numbers, we will generally treat them the same. Similarly we take a Bayesian approach and make no distinction between aleatoric uncertainty, arising from genuine randomness, and epistemic uncertainty which is a measure of our ignorance.

The input uncertainty can be thought of as the internal uncertainty within the simulator. The structural uncertainty is how our particular simulator relates to other simulators and more importantly to reality. This is discussed further below and in [5].

3   Quantifying Uncertainty

Before we consider structural uncertainty and the relationship between simulators and reality lets first consider the input uncertainty. We take a Bayesian approach to the problem, but it can be reformulated in terms of frequentist statistics [20]. The formulae generally stay the same but the justification is different.

We have some variables (y) that we are interested in and which we will call outputs. These are related to another set of variables (x), which we will call inputs. We can find the values of the outputs corresponding to the values of the inputs by running a complex computer program which we will denote by f. Mathematically we can write

$$y = f(x)$$

We will assume for now that f is deterministic, so if we run f with the same set of inputs we will obtain the same set of outputs. We assume that any numerical error is small enough to be ignored. Some or all of our inputs might be uncertain. This might arise from genuine randomness, which we call aleatoric uncertainty; or the uncertainty could be a result of lack of knowledge, called epistemic uncertainty. The distinction between these two forms on uncertainty is not as clear as might appear. For example consider a coin toss. While the coin is in the air our uncertainty is aleatoric, but once it has landed, but is hidden from us, the uncertainty becomes epistemic. Because of this interchangeability we treat all sources of uncertainty the same and describe them with probability density functions ($\pi(x)$). If we want to know the uncertainty on y given $\pi(x)$ we need to calculate the transformation of $\pi(x)$ induced by f. If f is linear this is an easy problem and can be solved analytically, but the complex codes we are interested in are non-linear and an analytical solution does not exist. The naive solution is to use Monte Carlo methods: draw a sample from $\pi(x)$, x_i, propagate this through the program to produce $y_i = f(x_i)$. The resulting y_i are then a sample from $\pi(y)$ from which we can estimate $\pi(y)$. Such methods are effective but large samples are needed, particularly for high dimensional x and y. This means that Monte Carlo methods are not viable for expensive computer codes.

Our solution to this problem is to build a fast approximation to the full numerical simulator. This is known as an *emulator*. We not only want fast approximators, we want fast approximators that estimate their own error. It cannot be stressed too much how useful it is having knowledge of the uncertainty in the emulator estimate. When we come to validate our emulators it is invaluable as it gives us a measure against which we can gauge how far the estimator may be from the truth. Similarly it is very helpful in sequential design, where we can put the next point at the most uncertain current point and when we come to estimate the uncertainty of the outputs we need to include the uncertainty arising from the process of emulation.

4   Gaussian Processes and Emulators

Our requirement that we have an emulator that is fast and contains an estimate of its own uncertainty is satisfied by the Gaussian process, [18]. Gaussian processes are very adaptable stochastic processes which can be used to fit non-linear data. A Gaussian process (GP) is the infinite dimensional analogue of a Gaussian distribution. It is defined by a mean function and a covariance function. The mean function gives the expected value of the GP at any point. The covariance function then gives the covariance between any two points. The form of the covariance function dictates how 'smooth' a realisation of the GP is. For example if we use the 'exponential' form of the covariance function

$$cov(x_1, x_2) = \sigma^2 e^{-|x_1 - x_2|} \tag{1}$$

we get Brownian motion and any realisation does not possess any derivatives. On the other hand if we use the squared exponential covariance

$$cov(x_1, x_2) = \sigma^2 e^{-(x_1 - x_2)^2} \tag{2}$$

then all derivatives exist and we get a very smooth set of realisations. There are other forms of the covariance function, see [18] for further details. For uncertainty quantification a limited selection of covariance functions tend to be used. We rarely believe our simulators do not possess any derivatives so we tend not use the exponential form. The squared exponential is probably the most used but this can give rise to numerical problems (see below). Other options include the Matern and generalised exponential form $\sigma^2 exp(-||x_1 - x_2||^\alpha)$.

The smoothness of the GP is dictated by its covariance function but its large scale properties come from the mean function. In general a linear model is used as the mean function

$$\mu(x) = h(x)^T \beta \qquad (3)$$

where the $h(x)$ are basis functions and the β's are coefficients. The $h(x)$ can be any basis functions but are usually taken to be monomials $\{1, x, x^2, \ldots\}$. However more complex functions such as Fourier bases could be used if for example our output was on a circle. There is some discussion within the emulation community on how much effort should be put into building a good mean, or regression, model and how much the GP can be allowed to fit the large and small scale variation in the data. For example [11] make the case for not including a linear model term at all. However most practitioners do include regression terms even if they are only low order polynomials.

To analyse our simulator we use a GP as a prior and combine it with the simulator runs to produce a posterior emulator. Our prior has a mean function as given above

$$\mu(x) = h(x)^T \beta \qquad (4)$$

where the β are parameters that will need to have their own prior specified. We then specify a covariance function

$$cov(x_1, x_2) = \sigma^2 c(x_1, x_2; C) \qquad (5)$$

where σ^2 is a variance term and C is a matrix of 'smoothing' terms for the correlation function, c. For example if we have a squared exponential correlation function (sometimes called the Gaussian correlation function for obvious reasons) C might be a diagonal matrix such that

$$c(x_1, x_2; C) = exp(-(x_1 - x_2)^T C(x_1 - x_2)) \qquad (6)$$

C does not have to be diagonal but it reduces the number of parameters to estimate and making this assumption does not appear to impact on our emulators.

5   Bayes Linear Methods

When we decided to use a Gaussian process as our emulator we made more assumptions than we needed to. The assumption that all the points on our stochastic process have a multivariate Normal distribution is not necessary. All we need assume is that second moments exist and we can then specify the mean and covariance functions without making any assumptions about the statistical distribution of the process.

Such methods that use Bayes theorem but do so in terms of first and second moments are known as Bayes Linear methods and are described in [6]. The clear advantage over a full Bayes solution is that the priors are also specified only in terms of first and second moments so we do not need to elicit full probability distributions from experts. The second advantage as we shall see is that the calculations are much simpler and faster. However nothing is free and without adding some distributional assumptions we cannot make realisations of the process or produce any form of probabilistic credibility or uncertainty limits. Because Normal distributions are defined by their first two moments there is often confusion between Bayes linear methods and making an assumption of Normality. A true Bayes Linear analysis refuses to make any distributional assumptions; although the results may look similar they are conceptually very different.

As with the GP emulator our basic form is

$$f(x) = \sum_j \beta_j h_j(x) + w(x) \tag{7}$$

The regression terms are identical but the $w(x)$ is not a Gaussian process but rather a general second order process, defined by its covariance function.

The equivalent to Bayes Theorem for Bayes Linear are the Bayes Linear update equations. If θ is a vector of our parameters (in our case this will be the β's, σ^2 and the length scales) and x are the results of our runs

$$E(\theta|x) = E(\theta) + Cov(\theta, x)V(x)^{-1}(x - \sum_j \beta_j h_j(x)) \tag{8}$$

for the adjusted expectation and by

$$V(\theta|x) = V(\theta) - Cov(\theta, x)V(x)^{-1}Cov(x, \theta) \tag{9}$$

The full equations for the posterior moments of the Bayes Linear emulator are rather more complicated and are given in full in the Core BL Emulator Thread of the MUCM toolkit [14]

6   MUCM and the Toolkit

The Managing Uncertainty in Complex Models (MUCM) consortium consists of five UK research institutions (University of Sheffield, University of Durham, University of Aston, London School of Economics and the National Oceanography Centre). As part of its research activities MUCM has set up an on-line toolkit. This consists of more than 300 pages describing most aspects of emulation. The toolkit is not a software package. The best analogy is a recipe book, one of those good recipe books that encourage you to experiment. There are worked examples so you can check that your code works and there are links to existing software packages. Part of an example page is shown in Fig. 1. The URL for the toolkit is http://www.mucm.ac.uk/toolkit.

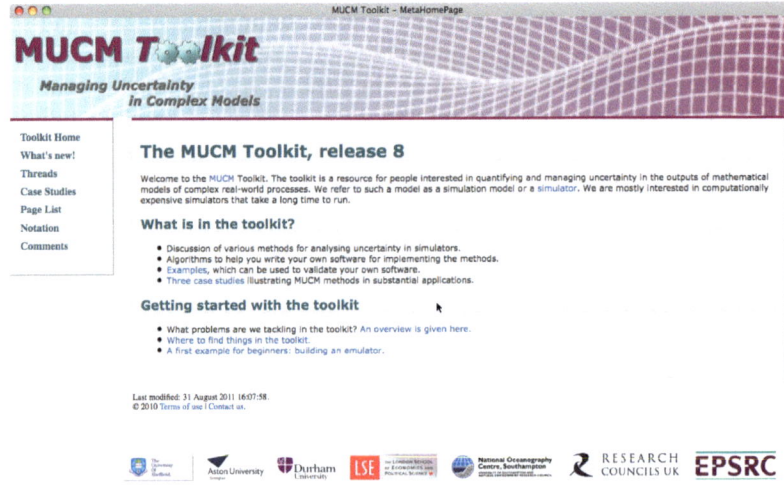

Fig. 1. The top page for the MUCM toolkit

The toolkit is arranged in thirteen threads plus three large case studies. The threads include two 'core' threads on emulating with Gaussian processes and Bayes linear emulation. Further threads explore how to extend these two core procedures when there are multiple outputs, dynamic emulators, two level emulators (one more complex than the other), combining multiple, independent emulators and using and obtaining derivative information. The issues of model discrepancy, history matching, calibration and sensitivity analysis are also covered as well as the important problems of experimental design and screening. In the near future we expect at least one additional thread on emulators for stochastic simulators. Within what is a relatively short paper it is impossible to cover the breadth of material in the toolkit so we will concentrate on a few highlights.

7   Building an Emulator

In the MUCM Toolkit we have two threads that cover what is described as the core problem. The core problem is in many ways the simplest application for an emulator. We have a single output of interest, we do have multivariate inputs but no information on the derivative of the output. There are no real world data to compare the simulator to and we are not concerned with making inferences about reality; we are only interested in the simulator. This is not as uncommon as might be thought. Simulators, particularly relatively simple ones, are often used to study an idealised version of the system rather than to make inferences about the real world.

The steps in building an Gaussian process emulator are:

1. Set up the initial GP model. Decide on the form of the regression terms and the correlation function.

2. Decide what priors we are going to use for the GP parameters.
3. Design and run the initial experiment. This is sometimes called the training experiment. Design is an important issue. Computer time is usually limited and each simulator run is expensive so every run must be made to count. Design has its own thread on the MUCM toolkit and is discussed further below.
4. Build the emulator.
5. Validate the emulator. This is a crucial step. Our first attempts at building good emulators are often failures and we need to establish, and possibly convince others, that we have built an emulator which we can have confidence in. There are two approaches. The first is to use a leave one out cross validation. Each point in the training set is left out in turn and an emulator is built using the remaining points. This reduced emulator is used to predict the left out point. The mean square error of the differences between these predictions and left out runs gives a global measure of the quality of the emulator, while a plot of the individual differences may reveal areas of input space where the emulator is performing badly. The advantage of leave one out is that we do not need to perform any more runs of the simulator. The disadvantage is that we are not testing the full emulator but emulators built using reduced datasets. Our carefully crafted space filling design is continually compromised by having one point at a time omitted.

   The alternative is to run a second experiment with an independent set of simulator runs. These are then used to test the emulator. This can be done by simply comparing the difference between the predicted values and the truth (scaled by the predicted uncertainty) similar to the leave out method or there are more rigorous methods based on regression diagnostics that take into account the correlation between points [1]. Once we are satisfied that we have an emulator that validates we can combine the training and validation runs and create a single emulator.

The procedure for a Bayes linear emulator is similar but the specification of the priors is in terms of moments rather than full prior distributions.

8 Design

One of the first steps in any experiment is the design. This is as true of computer experiments as it is of field trials. Our methods have been developed with expensive simulators in mind, so our designs need to minimise the number of runs required. However we also we wish to cover the complete input space. This seems like an impossible task. If we were to use a traditional factorial design with only 2 levels for each input the number of runs required would be 2^p where p is the number of inputs. Two levels of each input gives us very poor coverage and 2^p becomes unaffordable for even a small number of inputs.

We therefore have to look at other designs. There are two main families of design: those based on the Latin Hypercube and those based on Quasi-Monte Carlo sequences.

The traditional design is the Latin Hypercube [13]. In a Latin Hypercube design we first decide how many evaluations of the simulator we can afford in total, let this be n. We then divide the range for each input variable into n equal sections. The simulator is evaluated once, and only once, in each of these settings. This means that we have good marginal coverage of each of the variables. The Latin Hypercube design is now produced by permuting the numbers $1, ..., n$ for each variable separately. Note the randomisation is to produce a design rather than to randomise for external factors as it is used in field experiments. All Latin Hypercubes have good marginal properties but are not necessarily the space filling. There is no algorithm for an optimal space-filling Latin Hypercube. In general an additional space-filling criteria, maximising the minimum distance between points [9] or minimising the sum of inverse distances for example, is imposed on the design. Alternatively we can use Latin Hypercubes based on orthogonal arrays [17].

An alternative to the Latin Hypercube is to use a low discrepancy sequence to define the design points. Such sequences were originally devised to efficiently compute multidimensional integrals [15]. These sequences are space filling but for small numbers of simulator runs there can be problems with certain projections having 'holes' in them. Examples include the Halton sequences [7] and Niederreiter nets. For the design of computer experiments the most commonly used low discrepancy sequence is the Sobol sequence [22,3].

Current work in the design of computer experiments explored in the MUCM toolkit [14] includes sequential design where we use an initial space filling design to learn about the system and then use this information to guide us on where future runs should be carried out.

9   An Example

As a simple example consider a energy balance model of the Earth's climate. The Earth is reduced to a line of grid boxes all ocean with a single box below to represent the deep ocean. If it becomes cold enough in a box sea ice forms this has a different albedo to water so a different amount of radiation is reflected from sea ice. As the surface water becomes colder it becomes denser and can sink at a variable location in the North. Deep water upwells back to the surface at a fixed location in the South. Heat is transferred between the surface grid boxes by both advection and diffusion. The system is driven by incoming short wave radiation from the Sun. As in the real world this is greater at the equator than the poles. The total amount of incoming radiation is set by the Solar constant. As this is a simple example we only vary one input, the Solar constant, and we only use a single output, the mean surface temperature. Note as is often the case the output we emulate is a function of the simulator not a state variable.

A space-filling Latin Hypercube in 1-d is simply a set of evenly spaced points. Using 6 points we get the emulator shown in panel (a) in Fig. 2. The dashed line shows the expected value of the emulated and the shaded region gives 95% uncertainty bands. Because we do not have a nugget term in the emulator the uncertainty collapses to zero where we have simulator runs. This makes perfect sense as here we *know* the value of the simulator output. Note that because this simulator is very cheap to run we can run it across all of input space and plot the true simulator value across the range and this is shown by the solid line in the figure.

We now do a further three runs of the simulator. Fig. 2 shows the standardised residuals for each of these new points. This is the distance between the expected value from the emulator and the true value divided by emulator standard deviation. If the emulator were correct we expect these numbers to have Normal distributions (for simplicity we are ignoring correlation here; for a full solution see [1]). One of the points is outside the limits shown in the figure, indicating that we do not have a good emulator. We therefore combine our existing points, both the six original training points and the three new validation points, into a single dataset and build a new emulator. This is shown in Fig. 2 panel (c). We now need to validate this new emulator. A further three runs are carried out and this time all three points are within the bounds for the standard used residuals. The emulator in panel (c) therefore validates. However we now have three additional runs which we can use to build an even better emulator. This, our final emulator, is shown in panel (d) of Fig. 2.

10   Sensitivity Analysis

One important application of emulators is in sensitivity analysis. Although computer simulators may have large numbers of inputs, often the outputs are dependent on only a few. Formally we can look at sensitivity analysis as answering the question: if the inputs x are changed by a small amount δx what is the effect on the simulator output $f(x)$. One way of looking at this problem is to vary each input in turn, run the simulator and see what effect the change has on the output. This is known as one at a time sensitivity analysis. If we could guarantee that all the inputs were completely independent of each other it might not be a bad idea, but we are dealing with large, complex, non-linear simulators and it would be foolhardy to make such an assumption. Any approach to measuring the sensitivity of a simulator must acknowledge that there will interactions between the inputs and should at least estimate not only the effect of single inputs (main effects) but also at least the first order interactions.

Traditionally this has been tackled by looking at the derivatives of $f(.)$ w.r.to x. This gives the *local* sensitivity since $\partial f(x)/\partial x$ depends on x and may change radically as we move around input space. An alternative is to use variance based sensitivity analyses, see for example [19]. [16] extend these methods so that they can be used with emulators.

11   History Matching and Calibration

After sensitivity analysis, probably the most important application of emulators is in comparing simulators with data and calibrating the simulators. We make a

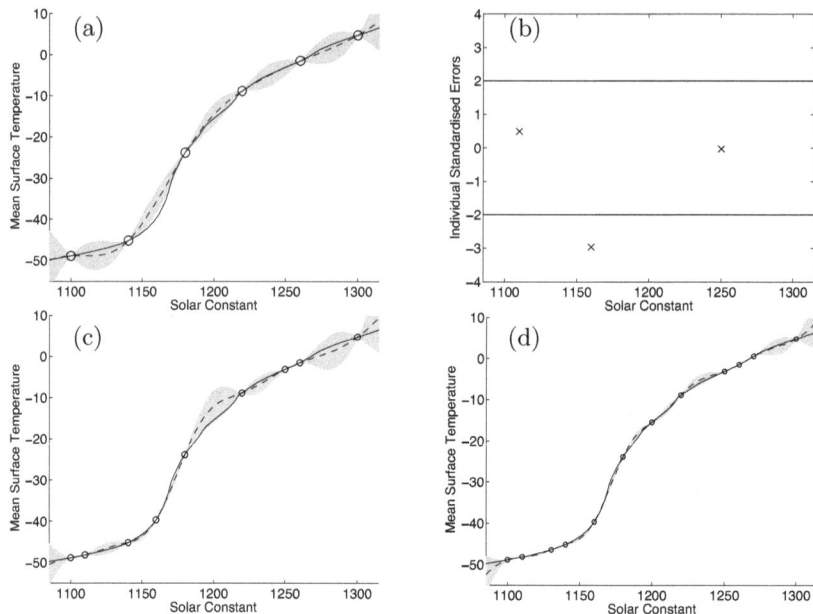

Fig. 2. Panel (a) shows the expected value of the emulator (dashed) and the true simulator based on six runs (o). The shaded region is a 95% uncertainty band around the expectation. Panel (b) shows the values for 3 additional validation runs with the emulator uncertainty. Note that one of the points is outside the uncertainty limits. Panel (c) shows the emulator built using all nine points (the original training set plus the three validation points). This emulator validates with an additional three points (not shown). The final emulator based on 12 points (6+3+3) is shown in panel (d)

distinction between simulator comparison, often referred to as history matching, where we simply rule out areas of input space that are incompatible with the data and calibration where we estimate the 'best' input values from the data.

Both rely on the concept of 'model discrepancy'. The simulator is attempting, in most cases, to simulate some property of the real world. But if we measure that property do we expect the simulator to give an exact fit to the data? The answer is almost certainly not. In building simulators we make assumptions, we parameterise processes and there are real world processes we do not include, or even no about. All these omissions and approximations mean that we do not expect our simulator to explain the data perfectly. We describe this difference between the real world and the simulator as the model discrepancy. Model discrepancy will change as the input values change and often will have to be elicited from experts rather than being estimated from the data. Statistically it is possible to think of model discrepancy in two ways: as a bias between the simulator/emulator and the data, or as an increase in variance around the simulator/emulator.

First consider the problem of history matching. Some data are collected and we wish to know which values if any of the inputs could have produced them.

Table 1. Table showing the reduction in Not Ruled Out Yet space by application of an implausibility measure to the Galform simulator of galaxy formation. For details see [23].

Wave	Runs	% Space
1	993	14.9%
2	1414	5.9 %
3	1620	1.6 %
4	2011	0.26 %

Translating this into the negative language of statisticians, who only ever reject hypotheses and never accept them, we are looking for input values that are *implausible* given the data. This is done via an implausibility measure

$$I_{mp} = \sqrt{\sum_{all\ x} \frac{(x - x_{emul})^2}{\sigma_{data}^2 + \sigma_{emul}^2 + \sigma_{discrep}^2}}$$

where x is a data point; x_{emul} is the emulator value corresponding to that data point; σ_{data}^2 is the variance of the data; σ_{emul}^2 is the variance of the emulator; and $\sigma_{discrep}^2$ is the model discrepancy expressed as a variance. Values of implausibility greater than 3 (or 5 depending on how conservative we want to be) are ruled to be implausible. Additional simulator runs are then done within the input space not yet deemed implausible (sometimes to as NROY, Not Ruled Out Yet, space). By refining the emulator and adding additional runs in waves within NROY space the volume that hasn't yet been deemed implausible falls rapidly. Table 1, taken from [23], shows the volume of NROY space at each wave of the experiment as well as the number of runs in each wave.

If we want to go beyond history matching to model calibration we need to reformulate the problem. We follow [10]. First we split the inputs into two: control inputs and calibration inputs. The calibration inputs are those inputs we are trying to calibrate. Control inputs on the other hand are inputs that control the simulation but which can't or won't be calibrated. For example in an environmental simulator we may have inputs that give the spatial position of the outputs, these would be control inputs. We now say that the simulator $f(x_{con}, x_{cal})$ is the sum of reality $y(x_{con})$ and discrepancy $d(x_{con})$. Note that in this formulation both reality and the discrepancy are functions only of the control inputs not the calibration inputs. We have some measurements of reality, $z(x_{con})$ which are given by

$$z(x_{con}) = y(x_{con}) + \epsilon \tag{10}$$

We can now build an equation that links the data (z) and the simulator.

$$z(x_{con}) = f(x_{con}, x_{cal}) + d(x_{con}) + \epsilon \tag{11}$$

We now use Gaussian processes not only to build an emulator for $f(x_{con}, x_{cal})$ but also for $d(x_{con})$. The posterior of x_{cal} is the calibrated distribution inputs. For full details see the toolkit [14] or [10].

12 Beyond the Core Problem

The core problem as discussed above gives us a basis on which to expand to more complex problems. Examples dealt with in the toolkit include:

1. Multiple outputs where we are interested in more than a single output from the simulator.
2. Dynamic emulators, where we are interested in an output that is itself changing over time.
3. Multiple level emulators, where we have a number of simulators ranging from a coarse, fast simulator through a hierarchy to a slow, but more accurate, simulator.
4. Derivatives. Often we have derivative information available from the simulator in the form of an adjoint model. We can use this information to improve our emulator. However there is a trade off between the extra expense in calculating the derivatives and producing more runs of the simulator without derivatives. In other cases we may not have derivative information from our simulator but we are interested in the form of the derivatives. For instance we may be interested in local sensitivity. Even with the use of automatic differentiating compilers [8] the production of an adjoint for any reasonably sized simulator is a major undertaking. An alternative is to produce an emulator for the derivative. This is relatively easy as the derivative of a Gaussian process is another Gaussian process. Validating the derivative emulator is more difficult if we do not have an adjoint to compare with.

13 Conclusions

I hope that I have managed to show in this short introduction that both Gaussian process and Bayes Linear emulators are powerful tools in the quantification of uncertainty. The MUCM Toolkit [14] is a good reference for these methods in sufficient detail to allow code to be developed. The toolkit is not static and is updated approximately quarterly. In the near future we expect it to be extended to cover stochastic simulators

Acknowledgements. This paper wouldn't have been possible without the help and support of the MUCM team, in particular Ioannis Andrianakis (NOC), Ian Vernon (U. Durham) and Hugo Maruri-Aguilar (QMUL). John Shepherd (U. Southampton) kindly provided the simple climate model used in section 9.

References

1. Bastos, L., O'Hagan, A.: Diagnostics for Gaussian Process Emulators. Technometrics 51(4), 425–438 (2009)
2. Bates, R., Kenett, R., Steinberg, D.: Achieving robust design from computer simulations. Quality Technology and Quantitative Mangement 3(2), 161–177 (2006)

3. Bratley, P., Fox, B.: ALGORITHM 659: implementing Sobol's quasirandom sequence generator. ACM Transactions on Mathematical Software (TOMS) 14(1), 88–100 (1988)
4. Collins, W., Bitz, C., Blackmon, M., Bonan, G., Bretherton, C., Carton, J., Chang, P., Doney, S., Hack, J., Henderson, T., Kiehl, J., Large, W., McKenna, D., Santer, B., Smith, R.: The Community Climate System Model: CCSM3. J. Climate 19, 2122–2143 (2006)
5. Goldstein, M.: Bayesian analysis for complex physical systems modelled by computer simulations: current status and future challenges. In: IFIP (2011)
6. Goldstein, M., Wooff, D.: Bayes Linear Statistics, Theory and Methods. Wiley (2007)
7. Halton, J.: On the efficiency of certain quasi-random sequences of points in evaluating multi-dimensional integrals. Numerische Mathematik 2, 84–90 (1960)
8. Hascoët, L., Greborio, R.M., Pascual, V.: Computing adjoints by automatic differentiation with Tapenade. In: Sportisse, B., LeDimet, F.X. (eds.) Ecole INRIA-CEA-EDF Problemes Non-lineaires Appliques. Springer (2005)
9. Johnson, M., Moore, L., Ylvisaker, D.: Minimax and maximin distance designs. Journal of Statistical Planning and Inference 26(2), 131–148 (1990)
10. Kennedy, M., O'Hagan, A.: Bayesian calibration of computer models. Journal of the Royal Statistical Society: B 63, 425–464 (2001)
11. Lim, Y., Sacks, J., Studden, W., Welch, W.: Design and analysis of computer experiments when the output is highly correlated over the input space. Canadian Journal of Statistics-Revue Canadienne De Statistique 30(1), 109–126 (2002)
12. Maltrud, M.E., McClean, J.: An eddy resolving global $1/10°$ ocean simulation. Ocean Modelling 8, 31–54 (2005)
13. McKay, M., Beckman, R., Conover, W.: A comparison of three methods for selecting values of input variables in the analysis of output from a computer code. Technometrics 21(2), 239–245 (1979)
14. MUCM: MUCM Toolkit, http://www.mucm.ac.uk/toolkit
15. Niederreiter, H.: Random number generation and quasi-Monte Carlo methods. In: CBMS-NSF Regional Conference Series in Applied Mathematics, vol. 63. SIAM, Philadelphia (1992)
16. Oakley, J., O'Hagan, A.: Probabilistic sensitivity analysis of complex models: a Bayesian approach. Journal of The Royal Statistical Society Series B-Statistical Methodology 66(3), 751–769 (2004)
17. Owen, A.: Orthogonal arrays for computer experiments, integration and visualization. Stat. Sinica 2(2), 439–452 (1992)
18. Rasmussen, C., Williams, C.: Gaussian processes for machine learning. MIT Press (2006)
19. Saltelli, A., Chan, K., Scott, E.: Sensitivity Analysis: Gauging the Worth of Scientific Models. Wiley (2000)
20. Santner, T., Williams, B., Notz, W.: The Design and Analysis of Computer Experiments. Springer, New York (2003)
21. Shakun, J., Carlson, A.: A global perspective on Last Glacial Maximum to Holocene climate change. Quaternary Science Reviews 29, 1801–1836 (2010)
22. Sobol, L.: On the distribution of points in a cube and the approximate evaluation of integrals. USSR Comput. Math. and Math. Phys. 7, 86–112 (1967)
23. Vernon, I., Goldstein, M., Bower, R.G.: Galaxy Formation: a Bayesian Uncertainty Analysis. Bayesian Analysis 5(4), 619–669 (2010)
24. Wilkinson, D.: Stochastic Modelling for Systems Biology. Chapman and Hall (2006)

Discussion

Speaker: Peter Challenor

Maurice Cox: This seems a very useful toolkit. I understand you are on your second grant (MUCM2). I would be interested in hearing plans for your maintenance of the toolkit beyond MUCM2.

Peter Challenor: That is a very pertinent question that is at the top of our agenda. The MUCM2 project runs until October 2012. Maintaining the toolkit frozen from that point would not be expensive and even further developments could be done for little cost and we are looking for funding to put the toolkit on a long term sustainable basis. We would like to encourage non-MUCM participants to contribute to the toolkit and make it a true community resource.

John Reid: You have said nothing about parallel programming. It strikes me that you scope for "embarrassingly parallel" (ideal) execution.

Peter Challenor: Because we rely on ensembles of simulator runs it is quite correct that have an embarrassingly parallel problem. However some of the simulators we are working with are so large that we do not have the computer resources to make use of this. Parallel computing brings up some interesting questions in the design of experiments. I mentioned sequential designs above. Traditionally sequential designs would involve additional single simulator runs, but if we have access to parallel computing it is much more efficient to use 'batch sequential' designs where we run n additional simulations at a time.

John Rice: Have you used your simulator to explore the possible sources of the unusual climate changes observed in historical data? For example, the "little ice age" that occurred a few centuries ago or the major climate change that occurred about 12000 years ago and which appears to have occurred several times earlier with a periodicity of (as I recall) about 125000 years? Historically speaking, we appear to be near the end of a long warm period if this cycle is persistent.

Peter Challenor: The simple Earth radiation balance simulator we use in the example is not suitable for reproducing the history of the Earth's climate. It is much too simple and is lacking too many process. For a good description of the Earth's climate over the last 20000 years see [21]

Will Welch: The MUCM Toolkit provides recipes for developing code. To test that code it would be useful to have test cases complete with data and results. What plans do you have to provide such test cases?

Peter Challenor: This is a very good point and it is our intention to supply such test cases with most, if not all, the pages. We currently have worked examples for some pages; the example in section 9 is taken from the coreGP pages. We currently have nine test cases plus the three large case studies. Over the next year I hope we will see many more being produced.

Measuring Uncertainty in Scientific Computation Using Numerica 21's Test Harness

Brian T. Smith

Numerica 21 Inc.
Angel Fire, NM, USA
carbess@swcp.com

Abstract. The test harness, TH, is a tool developed by Numerica 21 to facilitate the testing and evaluation of scientific software during the development and maintenance phases of such software. This paper describes how the tool can be used to measure uncertainty in scientific computations. It confirms that the actual behavior of the code when subjected to changes, typically small, in the code input data reflects formal analysis of the problem's sensitivity to its input. Although motivated by studying small changes in the input data, the test harness can measure the impact of any changes, including those that go beyond the formal analysis.

Keywords: testing, scientific application code, floating-point computation, data perturbation, computational sensitivity, test harness tool.

1 Introduction

An article on the website of the National Physical Laboratory on a Framework for Uncertainty in Measurement, June 5th, 2007 [1] makes the following statement:

> "A measurement is meaningless without a quantitative statement of its quality in the form of an uncertainty."

This statement is just as true about a scientific computation as it is about a physical measurement. Software is useless unless the uncertainty in the computed results due to changes in its input or instabilities in the way the results are computed are measured or analyzed. To believe computational results, it is essential to demonstrate that the sensitivity of computed results to changes in input data, precision of computation, and other key data for the software is consistent with what is predicted from the characteristics of the problem being solved. Ideally, a measure of the uncertainty of the computed results as a consequence of the uncertainty of data that the results depend upon needs to be obtained.

In many cases, such a measure of uncertainty is difficult to obtain analytically. However, uncertainty in software can still be measured by running the software with perturbed data values to see whether the computed solution changes as expected. With some thought, a measure can often be devised to indicate how

A. Dienstfrey and R.F. Boisvert (Eds.): WoCoUQ 2011, IFIP AICT 377, pp. 165–179, 2012.

the results change with small perturbations in the input and other key data. This paper is about a tool to help provide an assessment of uncertainty in scientific software and computation.

The tool is a general-purpose test harness modified to facilitate the measurement of changes in results due to changes in input data and other values key to the computation.

Presented in this paper is a brief description of the test harness, its design, its features, and how it was modified to facilitate the measurement of uncertainty of a computation with respect to changes in its input. A case study is provided to show how this tool has been used to measure the uncertainty of the solution of a 3-D magnetostatics computation for the vector potential and magnetic flux or field. The solution technique used for this magnetostatics case study is a boundary element package for 3-D magnetostatics problems from a software firm Accurate Solutions In Applied Physics [2].

2   Motivation – Measuring Uncertainty in Software

Software is at the end of the development chain, depending on mathematical models of physical problems that become the basis for the numerical computation. The solutions to these problems depend on the algorithms used and the data used to drive those algorithms.

Software is the final step. As such, we want to determine if the software is behaving as the mathematical and physical models are predicted to behave. The approach proposed here is to provide a tool that allows one to measure the sensitivity of computed results due to changes in input values or critical parameters in the models, algorithms, and software. Such a measurement of sensitivity indicates how uncertain the computational results are with respect to the uncertainty in the values of such critical data.

3   What Is the Test Harness?

The test harness is a tool to evaluate software. In its initial form, it was a change-detection tool that measured differences in results of two programs that were supposed to create the same results. The applications for such a tool are many: to give a few examples,

- the two programs may be actually the same program compiled by two different sets of compiler flags, such as optimization flags;
- the two programs may be the same but run on different machines;
- one program is an enhancement of one other, enhanced to improve performance but compute the same results, enhanced to use different data structures or organized differently, or enhanced to add some new feature but the developer wants to show that the other features remain unaffected by the enhancements.

The key to the test harness in its original form was to measure "significant" differences in results, that is, differences that represent errors and not differences that can be traced to reordering of operations, changing results of stable computations in minor ways. Therefore, it was essential to have the user provide both the criteria for the comparison (say, relative or absolute difference) and a threshold to indicate whether the difference was predicted and thus acceptable, or unpredicted and thus not acceptable, indicating something was wrong. Also, in scientific applications, arrays and other aggregates need to be compared and criteria for them are needed and have in some cases to be specified by the user.

The test harness is designed to support large scientific codes. As such, these codes involve large collections of data and with such programs, the writing, reading, and comparisons of large volumes of data can be costly. Consequently, the test harness allows the user to select which procedures are monitored, which variables are monitored, which parts of arrays are monitored, how often they are monitored or when the monitoring begins or ends. The test harness measures and reports the volume of data it is monitoring so that the user controls how much is monitored on a given run. It also gives execution counts and execution times for each procedure monitored.

Without going into all the details, the test harness can address all these issues, as described in a previous paper [3] and in its user's guide [4]. Enhancements made to the test harness to support uncertainty measurements are described below. The test harness is currently written for scientific codes in Fortran 77/90/95/2003.

4   The Design of the Test Harness

The test harness is a collection of modules containing input/output procedures to read and write the monitored data, a collection of generic INCLUDE files modified by tools to create application-specific INCLUDE files that include application specific source text into the application code, and a collection of procedures to perform data comparisons and report differences in results. Either the program terminates with the first significant difference or reports its results in a tabular form for an entire program execution. In addition, described in the next section, there are a series of tools that read and analyze the application code, determine default places to monitor results, and build the test harness into the application. The application code with the test harness installed into it can be run in one of two modes described below.

The modes for the application code are: **generate** mode and **check** mode. In **generate** mode, the application code runs to completion, creating data from the run to be compared in **check** mode with another version of the application code. In **check** mode, the data written into files in **generate** mode are read at the point where the corresponding data in the second program is computed and a comparison of the results is performed. In **check** mode, there is the option to terminate the execution at the end of the probe where the first unacceptable result (difference or evaluation that indicates a problem) is encountered or to tabulate the difference and continue execution until the application code completes. Upon completion in the latter case, a summary of the unacceptable results is printed.

Four types of probes or monitoring can be specified; three of the types, namely `input`, `output`, and `specific`, probe record data in a file when the test harness is in `generate` mode and read recorded data and perform comparisons with results recorded in `generate` mode. An `input` probe can be placed at any entry point to a procedure; an `output` probe can be placed just before any exit point from a procedure; and a probe of type `specific` can be placed at any place in the execution part of the application code. The probes are different in what they record; this enables them to make certain checks to ensure only corresponding data is being compared. The fourth type of probe is a `perturb` probe which perturbs specified data values in specified ways when the test harness is in `check` mode, allowing the other probes, if present, to read and compare results between the application code with the original data and results with perturbed data.

5   Building the Test Harness into an Application Code

Much like a debugger, monitoring probes must be placed into the application code. The emphasis though with the test harness is to facilitate the comparison and evaluation of results for floating point (although the test harness supports the monitoring of any intrinsic type or derived type object). Besides addressing the added complication of comparing floating point values, test harness must ensure that the data being compared between the `generate` and `check` modes are comparable values; to do this, it has to trace the execution flow by procedure and order the data so that comparable values (values of the same entities) are compared.

Tools have been created to accomplish these tasks and ensure the integrity of the comparisons. Fig. 1 shows the use of the tools to produce a source code file that represents the application code with the test harness build into it. The `analyzer` tool first analyzes the application code, providing a complete specification of all variables in all procedures in the application and performs a simple usage analysis of each variable to determine if it is referenced for its value before it is written into or is always written before it is referenced. Given this analysis and a list of procedures to monitor, the `installer` tool creates a file readable by the `builder` tool that specifies the input and output probes for each entry and exit point for the listed procedures. Also, given the results of the `analyzer` tool and a list of procedures, the probe tool creates a version of the application code with INCLUDE lines inserted into it. The INCLUDE lines will include source text that will be synthesized by the `builder` tool that represents the test harness built into the application code.

At this point, the user is expected to modify both the files created by the `installer` and `probe` tools. The reason is that these files specify default comparisons and thresholds, probably specify more probe variables than are appropriate for the goal of investigating the code, and may specify more probes than are desirable or appropriate (the reference/definition analysis is only approximate and in general includes variables that need not be monitored). In the case

The TH Tools

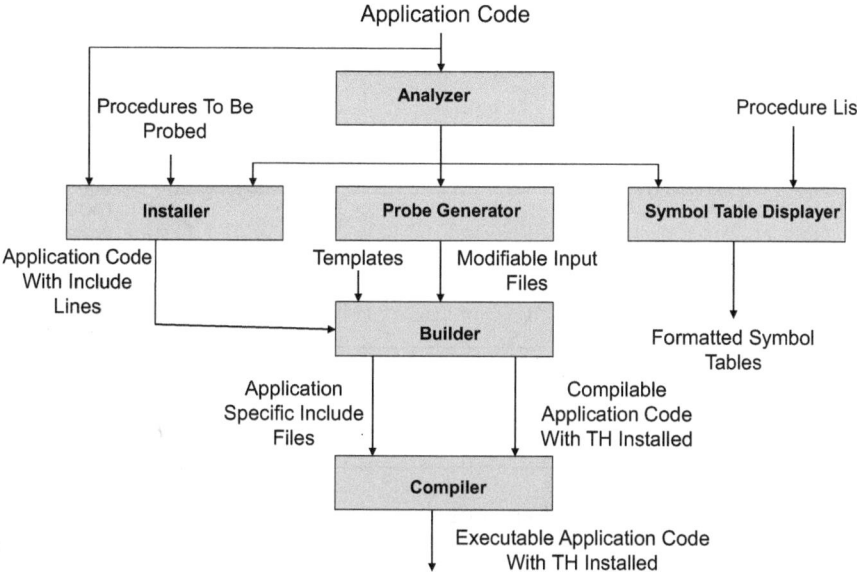

Fig. 1. The test harness tools showing their input and output connections to install the test harness into an application code

of the files created by the **probe** and **installer** tools, these can be modified and reused in subsequent runs without rerunning these tools. In the case of the **installer** file, the only expected modification is the deletion of INCLUDE lines that represent unwanted probes. One of the major modifications to the **builder** monitoring specification files (created by the **probe** tool) is also the deletion of inappropriate monitored variables; changing the default monitoring thresholds to appropriate thresholds for the computation is unavoidable until further tools are provided.

Once the builder input files are modified, the **builder** and **includer** tools with the modified files complete the installation of the test harness into the application code. The monitoring process proceeds by running the application code with the test harness installed and often involves revisiting the choice of thresholds and monitoring. The typical situation is that the **analyzer** and **probe** tools are not rerun while investigating the behavior of the code. The user can change what and how variables are monitored, even what procedures are monitored, without rerunning the **analyzer** and **probe** tools. If variables that are monitored are changed or their comparison criteria or thresholds are changed, the test harness must be rerun in **generate** mode (including measurements of uncertainty or code sensitivity to data).

Fig. 2 shows a typical scenario with the use of the test harness. The top line represents versions of the code that are run in **generate** mode. In this

Usage Scenario

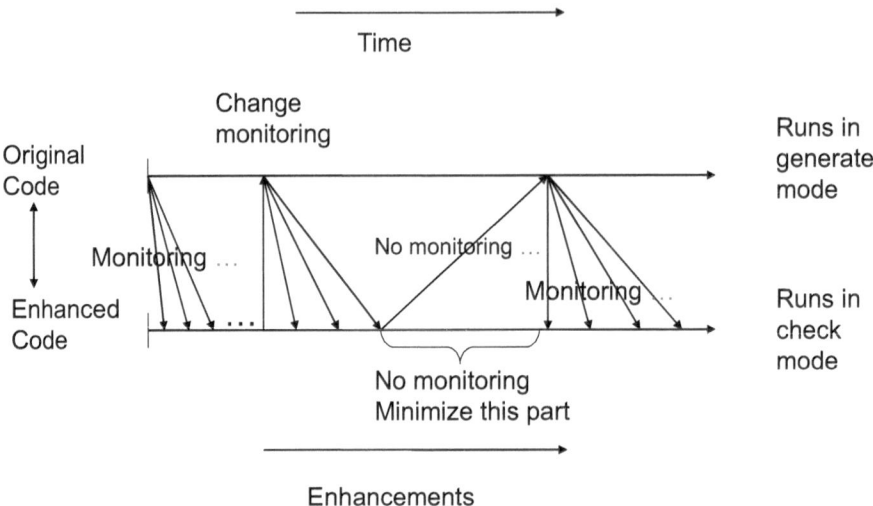

Fig. 2. A scenario for using the test harness to support code development. The top line represents the original version of the code compared with the modified version on the bottom line. The code is enhanced as time progresses to the right and the diagonal lines down represent comparisons made between the original and enhanced versions of the monitored variables. The line with the up arrow indicates the modified version becomes the production version at certain points and the monitoring is changed.

mode, values of variables in procedures specified in the `builder` file are recorded for later comparison. Many runs of the test harness in `check` mode (including perturbing variables) can be run and compared with the monitoring data created in the version run in generate mode.

The bottom line of Fig. 2 represents versions of the application with the test harness built in the application run in `check` mode. Each of these runs may have different criterion for comparison or different perturbations specified in the builder input files but require the `builder` and `installer` tools to be rerun to generate a version to run these different cases. Time progresses towards the right. The slanted lines down from the same "generate" version indicate different runs with typically different builder input files. Also, the application source files with the probes inserted may also be changed as long as the variables that are monitored or evaluated or the order in which they are generated are not changed. At some point, it is desirable to use the code as it has been changed or monitor different items. That process is indicated by the lines with the up arrows in Fig. 2 where the version from the lower line is run is `generate` mode to create anew the monitored data file. The process of monitoring and changing codes continues anew, until the user is satisfied with the results.

The diagonal line with an up arrow indicates a jump in versions where the code that is being monitored is changed in some substantial way and no testing of the changes is made. Hopefully, these kinds of changes do not often occur because they represent situations where code is enhanced and tests are not performed to ensure that the existing code still performs the way it used to before the changes.

For uncertainty measurements, the process is much simpler, essentially depicted only by the downward arrows and the code is never changed. However, the builder input files may be changed to measure the sensitivity of the computed results caused by changes in different variables or combinations of variables. Using the test harness for uncertainty measurements is described in more detail in Sect. 7.

6   Brief Summary of the Test Harness Features

When the execution of the application code with the test harness installed in it is executed, the test harness reads files which specify several options:

- The mode, either **generate** or **check**;
- Whether execution performance for each monitored procedure is to be measured;
- Whether the sizes of the files containing the monitor data for each procedure are recorded and printed at the completion of the application code, when in **generate** mode;
- Whether the application code terminates or continues on the first occurrence of a difference that violates the comparison threshold;
- Whether a summary report of all comparisons performed during the run upon completion of the application code is printed;
- The Fortran logical units for diagnostics and for debugging output;
- The Fortran logical unit for the monitored data;
- The maximum number of monitored procedures;
- The name of the main program;
- The maximum number of routines to be monitored; if it is not provided, the default is 100;
- The default value for the tabulate option. If it is not present, the default value is no tabulation.

The *builder* input files allow the following attributes to be provided; if not provided, a default value is set in all cases:

- The lower bound for array subscripts, when such lower bounds cannot be determined from the source code, for example, when the array is assumed-size. The default is a vector of 1's of the rank of the variable;
- The relative tolerance or threshold used to compare floating point data values. The default is zero, which implies the compared values must be identical. For types other than floating point, the relative tolerance test is not made;

- The norm type, either element or global; the default is element. Element means the comparisons for an array are element by element and for the relative threshold, relative to each element. Global means the relative comparisons are element by element but relative to the norm of the array;
- The absolute tolerance or threshold used to compare arithmetic data values. The default is zero, requiring the compared values to be identical;
- The scope of the comparison, either global for all variables of this name, or local to this procedure. The default is local;
- The section of the array to be compared. The default is the whole array. However, for assumed-size arrays, a section which specifies the final subscript value or range is required because the **analyzer** cannot determine this value;
- Whether to tabulate the comparison results and print the tables at the end of the execution of the application code when in **check** mode. The default is the **tabulate** option specified in the test harness input file;
- The specification of "how" the perturbation is to be performed. The options are a relative or absolute perturbation or by a specific value of the specified variable with the random perturbation at most the size of a specified value, selected from a uniform distribution. There is no default value; the option must be specified;
- The name of a procedure that is to perturb the variable. If provided, the procedure overrides all other specifications of how the variable is to be perturbed. The default is no procedure specified;
- The name of a procedure that is to perform the comparison of data values. The default is no procedure specified.

The output generated when "tabulate=yes" is specified is printed on standard output after the application completes execution. It is a large table with a collection of lines for each probe for each procedure that is monitored. For each procedure, there is a line for each variable. The information printed in the table is:

- The name of the procedure;
- The kind of probe (1 for input probe, 2 for output probe, 3 for a specific probe, and 4 for a perturb probe);
- The probe name;
- The variable name;
- The flag E or N; E indicates the threshold was exceeded; N indicates the threshold was never exceeded;
- The type of comparison, when an array; elemental or global;
- Two sub-tables, one for absolute comparisons and one for relative comparisons. In each sub-table, 2 or 3 columns are provided for the first, maximum, and average differences, indicating the value of the difference. Also provided is the procedure call count for the reported difference and with the linear position in the array, if an array, where the difference occurred.

7 How Uncertainty Evaluation of Software Is Performed with the Test Harness

First, the `builder` input file is modified to specify the variables to be monitored and perturbed. This includes how they are to be perturbed and how the results are to be compared. Then the application code with the test harness in `generate` mode is executed to create a collection of monitored data. Next, the application code is rerun with the test harness in `check mode`, tabulating the results. Just before the test harness closes in the `check` mode run, it prints the results, indicating how the perturbation changed the results.

8 A Case Study – Measuring Uncertainty by Perturbing Data

The case study to demonstrate how to use the test harness to study uncertainty of computed results due to changes in input is a package of double precision codes to solve 3-D magnetostatics problems for the vector potential and magnetic flux using the boundary element method developed by ASAP LLC [2]. The equations solved are the 3-D Laplace equations with boundary conditions specified over the surfaces of 3-D objects. The test problems use spheres, annular cylinders, cubes, and tori.

The boundary element methods solve the Laplace equation by integrating Green's functions over boundary elements to produce a relatively large linear system of equations. The size of the system is dependent on the number of boundary elements. The integrands are singular in many cases and are transformed in several ways to remove the singularities, but unless care is taken, the matrix of the resulting linear system of equations may be near singular; how singular depends on the shape of the object and the aspect ratio of the elements as well as the techniques used to avoid the near singular integrands.

The goal of the study is to measure the uncertainty of the computed boundary solutions for these test objects and to show that the uncertainty analysis could be extended to objects for which the solutions are not known. The perturbations of interest were to the boundary conditions and to the weights and points of the quadrature formulas used to perform the needed surface integrals. The applications for this software are in cases where the boundary conditions are likely known to a few digits (usually 3 digits), but we were interested to find some quantity in the computation that might indicate or measure the sensitivity of the solution to the boundary conditions that could be computed when the solution was not known.

The package of software is approximately 50K lines and over 300 procedures. The test harness has been installed in most of the computational components for the regular testing of the package but for this study, the test harness was used only in the solver routine and the procedures involved in the solution, representing approximately 15K lines and 100 procedures, of which only 24 procedures and 168 variables were monitored.

8.1   Measurements

For this study, only the results for the sphere, annular cylinder, and torus are reported here. In all cases the solution vector, consisting of either the Cartesian coordinates of the vector potential or the tangential components of magnetic flux at surface nodes, was examined to study its dependence on boundary data. Solution variation was measured using the maximum element norm of the difference between the vector solution computed with the unperturbed and perturbed data. The vectorial boundary condition data were perturbed by addition of uniform random variables to their Cartesian coordinates. The magnitude of the perturbations was scaled relative to coordinate value with a change up to 100 units in the last place of double precision, 1 unit in the last place in single precision, 10,000 units in the last place in single precision (roughly a change in the third digit), and 100,000 units (roughly a change in the second digit). The boundary conditions are different for each problem; for the sphere and torus, the boundary conditions were Dirichlet and for the annular cylinder, the boundary conditions were mixed Dirichlet and Neumann. The element shapes were quadrilaterals for the torus and annular cylinder and mixed quadrilaterals and triangles for the sphere.

Also, measured as part of the case study were perturbations in the Gaussian weights and points. For all formulas (formulas with more weights and points are used when the integrand is determined to be near singular), the weights and points were changed by random perturbations relative to themselves at the same levels of 100 units in the last place of double precision and 1, 10,000, and 100,000 units in the last place of single precision. Perturbations to both the boundary conditions and Gaussian quadrature parameters at the same time were not performed in the material for this demonstration although this is possible with the test harness tool.

As a general practice, the Laplace solvers estimate the condition number of the linear system [5]. The expectation was that the size of the perturbations of the computed results would depend on the condition number, larger for larger condition numbers of the linear system. The concern was that other commodities might contribute, like how close to singularity were the integrands or how often the higher order quadrature rules or the Telles transformations [6,7] were used to handle very singular integrands.

8.2   The Results

Fig. 3 to Fig. 7 plot the sizes of the perturbation of the solution with respect to the perturbations of the boundary conditions and the quadrature weights and points, tested separately. The plots show that the effect of perturbations of either of these quantities on the solution is relatively small in general, roughly of the size of the perturbation but roughly proportional to and dependent on the condition number. That is, for Fig. 3 (the sphere), the perturbations in the solution follow closely the perturbations in the "input"; for the sphere, the condition number of the linear system is approximately 250. Similarly, in Fig. 5

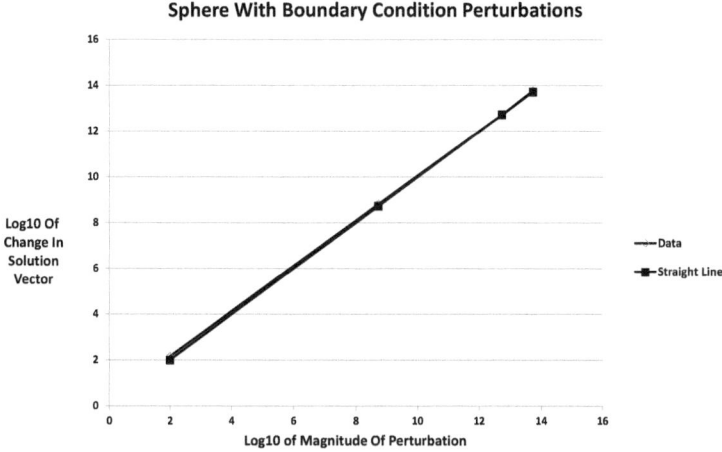

Fig. 3. For a sphere, measuring the perturbations of the solution where the boundary conditions are perturbed by a relative amount of approximately 10^2, 10^9, 10^{13}, and 10^{14} units in the last place of double precision. The solution is perturbed only slightly more than the perturbation in the boundary conditions (that is, the lines are on top of one another). The condition number of the linear system is approximately 250.

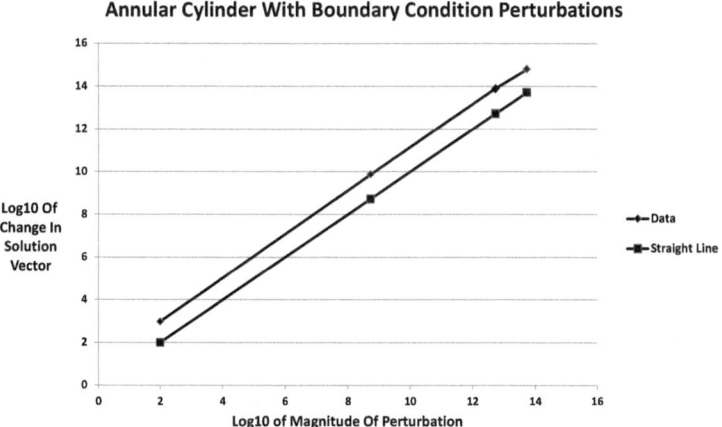

Fig. 4. For an annular cylinder, measuring the perturbation of the solution where the boundary conditions are perturbed by a relative amount of approximately 10^2, 10^9, 10^{13}, and 10^{14} units in the last place of double precision. The solution is perturbed more than the perturbation in the boundary conditions. The condition number of the linear system is approximately 6000.

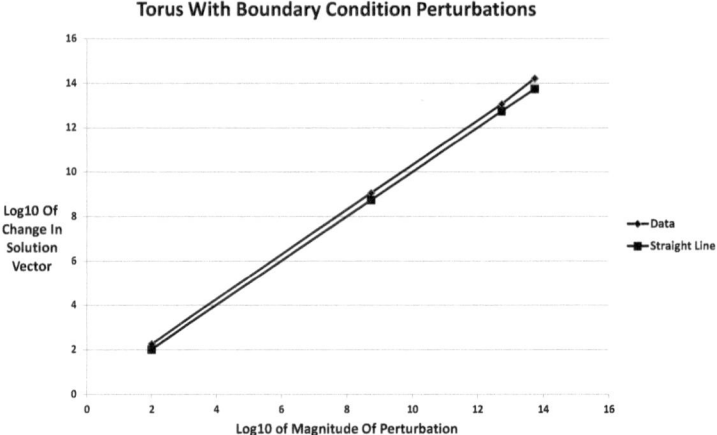

Fig. 5. For a torus, measuring the perturbations of the solution where the boundary conditions are perturbed by a relative amount of approximately 10^2, 10^9, 10^{13}, and 10^{14} units in the last place of double precision. The solution is perturbed more than the perturbation in the boundary conditions. The condition number of the linear system is approximately 170.

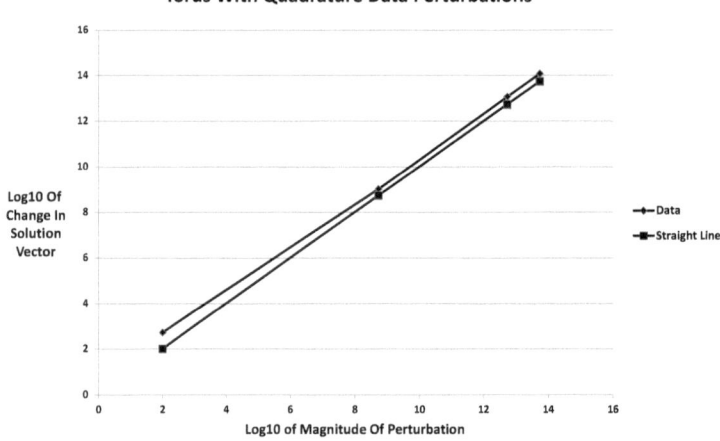

Fig. 6. For a torus, measuring the perturbations of the solution where the Gaussian quadrature weights and points are perturbed by a relative amount of approximately 10^2, 10^9, 10^{13}, and 10^{14} units in the last place of double precision. The solution is perturbed more than the perturbation of the Gaussian parameters and is more sensitive to weights/points perturbations than boundary condition perturbations. The condition number of the linear system is approximately 170.

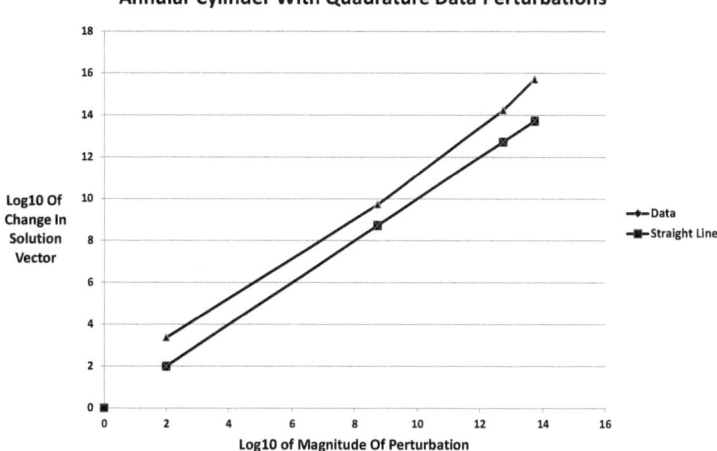

Fig. 7. For an annular cylinder, measuring the perturbations of the solution where the Gaussian quadrature weights and points are perturbed by a relative amount of approximately 10^2, 10^9, 10^{13}, and 10^{14} units in the last place of double precision. The solution is perturbed more than the perturbation of the Gaussian parameters and is more sensitive to weights/points perturbations than boundary condition perturbations. The condition number of the linear system is approximately 6000.

(the torus), the perturbations follow the perturbations in the input; the condition number approximately 170. However, in Figure 4 (the annular cylinder), the perturbations in the solution magnify those of the boundary conditions but still follow the perturbations in the boundary condition; the condition number of the linear system for the annular cylinder is approximately 6000, 25 to 40 times larger than that for the sphere and torus, but the perturbation in the solution is approximately 10 times larger, slightly smaller for the small perturbations and the factor increasing to slightly more than 10 times for the larger perturbations. The line labeled "Straight Line" (using square points) plots the perturbation in the solution as if the perturbation of the input data created the same perturbation in the solution. The line labeled "Data" (using diamond points) plots the measured perturbation in the solution.

Fig. 6 and Fig. 7 show similar behavior as a consequence of perturbations in the quadrature parameters for the torus and cylinder, with the magnification of the perturbations in the solutions being smaller for the torus where the condition number is smaller than that for the annular cylinder by a factor of approximately 40.

9   Conclusions

For this case study, it was relatively straightforward to make perturbations in the data and to measure the changes in the computed solution. The results of these measurements are consistent with the conjecture that the condition number of the linear system will indicate the sensitivity of the solution to changes in the input boundary conditions. Further experiments not reported here continue to confirm the significance of the condition number of the linear system when other boundary conditions are selected.

The test harness has provided a convenient tool to measure uncertainty due to data changes. It is conjectured that by changing the application code so that a version run in **generate** mode uses a slightly different model than the application code run in **check** mode would allow measurements to be made of the uncertainty in the solution caused by using a different model. The only requirement is that the perturbations in the model can be computed by specifying a user-supplied procedure to make the perturbations and a second user-supplied procedure can be written to measure the effect of the perturbations on the computed results.

Acknowledgements. The design of the test harness was supported in part by a DOE Phase 1 SBIR award DE-FG-02-04ER84028, July 14, 2004 – April 12, 2006.

References

1. National Physical Laboratory, UK: A Framework for Uncertainty in Measurement (2010), `http://www.npl.co.uk/mathematics-scientific-computing/` `mathematics-and-modelling-for-metrology/measurement-uncertainty-` `framework/a-framework-for-uncertainty-in-measurement` (accessed November 17, 2011)
2. Accurate Solutions In Applied Physics LLC, Albuquerque, New Mexico (2011), `http://www.manta.com/c/mtvfjt3/` `accurate-solutions-in-applied-physics-llc` (accessed November 17, 2011)
3. Smith, B.T.: A Test Harness TH For Numerical Applications and Libraries. In: Gaffney, P.W., Pool, J.C.T. (eds.) Grid-Based Problem Solving Environments. IFIP, vol. 239, pp. 227–241. Springer, Boston (2007)
4. Smith, B.T.: The Test Harness User's Guide, Version 0.6.9, Numerica 21 Incorporated (2010)
5. Anderson, E., et al.: LAPACK User's Guide, 3rd edn. SIAM, Philadelphia (1999), `http://www.netlib.org/lapack/lug`
6. Telles, J.C.F., Oliveira, R.F.: Third Degree Polynomial Transformation for Boundary Element Integrals: Further improvements. Eng. Anal. BEM 13, 135–141 (1994)
7. Baltz, B., Mammoli, A.A., Ingber, M.S.: Incremental Improvements to the Telles Third Degree Polynomial Transformation for the Evaluation of Nearly Singular Boundary Integrals. In: Chen, C.S., Brebbia, C.A., Pepper, D.W. (eds.) Boundary Element Technology XII, pp. 459–473. WIT Press, Southampton (1999)

Discussion

Speaker: Brian Smith

William Kahan: What if the tool's INCLUDEs insert statements that should not, but do, change the arithmetic because of over-agressive compiler optimizations triggered or inhibited by the INCLUDEs?

Brian Smith: The short answer is that this situation of the included text disturbing the optimization is very unfortunate but certainly possible. I have tried to minimize the likelihood of this happening for the default probe insertion locations where I have some expectation that the included text will not disturb the optimization. The included text is, in most cases, a CALL statement and typically the impact of a CALL statement on compiler optimization is predictable at an entry point or exit point of a procedure. However, when the user inserts a probe to monitor or change program variable values, the user must be aware of what impact the probe is having on optimization; the documentation in the users guide warns the user of this issue. For example, placing a probe in the middle of a loop will likely change the optimization and the user needs to be aware of this impact and how it affects the computed results.

John Reid: It looks as if you are working in Fortran 95 and do not support nested procedures. Is this true?

Brian Smith: No. The test harness supports nested procedures as restricted by Fortran 90/95/2003; that is, these versions of Fortran prohibit internal procedures nested in internal procedures so that the installation of a monitoring, evaluation, or uncertainty probe in an internal procedure must not create an internal procedure within an internal procedure. This is mainly an annoyance and inconvenience in that INCLUDE lines representing a probe are replaced by a CALL statement to an internal procedure that is created and inserted at the end of the procedure unit in all procedures except an internal procedure. For an internal procedure, the INCLUDE line is replaced by many lines of code that implements the probe.

If a version of Fortran removes this restriction, then the test harness code installer will be modestly modified to replace the INCLUDE line for a probe with a CALL statement to an internal procedure in all cases in the same way it treats all other procedures.

So the bottom line is the test harness currently supports nested procedures in all ways allowed by the Fortran 90/95/2003 standards.

Numerical Aspects in the Evaluation of Measurement Uncertainty

Maurice Cox, Alistair Forbes, Peter Harris, and Clare Matthews

National Physical Laboratory
Teddington, United Kingdom
maurice.cox@npl.co.uk

Abstract. Numerical quantification of the results from a measurement uncertainty computation is considered in terms of the inputs to that computation. The primary output is often an approximation to the PDF (probability density function) for the univariate or multivariate measurand (the quantity intended to be measured). All results of interest can be derived from this PDF. We consider uncertainty elicitation, propagation of distributions through a computational model, Bayes' rule and its implementation and other numerical considerations, representation of the PDF for the measurand, and sensitivities of the numerical results with respect to the inputs to the computation. Speculations are made regarding future requirements in the area and relationships to problems in uncertainty quantification for scientific computing.

Keywords: Measurement uncertainty, uncertainty quantification, probability density function, Monte Carlo method, sensitivity measure.

1 Introduction

Metrologists at National Metrology Institutes (NMIs) and industrial laboratories routinely propagate uncertainties related to input quantities through mathematical models of measurement to provide uncertainties related to output quantities. A traditional approach uses model linearization and normality assumptions. Relevant guidance is available and supporting software exists.

The Joint Committee for Guides in Metrology (JCGM) is responsible for the GUM, the Guide to the expression of uncertainty in measurement [3], and for supporting documents, e.g., references [2,6]. The JCGM is aware of limitations of the traditional approach. To overcome such limitations, current JCGM activity [4,5] characterizes input quantities by probability density functions (PDFs), which are propagated through the model to obtain a joint PDF for the output quantities. Best estimates, covariance matrices and coverage intervals and coverage regions for the output quantities (or measurand), all used by metrologists, can then be obtained.

Models with PDFs for the input quantities can be seen in a broader setting, where possibly many inputs do not simply relate to measurement. The main consideration here, though, remains measurement uncertainty (MU) quantification,

A. Dienstfrey and R.F. Boisvert (Eds.): WoCoUQ 2011, IFIP AICT 377, pp. 180–194, 2012.

although many of the principles apply more widely. We observe that the result of a computation represents the effect of uncertainty from all sources considered. Numerical methods of solution, especially Monte Carlo (MC) methods, are used.

We regard a numerical calculation as modeled by $Y = f(X)$, sometimes known as a *measurement equation* (ME) [22], where X denotes N input quantities, Y the output quantities, and f is a given function, specified by a computational model, that transforms X to Y. When X is uncertain, Y is uncertain. Given knowledge about X, knowledge is required about Y. Prior knowledge of Y may be available. The components of X are characterized by random variables, and we encode available knowledge about X as a PDF.

$p(Z)$ denotes the PDF for a quantity Z. z denotes an estimate of Z and U_z the associated covariance matrix (sometimes called uncertainty matrix [4]), taken respectively as the expectation $E(Z)$ and covariance $V(Z)$. $u(z_i)$ denotes the so-called standard uncertainty associated with the ith component of z. $U(z_i)$ denotes an expanded uncertainty corresponding to a stipulated coverage probability p.

Consider so-called *aleatory uncertainties* (due to random effects) and *epistemic uncertainties* (due to other effects). Some authors treat aleatory uncertainties as random variables having PDFs, and epistemic uncertainties as intervals with no assumed PDFs. In metrology we encode knowledge of all quantities by PDFs, a view fully consistent with the GUM [3]. The rules of probability calculus can then be employed. In contrast, in some references, e.g., [24], the two types of uncertainty are propagated separately and the results combined.

Numerical analysis has a long history in uncertainty quantification (UQ) when computing in finite-precision arithmetic. Two principal techniques for error propagation are interval analysis [20] and floating-point (FP) error analysis [30]. We will not consider these techniques here, although we recognize the value of FP error analysis, especially in making statements about the numerical stability of algorithms used within the computational model. We strongly distinguish between errors and uncertainties: an error is the difference between the value of a quantity and the true value of that quantity. An uncertainty is a measure of dispersion (such as the standard deviation determined from the PDF) for that quantity.

This paper reviews numerical tools currently used in MU evaluation. Section 2 makes remarks on the process of assembling and using information about the input quantities, that is, uncertainty elicitation. Section 3 reviews the propagation of distributions through a computational model. Section 4 considers the use of Bayes' rule and Monte Carlo Markov chain (MCMC) methods. Section 5 treats numerical considerations in generating Monte Carlo results and Section 6 considers the representation of such results. Section 7 considers sensitivity issues. Section 8 gives concluding remarks and speculations on tools needed for MU in the future in treating more complicated computational models in metrology.

2   Elicitation

Elicitation is the process of obtaining knowledge of an input quantity and transforming that knowledge to a PDF for that quantity. In metrology if the only knowledge about a component X_i of $\boldsymbol{X}$ is the endpoints of an interval for X_i, we use MaxEnt, the maximum entropy principle, to assign a uniform PDF to X_i [4]. If the only knowledge about $\boldsymbol{X}$ is its expectation $\boldsymbol{E}(\boldsymbol{X})$ and covariance $\boldsymbol{V}(\boldsymbol{X})$, we again use MaxEnt to characterize $\boldsymbol{X}$ by a multinormal distribution. Sometimes Bayes' rule is used to assign a PDF, such as when repeated observations of a quantity are available. Reference [4, subclause 6.4] gives PDF assignments for common circumstances in metrology.

Hibbert et al. [12] applied MaxEnt and Bayesian model selection to a fascinating class of decision-making problems in suspected cases of horse-doping. From a large mass of historical data they constructed PDFs for total carbon dioxide concentration in pre-race samples of plasma. Separate PDFs were obtained for 'clean' horses and horses that were subsequently tested positive. Using q leading moments of the data, MaxEnt delivers a PDF based on a set of Lagrangian parameters [19]. Bayesian model selection was used to obtain that value of q that maximized the Bayesian model probability p, thus avoiding model over-fitting. For clean horses, values of $-\log_{10}(p)$ for $q = 2, \ldots, 7$ were approximately $11, 13, 12, 14, 0, 2$, respectively, leading to the conclusion that Bayesian model selection has strongly settled for a moderately complex model of the form $\exp(a_1 X + \cdots + a_6 X^6)$. This model was chosen in preference to a simple model such as $\exp(a_1 X + a_2 X^2)$, which, for $a_2 < 0$, is Gaussian. Measured data for a further horse can be compared with these PDFs and a decision made on whether the horse has been subjected to doping.

3   Propagation of Distributions

Obtaining the joint PDF $p(\boldsymbol{Y})$ for $\boldsymbol{Y}$ given the joint PDF $p(\boldsymbol{X})$ for $\boldsymbol{X}$ is known as 'propagation of distributions' [4]. Formally, it constitutes an application of Markov's theorem [9],

$$p(\boldsymbol{Y}) = \int_{-\infty}^{\infty} \cdots \int_{-\infty}^{\infty} p(\boldsymbol{X})\delta(\boldsymbol{Y} - \boldsymbol{f}(\boldsymbol{X}))\,\mathrm{d}\boldsymbol{X}, \tag{1}$$

where $\delta(\cdot)$ denotes the Dirac delta function. Figure 1 illustrates the principle for a univariate model with $N = 3$ input quantities. $p(Y)$ is indicated as being asymmetric, as generally arises for non-linear models or asymmetric $p(X_i)$.

A quadrature rule can be used, albeit inefficiently, to evaluate the integral (1), so as to provide an approximation $\widehat{p}(\boldsymbol{Y})$ to $p(\boldsymbol{Y})$. $\widehat{p}(\boldsymbol{Y})$ is often obtained in metrology using an MC method [4]. Random draws are made from $p(\boldsymbol{X})$, $\boldsymbol{f}$ evaluated in each case, and the resulting set of values used to form $\widehat{p}(\boldsymbol{Y})$.

See GUM Supplement 1 [4] for details of an MC method that implements the propagation of distributions. These considerations apply when $\boldsymbol{X}$ does not depend on the measurand.

4   Bayes and MCMC

When observations of a component X_i of $\boldsymbol{X}$ that depend on the value of the measurand are available, an *observation equation* (OE) approach is appropriate.

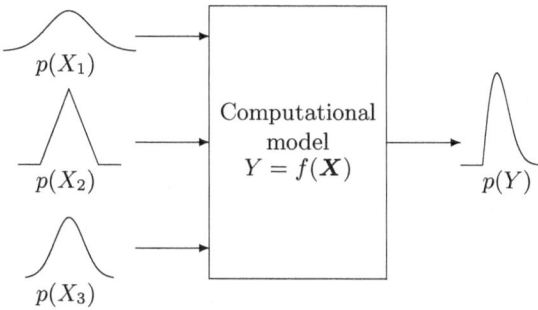

Fig. 1. Propagation of distributions for $N = 3$ independent input quantities

Then Bayes' rule can be used to determine $p(\boldsymbol{Y})$. Let W, one of the X_i, be directly observable. Let $\boldsymbol{X}$ now denote the original $\boldsymbol{X}$ *less* W. Re-express the ME as $\boldsymbol{Y} = \boldsymbol{f}(W, \boldsymbol{X})$ and consider the corresponding OE $W = \phi(\boldsymbol{Y}, \boldsymbol{X})$ [11,22] and observations $W_i \in \mathrm{N}(W, \sigma^2)$. Bayes' rule can be used to update prior knowledge of $\boldsymbol{Y}$, $\boldsymbol{X}$ and σ^2 (regarded as random variables) with observations W_i to give a posterior distribution for these variable, with $p(\boldsymbol{Y})$ obtained by marginalization.

An MCMC algorithm can be used to obtain $p(\boldsymbol{Y})$. MCMC generates a sequence $\{\boldsymbol{y}_k\}$, in which $\boldsymbol{y}_k$ is obtained from $\boldsymbol{y}_{k-1}$ under an iterative operation. It asymptotically generates draws from $p(\boldsymbol{Y})$. A Metropolis-Hastings algorithm [23, Chapter 6] is an MCMC algorithm that allows $p(\boldsymbol{Y})$ to be specified straightforwardly. One variant, an 'independence chain', can be applied when we have an approximating PDF $\widetilde{p}(\boldsymbol{Y})$ that is easily sampled. Given $\boldsymbol{y}_{k-1}$, $\boldsymbol{y}^*$ is drawn at random from $\widetilde{p}(\boldsymbol{Y})$. Then $\boldsymbol{y}_k$ is set to $\boldsymbol{y}^*$ with acceptance probability

$$\min\{1, r_k\}, \qquad r_k = \frac{p(\boldsymbol{y}^*)\widetilde{p}(\boldsymbol{y}_{k-1})}{p(\boldsymbol{y}_{k-1})\widetilde{p}(\boldsymbol{y}^*)}. \tag{2}$$

To implement the acceptance step, draw v_k from the uniform distribution $\mathrm{R}(0,1)$ and set $\boldsymbol{y}_k = \boldsymbol{y}^*$ if $v_k < r_k$; otherwise set $\boldsymbol{y}_k = \boldsymbol{y}_{k-1}$. This scheme replicates samples when $\widetilde{p}(\boldsymbol{Y})$ under-represents $p(\boldsymbol{Y})$ at the expense of rejecting samples when $\widetilde{p}(\boldsymbol{Y})$ over-represents $p(\boldsymbol{Y})$. On convergence of the Markov chain, the $\{\boldsymbol{y}_k\}$ are draws from $p(\boldsymbol{Y})$. Implementations typically involve repeat runs (chains) with different initial samples to gauge convergence.

The ME approach (propagation of distributions with W assigned a PDF based on the sampling distribution) can be analyzed in terms of a Bayesian approach [10]. Let ζ denote an observation. If the OE approach is implemented with prior $p_{\mathrm{OE}}(\boldsymbol{Y}, \boldsymbol{X}) \propto p(\boldsymbol{X})$ to produce posterior distribution $p_{\mathrm{OE}}(\boldsymbol{Y}, \boldsymbol{X}|\zeta)$, then, with $\partial\phi(\boldsymbol{Y}, \boldsymbol{X})/\partial\boldsymbol{Y}$ denoted by $\dot{\phi}(\boldsymbol{Y}, \boldsymbol{X})$, the posterior distribution for the ME approach is [15]

$$p_{\mathrm{ME}}(\boldsymbol{Y}, \boldsymbol{X}|\zeta) \propto |\dot{\phi}(\boldsymbol{Y}, \boldsymbol{X})|p_{\mathrm{OE}}(\boldsymbol{Y}, \boldsymbol{X}|\zeta),$$

based on a prior distribution

$$p_{\mathrm{ME}}(\boldsymbol{Y}, \boldsymbol{X}) \propto |\dot{\phi}(\boldsymbol{Y}, \boldsymbol{X})|p(\boldsymbol{X}).$$

There will be little difference between the ME and OE approaches if $|\dot{\phi}(\boldsymbol{Y}, \boldsymbol{X})|$ is approximately constant throughout the region of interest.

Consider a Metropolis-Hastings independence chain with $p_{\mathrm{OE}}(\boldsymbol{Y}, \boldsymbol{X}|\zeta)$ playing the role of $p(\boldsymbol{Y})$ and $p_{\mathrm{ME}}(\boldsymbol{Y}, \boldsymbol{X}|\zeta)$ that of $\widetilde{p}(\boldsymbol{Y})$. Expressions (2) then become

$$\min\{1, r_k\}, \qquad r_k = \left| \frac{\dot{\phi}(\boldsymbol{y}_{k-1}, \boldsymbol{x}_{k-1})}{\dot{\phi}(\boldsymbol{y}^*, \boldsymbol{x}^*)} \right|.$$

For models in which $\dot{\phi}(\boldsymbol{Y}, \boldsymbol{X}) \approx$ constant over the region of interest, $r_k \approx 1$ ensuring a high acceptance rate, reflecting the fact that little adjustment has to be made to the sample generated by the ME approach. The chain $\{\boldsymbol{y}_k\}$ will then have low autocorrelation so that sample statistics such as means and standard deviations will converge to their distribution counterparts at rates similar to that for the MC method of GUM Supplement 1.

5   Numerical Considerations in Generating Monte Carlo Results

5.1   Random Number Generation

In metrology the PDF $p(\boldsymbol{X})$ can often be decomposed into univariate PDFs or joint PDFs involving a smaller number of variables. Procedures for drawing randomly from a variety of PDFs commonly occurring in metrology such as normal, multinormal, t and arcsine are summarized in GUM Supplement 1 [4]. These procedures depend on the quality of an underpinning uniform random number generator (RNG).

An extensive test of the statistical properties of uniform RNGs is carried out by TestU01 [13], a suite containing many individual tests including the so-called Big Crush. Several RNGs passing the Big Crush test are listed by Wichmann and Hill [29], who also considered RNGs on distributed computing systems.

5.2   Practicalities

Rather than carrying out all M Monte Carlo trials and then processing them to obtain a histogram, the following procedure (where an MC trial constitutes making a draw from the input joint PDF and providing the corresponding model value) has benefits in terms of time and memory:

1. Perform a modest number of MC trials, $M_0 = 10^4$, say;
2. Establish a set of bins based on these M trials;
3. Carry out a further $M - M_0$ trials, allocating values to bins or, if a value lies outside the set of bins, storing it *individually*.

The bins, bin frequencies and additional values are used to obtain the required results. This approach is advantageous when computing coverage intervals and regions, which typically involve values in the tails of $p(\boldsymbol{Y})$. Further details are available [8, Annex E].

5.3   Monte Carlo Convergence

Suppose M random draws are made from $p(\boldsymbol{X})$ and the corresponding output quantity values are calculated. For any j, the average of these values for the jth component Y_j is a realization of a random variable with expectation $E(Y_j)$ and variance $V(Y_j)/M$. The closeness of agreement between this average and $E(Y_j)$ can be expected to be proportional to $M^{-1/2}$. This 'convergence rate' can be improved using schemes such as Latin Hypercube sampling (LHS) [18] for certain classes of problem.

5.4   Adaptive Schemes

The approach in Section 5.2 necessitates specifying M in advance. Thus, the numerical accuracy of the results obtained is unknown *a priori*. The aim of an adaptive scheme is to provide (a) an estimate $\boldsymbol{y}$ of $\boldsymbol{Y}$, (b) an associated covariance matrix $\boldsymbol{V}_{\boldsymbol{y}}$, and (c) a coverage region for $\boldsymbol{Y}$ for a stipulated coverage probability (or their univariate counterparts for an uncertainty evaluation problem with a single measurand), so as to meet a specified numerical tolerance.

An approach, involving carrying out a sequence of applications of an MC method, is detailed in references [4,5]. It operates in terms of a specified numerical tolerance δ used to assess the 'degree of approximation' required in the elements of the covariance matrix $\boldsymbol{V}_{\boldsymbol{y}}$. The approach, which utilizes a sequence of batches of, say, $M_0 = 10^4$ MC trials, consists of the following steps:

1. Carry out a batch of MC trials and use the model values to calculate batch results (averages, standard deviations, etc.);
2. Use updating techniques to calculate results for all batches;
3. Regard the computation as having stabilized when the standard deviations of the averages of the batch results is no greater than δ.

The test in step 3 regards the averages as realizations of variables distributed as Student's t, and corresponds to a coverage probability of 95 % [32]. This test is superior to that in reference [4], which is based on regarding the averages as realizations of Gaussian variables.

Adaptive schemes such as that above can be tailored to other sampling procedures such as LHS.

6   Representation of MC Results

6.1   General

Suitable representations of MC results are required for (a) visualization purposes, and (b) subsequent use. In terms of (b) the output of one MU evaluation should be transferable, that is, usable as an input to another evaluation [3]. In particular it is not always convenient to retain the $M = 10^6$, say, (vector) values produced by MC and use them subsequently. But, balanced with this, the MC output is ideal in that it automatically conveys covariance information. Methods such as kernel density approximation (KDA) can be used for (a) and (b).

6.2   Density Approximation

Kernel density estimation (or better approximation) can be used to approximate a PDF from sampled values from that PDF, but it is not greatly used as yet in metrological applications. A KDA to a univariate PDF p at the point Y is

$$\widehat{p}_h(Y) = \frac{1}{Mh} \sum_{r=1}^{M} K \left(\frac{Y - y_r}{h} \right). \tag{3}$$

In expression (3), $y_1, \ldots, y_M$ are sampled values with underlying density p, and K is a kernel function with unit area. Common kernel functions are Gaussian and B-spline [27,28]. B-splines have appreciable speed advantages when sampling from $\widehat{p}_h(Y)$ because of their compact support property [7].

In expression (3) h is a smoothing parameter known as the bandwidth and plays a similar role to that of bin width in a histogram. Too small a value of h results in spurious oscillatory behavior. Too large a value results in over-smoothing, losing local detail. A number of methods are used to determine h given the y_r [27,28]. Figure 2 illustrates such a representation. The bandwidth in the kernel density approximation and the bin width of the histogram are identical there.

A (conventional) KDA (3) has the same information content as the data it represents. With M often of $O(10^6)$, it is possible to produce KDAs with many fewer terms. It is also possible to describe the sampled values by some parametric form with adjustable parameters.

If a complete PDF is not required Willink [31] suggests summarizing the PDF in terms of a model for its quantile function (the inverse of the distribution function). Willink proposes an asymmetric form of the 'lambda distribution' for

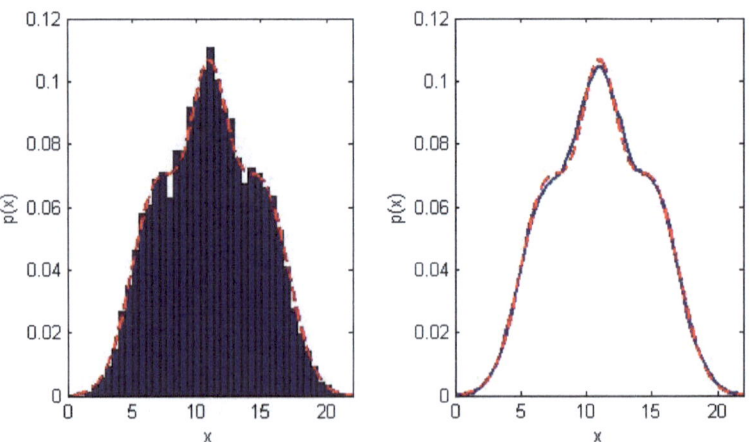

Fig. 2. Representation of a set of observations by a histogram and (right) a KDA, the broken red line showing true density

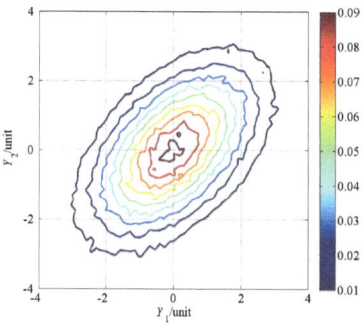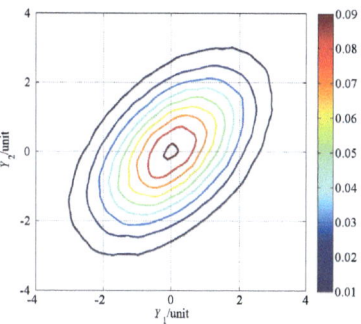

Fig. 3. Contours of the PDF, based on MC results, for a bivariate output quantity and (right) with contour smoothing

this purpose. The quantile function for this distribution has four free parameters which may be related by non-linear transformations either to the first four moments of the PDF obtained by Monte Carlo analysis, or, if desired its quantile information. As the extended lambda distribution is defined with respect to its quantile function, drawing random samples from the resulting model approximation is straightforward. Should this approximation be inadequate in any particular case, the more complete KDA can be used.

A bivariate PDF is sometimes represented by a set of contour lines. The contour lines should be faithfully reproduced: as $M \to \infty$, they should converge to the contours of the corresponding PDF. Doing so requires appropriate smoothing [25,28]. Some contour diagrams can be computed directly from a KDA to the corresponding PDF. For others an appropriate smoothing algorithm can be applied to the MC results and the resulting smoothed contours drawn. Figure 3 illustrates the effect of a simple smoothing algorithm.

6.3   Coverage Regions

In metrology, coverage intervals and regions are often required to accompany measurement results. A procedure included in reference [5], based on work of Possolo [21], for an approximation to the smallest $100p\%$ coverage region is as follows:

1. Construct a (hyper-)rectangular region in the space of the output quantities;
2. Subdivide this rectangular region into a mesh of small rectangles;
3. Assign each output quantity value to the small rectangle containing it;
4. Use the fraction of output quantity values assigned to each rectangle as the approximate probability that Y lies in that rectangle;
5. List the rectangles in terms of decreasing probability;
6. Form the cumulative sum of probabilities for these listed rectangles, stopping when this sum is no smaller than p;
7. Take the selected rectangles as defining the smallest coverage region.

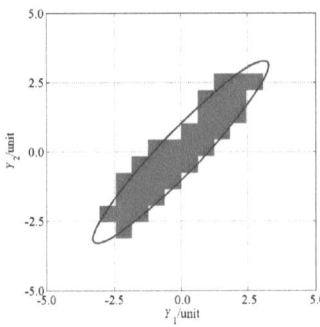 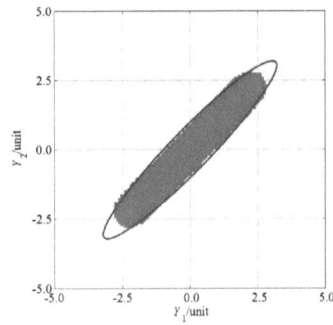

Fig. 4. Coverage regions based on a 10×10 mesh and 1 000 points drawn from the PDF for Y, and (right) for a 100×100 mesh and 1 000 000 points

A coverage region that is less disjointed and having a smoother boundary would be expected to be obtained if step 4 were replaced by the use of a more sophisticated approximation to the probability density [26]. In the bivariate case, steps 1 to 4 are also carried out in the initial stages of a typical contouring algorithm used in visualizing an approximation to the PDF for Y.

As a simple example consider the linear bivariate model $Y_1 = X_1 + X_3$, $Y_2 = X_2 + X_3$, with mutually independent $X_1 \sim \mathrm{N}(0, 0.1)$, $X_2 \sim \mathrm{N}(0, 0.1)$ and $X_3 \sim \mathrm{R}(-(5.7)^{1/2}, (5.7)^{1/2})$. Figure 4 shows approximations to the smallest $95\,\%$ coverage region, obtained using the above procedure, based on a set of rectangles forming a 10×10 and a 100×100 mesh. For comparison the $95\,\%$ elliptical coverage region for Y based on Gaussian parameters estimated from the model values is shown by a solid line. The assumption of normality yields a conservative coverage region in this example.

7   Sensitivity

7.1   First-Order Sensitivity Coefficients

Consider a univariate computational model $Y = f(X)$. A first-order sensitivity coefficient is $c_i = \partial f / \partial X_i$ evaluated at x. Then $u_i(y) = |c_i| u(x_i)$ is the corresponding first-order contribution to $u(y)$. It can readily be obtained using the *complex-step method* [17], which in our opinion deserves greater recognition, particularly when f is complicated. This method is applicable when the real types in the software that implements the model can be replaced by complex types.

The complex-step method is similar to the use of finite differences but complex arithmetic is used to obtain first derivatives. It uses the Taylor expansion of a function f of a complex variable:

$$f(z + w) = \sum_{r=0}^{\infty} \frac{w^r}{r!} f^{(r)}(z),$$

where z and w are complex. Setting $z = x$ and $w = ih$ where x is real and h is real and small, and taking real and imaginary parts, we have

$$\Re f(x+ih) = f(x) - \frac{h^2}{2} f''(x) + O(h^4), \qquad \Im f(x+ih) = hf'(x) - \frac{h^3}{6} f'''(x) + O(h^5),$$

from which

$$f(x) = \Re f(x + ih), \qquad f'(x) = \frac{1}{h} \Im f(x + ih),$$

with truncation errors of order h^2. Unlike the use of a finite-difference formula for $f'(x)$, h is chosen to be *very* small with little concern about the loss of significant digits through subtraction cancellation. NPL routinely applies the method with $h = 10^{-100}$ [1], suitable for all but pathologically-scaled problems.

7.2   The Use of Monte Carlo

MC can be adapted to provide a non-linear counterpart of a sensitivity coefficient. Hold all input quantities but one, say X_k, at their estimates. Make random draws from $p(X_k)$ and determine an approximation to the PDF for Y that depends only on X_k. The according standard deviation $\tilde{u}_k(y)$ is an approximation to the component of the standard uncertainty corresponding to X_k.

The use of 'non-linear' sensitivity coefficients in place of 'linear' sensitivity coefficients permits individual non-linear effects to be taken into account. A 'non-linear' sensitivity coefficient $\tilde{c}_k$ is defined by $\tilde{c}_k = \tilde{u}_k(y)/u(x_k)$. It will be equal to the magnitude of the 'linear' sensitivity coefficient c_k for f linear in X_k. The deviation of $\tilde{u}_k(y)$ from $u_k(y) = |c_k|u(x_k)$ is a measure of the influence of the non-linearity of f with regards to X_k alone. This measure does not account for interaction (i.e., non-additive) terms in f.

8   Concluding Remarks and Forward Look

8.1   Intensive Computation Beyond GUM

There is in place an infrastructure to deal with MU when expressed as a standard uncertainty or an expanded uncertainty for some coverage probability. The GUM goals of having a universal, internally consistent, and transferable framework are then largely achieved. However, when uncertainty is expressed using PDFs, MU evaluation then relies on (possibly intensive) numerical computation, and generates data that need to be represented in some way. Furthermore, measurement models are becoming more complicated and can only be treated numerically. An example is a radiation-transport calculation, which is itself an MC calculation.

8.2   Need for Efficiency

With MU evaluation and uncertainty quantification for scientific computing (UQ) in general there is a need for more efficient techniques for propagating PDFs through computational models. MC is naturally highly parallelizable; NPL uses a grid of PCs to treat the more complex computational models. As all such techniques are based on MC, it is a matter of tuning those techniques appropriately. For some problems the basic technique can hardly be bettered. Approaches such as LHS can give appreciable gains for certain classes of problem.

8.3   Embracing Uncertainty Quantification

In the future, MU evaluation will embrace more closely concepts used in UQ. Model uncertainty is recognized, being termed *definitional uncertainty* [6]. Elicitation uncertainty is already treated in a small way [3,4]. Numerical uncertainty is considered when computational models constitute FE solvers, for example [16], or in using adaptive schemes [4]. Within general UQ a main aim is to provide probabilistic statements about quantities of interest to inform decision makers. A politician, lawyer or manufacturing production manager needs to consider available evidence and decide a course of action to achieve some goal. Decisions might relate to whether a process remains under statistical control or an athlete is regarded as using a banned substance. Such aspects imply that there are (further) software tools requiring development for MU evaluation.

8.4   Use of MaxEnt

Other metrological applications could benefit from the approach to the horse-doping problem in Section 2. Numerical difficulties, though, can give rise to ill-determined PDFs or prevent a PDF from being obtained at all when MaxEnt is applied to moments as in Section 2 [14]. These difficulties arise when a metrological problem is unreasonable in that it relates to an inadequate measurement of a quantity. It would be useful to have in place a characterization of such problems.

8.5   Efficiency of Forms for Kernel Density Approximations

A compact form for an approximation $\widehat{p}(Y)$ to the PDF $p(Y)$ is desirable when $\widehat{p}(Y)$ is to be used as input to a subsequent MU evaluation. The use of kernel density approximations (KDAs) based on B-splines have advantages, delivering computationally efficient representations because of the many properties of B-splines [7]. There is a choice here: either (a) assemble MC results as a histogram and use a KDA, or (b) represent the *ordered* MC results by a suitable approximating cumulative density function (CDF), incorporating monotonicity conditions because of the non-decreasing property of CDFs. For (a), evaluation of the inverse CDF (quantile function), when generating draws from the distribution using a B-spline basis, is very efficient because of the compact support of the B-splines. For (b), the CDF can be differentiated to form the corresponding PDF.

In metrology, the question is raised, and is yet to be addressed, of how calibration certificates of the future can usefully convey results from an MC calculation, particularly in the presence of an asymmetric PDF.

Acknowledgements. The National Measurement Office, an Executive Agency of the UK Department for Business, Innovation and Skills, supported this work through the Mathematics and Modelling for Metrology programme.

References

1. Al-Mohy, A.H., Higham, N.J.: The complex step approximation to the Fréchet derivative of a matrix function. Numer. Algor. 53, 133–148 (2010)
2. BIPM, IEC, IFCC, ILAC, ISO, IUPAC, IUPAP, OIML: Evaluation of measurement data — An introduction to the "Guide to the expression of uncertainty in measurement" and related documents. Joint Committee for Guides in Metrology, JCGM 104:2009 (2009),
   http://www.bipm.org/utils/common/documents/jcgm/JCGM_104_2009_E.pdf
3. BIPM, IEC, IFCC, ILAC, ISO, IUPAC, IUPAP, OIML: Evaluation of measurement data — Guide to the expression of uncertainty in measurement. Joint Committee for Guides in Metrology, JCGM 100:2008 (2008),
   http://www.bipm.org/utils/common/documents/jcgm/JCGM_100_2008_E.pdf
4. BIPM, IEC, IFCC, ILAC, ISO, IUPAC, IUPAP, OIML: Evaluation of measurement data — Supplement 1 to the "Guide to the expression of uncertainty in measurement" — Propagation of distributions using a Monte Carlo method. Joint Committee for Guides in Metrology, JCGM 101:2008 (2008),
   http://www.bipm.org/utils/common/documents/jcgm/JCGM_101_2008_E.pdf
5. BIPM, IEC, IFCC, ILAC, ISO, IUPAC, IUPAP, OIML: Evaluation of measurement data — Supplement 2 to the "Guide to the expression of uncertainty in measurement" — Models with any number of output quantities. Joint Committee for Guides in Metrology, JCGM 102:2011 (2011),
   http://www.bipm.org/utils/common/documents/jcgm/JCGM_102_2011_E.pdf
6. BIPM, IEC, IFCC, ILAC, ISO, IUPAC, IUPAP, OIML: International vocabulary of metrology — basic and general concepts and associated terms. Joint Committee for Guides in Metrology, JCGM 200:2008 (2008),
   http://www.bipm.org/utils/common/documents/jcgm/JCGM_200_2008.pdf
7. Cox, M.G.: The numerical evaluation of B-splines. J. Inst. Math. Appl. 10, 134–149 (1972)
8. Cox, M.G., Harris, P.M., Smith, I.M.: Software specifications for uncertainty evaluation. Tech. Rep. MS 7, National Physical Laboratory, Teddington, UK (2010)
9. Cox, M.G., Siebert, B.R.L.: The use of a Monte Carlo method for evaluating uncertainty and expanded uncertainty. Metrologia 43, S178–S188 (2006)
10. Elster, C., Toman, B.: Bayesian uncertainty analysis under prior ignorance of the measurand versus analysis using the Supplement 1 to the Guide: a comparison. Metrologia 46, 261–266 (2009),
    http://stacks.iop.org/0026-1394/46/i=3/a=013
11. Forbes, A.B., Sousa, J.A.: The GUM, Bayesian inference and the observation and measurement equations. Measurement 44, 1422–1435 (2011)
12. Hibbert, D.B., Armstrong, N., Vine, J.H.: Total CO_2 measurements in horses: where to draw the line. Accred. Qual. Assur. 16, 339–345 (2011)
13. L'Ecuyer, P., Simard, R.: TestU01: A software library in ANSI C for empirical testing of random number generators,
    http://www.iro.umontreal.ca/~simardr/testu01/tu01.html

14. Lira, I.: Evaluating the Uncertainty of Measurement. Fundamentals and Practical Guidance. Institute of Physics, Bristol (2002)
15. Lira, I., Grientschnig, D.: Bayesian assessment of uncertainty in metrology: a tutorial. Metrologia 47, R1–R14 (2010), http://stacks.iop.org/0026-1394/47/i=3/a=R01
16. Lord, G., Wright, L.: Uncertainty evaluation in continuous modelling. Tech. Rep. CMSC 31/03, National Physical Laboratory, Teddington, UK (2003)
17. Lyness, J.N., Moler, C.B.: Numerical differentiation of analytic functions. SIAM J. Numer. Anal. 4, 202–210 (1967)
18. McKay, M., Conover, W.J., Beckman, R.J.: A comparison of three methods for selecting values of input variables in the analysis of output from a computer code. Technometrics 42, 55–61 (2000)
19. Mead, L.R., Papanicolaou, N.: Maximum entropy in the problem of moments. J. Math. Phys., 2404–2417 (1984)
20. Moore, R.E.: Interval Analysis. Prentice-Hall, New Jersey (1966)
21. Possolo, A.: Copulas for uncertainty analysis. Metrologia 47, 262–271 (2010)
22. Possolo, A., Toman, B.: Assessment of measurement uncertainty via observation equations. Metrologia 44, 464–475 (2007)
23. Robert, C.P., Casella, G.: Monte Carlo Statistical Methods. Springer, New York (1999)
24. Roy, C.J., Oberkampf, W.L.: A comprehensive framework for verification, validation, and uncertainty quantification in scientific computing. Comput. Methods Appl. Mech. Engrg. 200, 2131–2144 (2011)
25. Scott, D.W.: Multivariate Density Estimation: Theory, Practice, and Visualization. John Wiley & Sons, New York (1999)
26. Scott, D.W., Sain, S.R.: Multi-dimensional density estimation. In: Rao, C., Wegman, E. (eds.) Handbook of Statistics. Data Mining and Computational Statistics, vol. 23, pp. 229–261. Elsevier, Amsterdam (2004)
27. Sheather, S.J.: Density estimation. Statist. Sci. 19, 588–597 (2004)
28. Silverman, B.W.: Density Estimation. Chapman and Hall, London (1986)
29. Wichmann, B.A., Hill, I.D.: Generating good pseudo-random numbers. Computational Statistics and Data Analysis 51, 1614–1622 (2006)
30. Wilkinson, J.H.: Rounding Errors in Algebraic Processes. Notes in Applied Science, vol. 32. Her Majesty's Stationery Office, London (1963)
31. Willink, R.: Representing Monte Carlo output distributions for transferability in uncertainty analysis: modelling with quantile functions. Metrologia 46, 154–166 (2009), http://stacks.iop.org/0026-1394/46/i=3/a=002
32. Wübbeler, G., Harris, P.M., Cox, M.G., Elster, C.: A two-stage procedure for determining the number of trials in the application of a Monte Carlo method for uncertainty evaluation. Metrologia 47, 317–324 (2010), http://stacks.iop.org/0026-1394/47/i=3/a=023

Discussion

Speaker: Maurice Cox

Mark Campanelli: Has there been any discussion on a "bias discovery" methodology for metrology?

Maurice Cox: Bias is an important concept in measurement. In the *Guide to the expression of uncertainty in measurement* (GUM) [3], concentration is given to *systematic error*, and its characterization by a probability density function (PDF). This distribution is obtained by accounting for available knowledge of the relevant quantity. Bias is regarded as an estimate of a systematic error. In an interlaboratory comparison, nominally the same measurand is measured by participating laboratories and a consensus value determined. The deviations of the measured values provided by the laboratories from the consensus value can be regarded as laboratory biases.

Mark Campanelli: How can the reliability of the Monte Carlo method be checked for computing 99 % coverage intervals for high-dimensional problems?

Maurice Cox: For multivariate problems an adaptive method is given in Supplement 2 to the GUM [6], as outlined in our paper. It constitutes a simple extension of the adaptive scheme in Supplement 1 to the GUM [4]. Further experience needs to be gained before its performance can be judged in obtaining 99 % coverage intervals and 99 % coverage regions. At the moment these guidance documents apply to simple Monte Carlo sampling; different considerations might apply to Latin Hypercube sampling or importance sampling.

Richard Hanson: The use of pure complex "divided differences" requires an algorithm for complex continuation of the function. Is this the case for your application?

Maurice Cox: In metrology, the measurement function, when expressed in terms of complex variables, is usually analytic, that is, the Cauchy-Riemann equations apply. In such a case, the replacement of real type by complex types is virtually all that needs to be done. When, rarely, the measurement function is specified in terms of functions that are not analytic (such as the absolute value function, which is not analytic at the origin), the measurement function should be re-expressed so that the Cauchy-Riemann equations are satisfied.

Jeffrey Fong: For problems involving controls based on "uncooked" models (i.e., models without a history of baseline data and experimental validation), and with small M ($\sim 10^2 - 10^3$) do you have any advice on using the approach you outlined in your talk?

Maurice Cox: The Monte Carlo method given in guidance documents [4,6] is simple and is not intended for small values of M. If the model is complicated, perhaps involving the solution of a large finite-element model, because of large computing times it may not be possible to use a sufficiently large value of M to obtain adequate distributional knowledge of the output quantity. In such a case we give very elementary advice. It involves the use of an approximate approach in which the PDF for the output quantity is regarded as Gaussian (as in the GUM). A value of M would be selected that is economically possible. The average and covariance matrix of the resulting M values of the (vector) output quantity would be taken respectively as the best estimate and covariance matrix of that quantity. A Gaussian PDF with these parameters would be used to characterize the output quantity and a desired coverage region calculated accordingly. Although this use of a small value of M is inevitably less reliable than that of a large value in that it does not provide an approximation to the PDF for the output quantity, it does take account of model non-linearity.

Tony O'Hagan: I suggest that a variance-based sensitivity analysis would be much more useful than derivatives. This is discussed, and efficient computation with GP emulators developed in: Oakley, J. and O'Hagan, A. (2004), Probabilistic sensitivity analysis of complex models : a Bayesian approach, *J. Royal Statistical Soc., series B*, vol 66, 751–769.

Maurice Cox: We do not use derivatives alone, but in conjunction with the input standard deviations. Thus, we consider, in the notation of our paper, the contributions $u_i(y) = |c_i|u(x_i)$. These terms are the first-order contributions to the standard deviation of the output quantity, and do not account for interaction effects. They are used by metrologists as indications of the input quantities to which most attention should be paid when there is a need to meet a target uncertainty. Nevertheless, we intend to examine variance-based sensitivity analysis.

Jon Helton: Is appropriate consideration given to the problem/inconsistencies that are present when variables with bounded ranges are represented with distributions that have infinite ranges (e.g., normal distributions)?

Maurice Cox: Attention is paid to the input quantities by characterizing them by PDFs that account for knowledge of those quantities. In my opinion too much reliance is placed on the central limit theorem in representing the output quantity by a Gaussian distribution. Inconsistencies arise when it it is known that the output quantity is bounded, such as a mass concentration lying between $0\,\%$ and $100\,\%$, and a coverage interval extends beyond one of these limits. The Joint Committee for Guides in Metrology, in its work on revising the GUM, will take account of such knowledge in recommending methods based on forming a suitable prior distribution and applying Bayes' theorem rather than using the propagation of distributions in such a case.

Jon Helton: Has the use of importance sampling been considered for use in the determination of the extreme tails of distributions?

Maurice Cox: The relevance of importance sampling is recognized, and it is intended that, when sufficient experience on metrology problems has been gained, appropriate guidance will be given.

Model-Based Interpolation, Prediction, and Approximation

Antonio Possolo

Statistical Engineering Division
Information Technology Laboratory
National Institute of Standards and Technology
Gaithersburg, Maryland, USA
`antonio.possolo@nist.gov`
`http://www.nist.gov/itl/sed/possolo.cfm`

Abstract. Model-based interpolation, prediction, and approximation are contingent on the choice of model: since multiple alternative models typically can reasonably be entertained for each of these tasks, and the results are correspondingly varied, this often is a considerable source of uncertainty. Several statistical methods are illustrated that can be used to assess the contribution that this uncertainty component makes to the uncertainty budget: when interpolating concentrations of greenhouse gases over Indianapolis, predicting the viral load in a patient infected with influenza A, and approximating the solution of the kinetic equations that model the progression of the infection.

Keywords: interpolation, prediction, approximation, uncertainty, influenza, greenhouse gases, projection pursuit.

1 Introduction

Three examples are described that illustrate the application of statistical methods to assess the contribution that model uncertainty makes to the overall uncertainty when doing interpolations, producing predictions, or building approximations.

The first example, described in §2, relates to the interpolation of concentrations of CO_2 measured in the course of a flight over Indianapolis: two interpolants are considered, local regression and kriging, among many others that could reasonably be entertained, and this suffices to make the point that model uncertainty can make a sizeable contribution to the overall uncertainty budget.

The second example, reviewed in §3, concerns a measurand (the quantity intended to be measured [15, 2.3], here the time that elapses after infection with the influenza A virus, until the viral load peaks in a particular patient) whose measurement involves the solution of a set of simultaneous differential equations. Since this depends on several parameters, each set of values assigned to them in fact defines a particular model. The example characterizes the dispersion of values that correspond to these multiple, alternative choices.

A. Dienstfrey and R.F. Boisvert (Eds.): WoCoUQ 2011, IFIP AICT 377, pp. 195–211, 2012.

The third example, presented in §4, is about the construction of a data-driven approximant to an unknown function of several variables. The quantity of interest is the same that is considered in the example of §3, the time that elapses after infection with the influenza A virus until the viral load peaks, but here regarded as a function of four parameters. The approximants considered all belong to the class of projection pursuit regressions. In this case, the alternative models under consideration correspond to the different numbers of "ridge functions" used to define the structure of the approximant.

These examples were all worked out using the facilities for statistical computing available in the R programming environment for statistical computing, data analysis, and graphics [22], which is free and open source. Writing for the *New York Times* of January 6, 2009, under the title "Data Analysts Captivated by R's Power", Ashlee Vance explains that R is a popular programming language used by a growing number of data analysts, and suggests that "it is becoming their lingua franca".

Some commercial instruments are identified in this paper accurately to represent the sources of some of the data that is used. Neither does such identification imply recommendation or endorsement by the National Institute of Standards and Technology, nor does it imply that the equipment identified is necessarily the best available for the purpose, or that it has been properly calibrated or used.

2   Interpolation

2.1   INFLUX Experiment

The Indianapolis Flux Experiment (INFLUX, http://influx.psu.edu/, [24]) serves to develop and assess methods for quantifying greenhouse gas emissions on an urban scale: it is a joint undertaking of Arizona State University, the Cooperative Institute for Research in Environmental Sciences (National Oceanic and Atmospheric Administration and the University of Colorado at Boulder), the Earth Science Research Laboratory (National Oceanic and Atmospheric Administration), the National Institute of Standards and Technology, Penn State University, and Purdue University.

Figure 1 shows the flight path of a small aircraft, instrumented with a Picarro greenhouse gas analyzer (http://www.picarro.com/gas_analyzers), which flew from Purdue University to Indianapolis on June 1, 2011, and then took measurements of CO_2 concentrations at about 17 000 points lying approximately on a vertical curtain about 80 km wide and 1.2 km tall, as illustrated in Figure 2.

We consider the problem of interpolating the observations to create a map of those concentrations covering the whole vertical flight curtain, which subsequently will be used to compute vertical CO_2 flux. Since such interpolation can reasonably be done in any of several alternative ways, the corresponding dispersion of results ought to be reflected in the map's uncertainty budget.

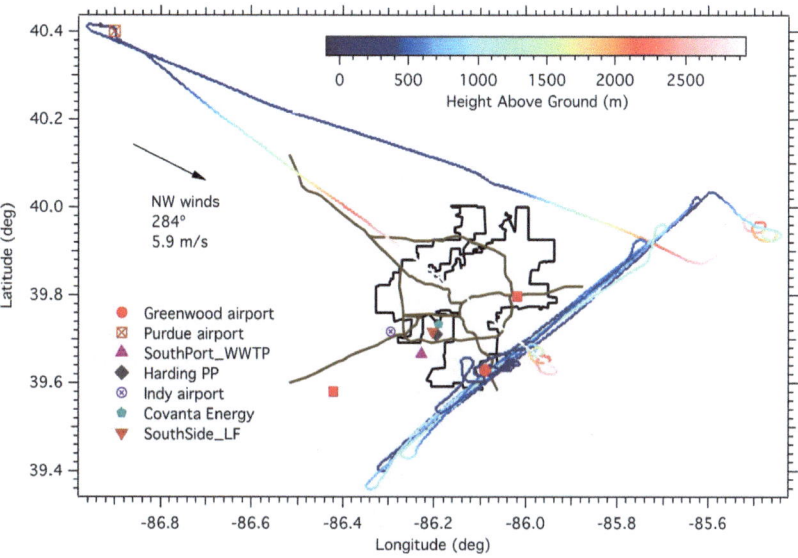

Fig. 1. Flight path of a small aircraft equipped with a Picarro greenhouse gas analyzer that made the measurements depicted in Figure 2: map drawn by Paul B. Shepson and M. Obiminda Cambaliza (Department of Chemistry, Purdue University) [24]

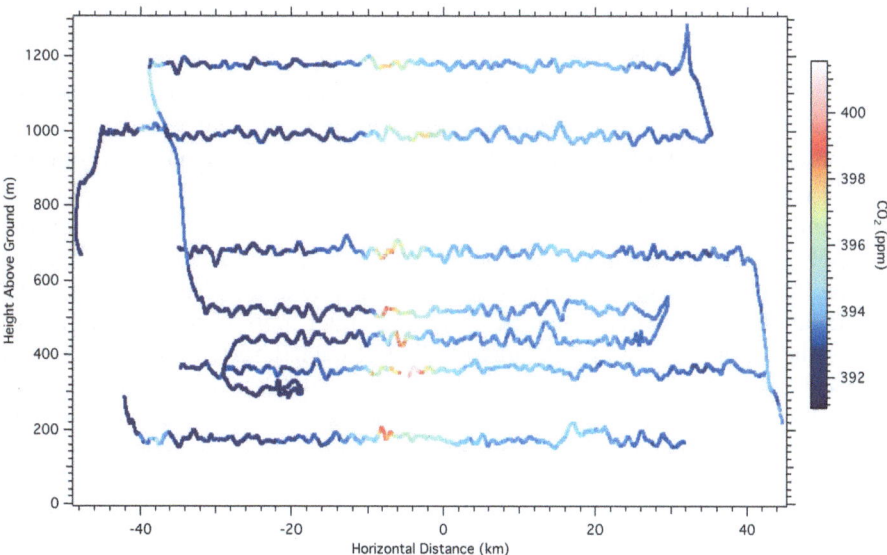

Fig. 2. Locations on the vertical curtain where concentrations of CO_2 concentrations where measured on June 1, 2011, using a Picarro greenhouse gas analyzer, with colors indicating the measured values: drawn by Paul B. Shepson and M. Obiminda Cambaliza (Department of Chemistry, Purdue University).

2.2   Local Regression and Kriging

The data are a set of values $y_1, \ldots, y_m$ of a real-valued, albeit unknown function θ, which were measured at points $x_1, \ldots, x_m$ in a space $\mathcal{X}$ on which there is defined a distance metric. The objective is to produce an estimate y of $\theta(x)$, for any x "in the middle" of the $\{x_i\}$, typically but not necessarily different from these $\{x_i\}$, and to characterize the uncertainty $u(y)$ associated with the estimate, conceived as an indication of how close to $\theta(x)$ one believes y to be.

According to the *Guide to the expression of uncertainty in measurement* (GUM) [13, 3.3.1] measurement uncertainty "reflects the lack of exact knowledge of the value of the measurand" (the value tlike $\theta(x)$, this paramater will be the standard deviation of a probability distribution that describes the dispersion of values that may reasonably be assigned to y.

The goal of estimating $\theta(x)$ is approached by modeling the observations $\{y_i\}$ probabilistically, so that the interpolation problem becomes a statistical estimation problem. In this conformity we represent the data as $y_i = \theta(x_i) + \epsilon_i$ for $i = 1, \ldots, m$, where the $\{\epsilon_i\}$ are regarded as realized values of non-observable random variables (*measurement errors*). The objective, then, is to interpolate the signal $\theta(x_i)$, not the signal plus the noise, $\theta(x_i) + \epsilon_i$. (The assumption that the signal and the noise combine additively can often be satisfied by suitably re-expressing, or transforming, the data before the statistical analysis.)

This goal can be achieved in any one of many different ways, several of which are comparably reasonable under mild assumptions about θ and about the probability distribution of the measurement errors. We choose to illustrate and compare the results of two well-known, widely used procedures: local regression [2,3,17] and kriging [6,16,27]. In both, a model needs to be chosen for θ, which next needs to be calibrated in light of the data (that is, its adjustable parameters need to be estimated), the resulting $\widehat{\theta}$ finally being used to compute the interpolated value as $y = \widehat{\theta}(x)$.

To apply local regression we assume that θ is continuous and locally quadratic. To apply kriging we assume that θ is a realized value of a stationary Gaussian random function Θ whose covariance function is a member of a particular parametric family. Both methods can be employed under more general assumptions, but neither of these assumptions already is particularly restrictive. In fact, and for local regression, assuming that θ is locally constant, linear, or quadratic makes little difference in most practical applications; and for kriging, the collection of realizations of a Gaussian random function, even a stationary one, is sufficiently varied to be able to mimic the majority of continuous functions likely to be encountered in practice.

The methods available to build these interpolants from empirical data are amply documented in the literature: for example in [4], which discusses interpolation of spatial data at great length. We used function `locfit` from the R package with the same name [18] to fit the local regression model, and function `automap` from the R package `intamap` [21] to fit the kriging model. Before discussing the results for the CO_2 data, we illustrate them in the context of a very simple example.

Figure 3 shows differences between a local regression model and a kriging model fitted to the same data, for a set of simulated data where $\mathcal{X}$ is the real line and θ is a real-valued function of a real argument. The kriging model is defined by a constant mean $\mu = \mathbb{E}(\Theta(x))$ for all x, and by a covariance function γ such that $\gamma(h) = \mathbb{E}[(\Theta(x+h) - \mu)(\Theta(x) - \mu)]$.

The assumption is made that γ depends only on h and not on x (where the increment h is a scalar when $\mathcal{X}$ is the real line, and it is a vector when $\mathcal{X}$ is the Euclidean plane), and that it belongs to the class of covariance functions introduced by Matérn [19], involving three adjustable parameters. This model was fitted to the data depicted in Figure 3 using facilities provided by the R package geoR [23].

Both the local regression and the kriging interpolants do some smoothing because the goal is to interpolate the signal — that is, values of $\theta(x)$ — not the signal plus the noise, $\theta(x) + \epsilon$. In practice, of course, since neither θ nor ϵ are known, a statistical procedure needs to be employed that "guesses" the most appropriate extent of smoothing: one such procedure is *cross-validation*, which will be discussed below.

The results obtained by the two methods are similar but different. Since either model could have reasonably been selected for the task at hand, these differences reflect an uncertainty component that is attributable to model selection. Another uncertainty component relates to model calibration (that is, estimation of model parameters), and derives from the fact that this calibration is based on only a finite amount of data.

Kriging is often heralded as providing assessments of uncertainty of its interpolations automatically. However, in many instances of application, kriging's built-in assessments *underestimate* uncertainty because one pretends that the estimate $\widehat{\gamma}$ of the covariance function is identical to the covariance function γ itself. Bayesian kriging provides means to account for this often neglected component of uncertainty [6]. Here, we will rely on cross-validation to evaluate the corresponding uncertainty component.

2.3 Cross-Validation and Model Uncertainty

Cross-validation [11,20] is an established procedure to assess the performance of statistical procedures realistically. In the context of interpolation, the idea is to partition the data into two subsets, use one (*training subset*) to develop and calibrate the interpolant, and the other (*testing subset*) to gauge its performance, by comparing the interpolation results with the observations at those locations present in the testing subset. The process may be repeated multiple times, over many different partitions, and some figure of merit (for example, root mean square of the differences between interpolated and actual values) averaged.

The partitioning may be done at random, or it may include consideration for the particulars of the situation. For the CO_2 data whose measurement locations are depicted in Figure 2, a random partition may lead to overoptimistic assessments. Given the fairly high rate at which the measurements were made, the measurement locations are very closely spaced along the horizontal legs of

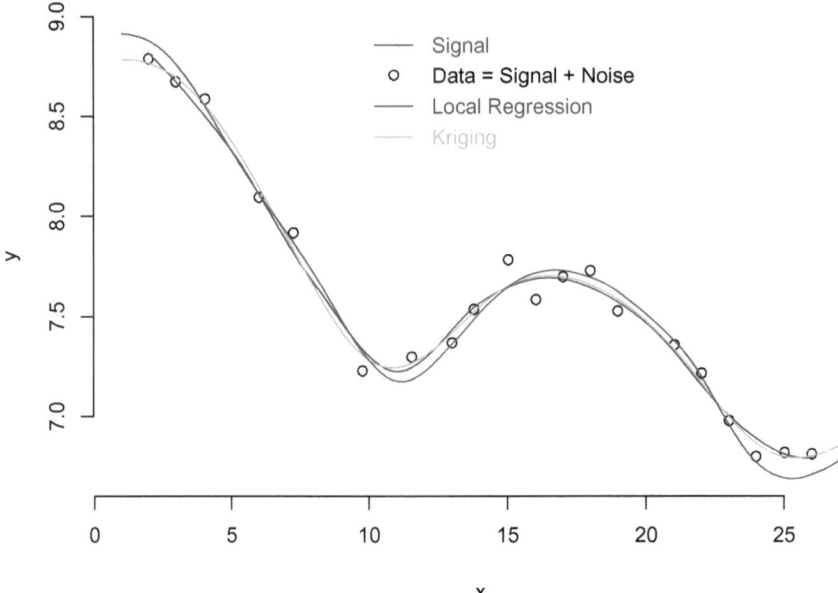

Fig. 3. Illustrative example of differences between a local regression model and a kriging model fitted to the same data. The red line, marked "Signal" in the legend, depicts the "true" function θ: this is known in this case because the data are simulated.

the curtain portion of the flight: for this reason, if the testing subset had been obtained by selecting, say, about one third of the measurement locations chosen uniformly at random, then there would have been data in the training subset very close to each of the locations in the testing subset, and the interpolation accuracy would misleadingly appear to be very high.

The interpolation challenge in this case, however, is mostly in the vertical direction, at points whose height lies between the heights of any pair of adjacent, horizontal legs of the flight. A suitable testing subset should then be one of these horizontal legs (Figure 4), with the rest of the observations being used to build an interpolant to estimate concentrations of CO_2 at all the points in the leg that will have been set aside. And then to repeat this procedure for each leg in turn.

2.4 INFLUX Experiment – Uncertainty Budget

Figure 5 shows the results of local regression and kriging interpolations for the INFLUX data. The differences are generally small, being most pronounced along the edges of the CO_2 plume. The cross-validation assessment of their respective performance indicates that they are quite comparable, and there is hardly any reason to prefer one over the other.

The uncertainty budget in Table 1 lists the recognized uncertainty components and the contributions they make to the expanded combined measurement un-

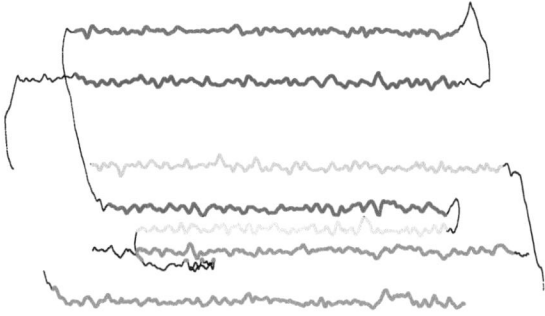

Fig. 4. Locations on the flight curtain where measurements of CO_2 were made, also depicted in Figure 2. Here the different colors indicate the testing subsets used when cross-validating the performance of the two interpolation procedures.

Table 1. Uncertainty budget for the map of CO_2 concentrations over the curtain flight plane. To within the single significant digit listed, the interpolation uncertainty is the same for the local regression and kriging models. The evaluations marked MANUF were drawn from the specification sheet for the Picarro CRDS Analyzer model G2301-m (http://www.picarro.com/gas_analyzers/flight_co2_ch4_h2o); CV indicates evaluation via cross-validation; and LAB+CERT indicates that the evaluation expresses laboratory calibration uncertainty and the uncertainty in the calibration standards.

SOURCE	EVALUATION STD.	UNCERT. mg/m^3 (ppmv)
Model selection	CV	0.36
Interpolation	CV	0.9
Instrument calibration	LAB+CERT	0.034
Instrument repeatability	MANUF	0.2
Instrument drift	MANUF	0.2
Atmospheric temperature	MANUF	0.0075
Atmospheric pressure	MANUF	0.7
Expanded Uncertainty	$U_{95\%} = 2.5\,\mathrm{mg/m^3}$ (ppmv)	

certainty in the last line of the table, computed as $U_{95\%} = 2\sqrt{0.36^2 + \cdots + 0.7^2}$ $= 2.5\,\mathrm{mg/m^3}$ (ppmv), meaning that, with approximate 95 % probability, the true CO_2 concentrations are within $\pm U_{95\%}$ of the values shown on the map.

3 Prediction

3.1 Viral Load in Influenza a Infection

[1] describe several mathematical models for the progression of influenza A infection within an individual patient. These models capture the principal features

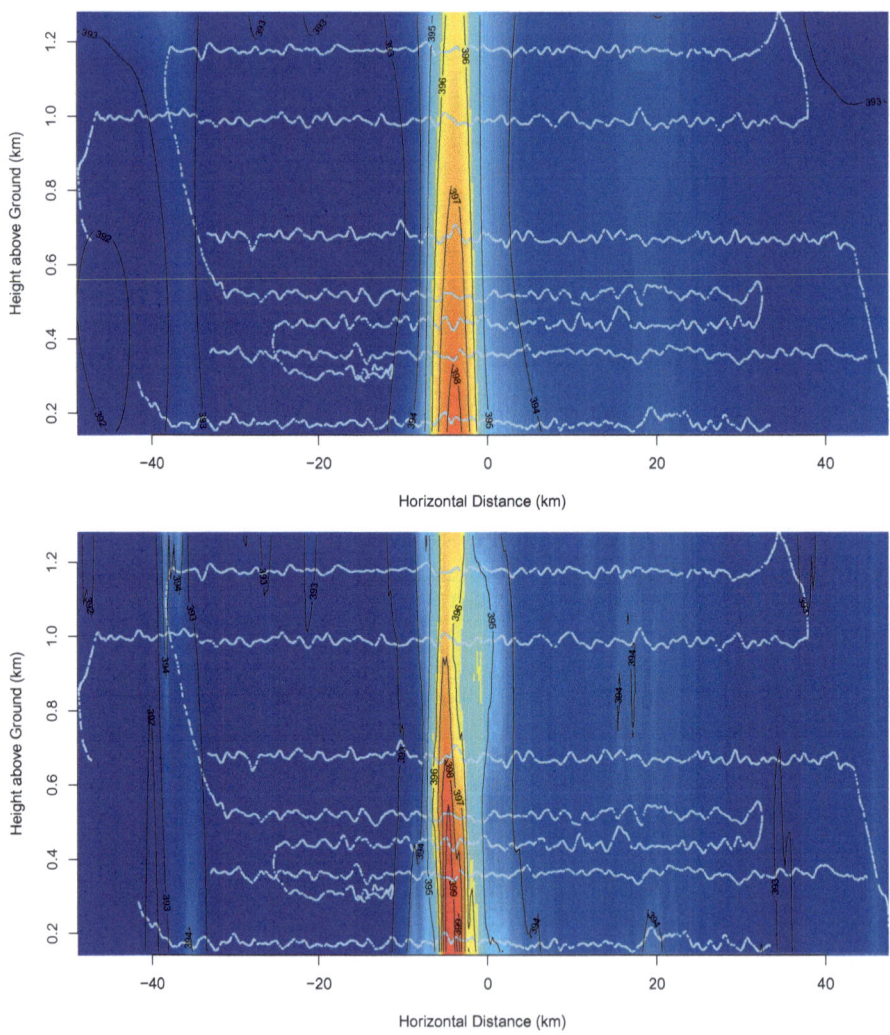

Fig. 5. Results of interpolating the CO_2 measurements over a regular grid, then colored and contoured, using local regression (top panel) and kriging (bottom panel)

of the underlying viral kinetics: initially the viral load grows exponentially fast, peaks 2 to 3 days post-infection, and finally decreases, also at an exponental rate, to undetectable levels within 6 to 8 days post-infection.

Here we will consider the simplest model that [1] describe. The infection is assumed to be limited by the availability of susceptible epithelial cells rather than by the patient's immune response. The number T of uninfected target cells, the number I of productively infected cells, and the viral load V satisfy the following system of differential equations, where β is the infection rate, $1/\delta$ is the expected lifespan of an infected cell, ρ is the increment to viral load contributed by each infected cell, and γ is the viral clearance rate:

$$\frac{dT}{dt} = -\beta TV, \quad \frac{dI}{dt} = \beta TV - \delta I, \quad \frac{dV}{dt} = \rho I - \gamma V. \tag{1}$$

Figure 6 shows the values of the viral load in patient 4 (from Table 1 of [1]) as measured on seven consecutive post-infection days, and the fit this model achieves assuming that $\log_{10} V$ is Gaussian, and that V satisfies equations (1). The differential equations were solved by application of the Livermore Solver for Ordinary Differential Equations (LSODA) [12], as implemented in R package deSolve [26]. The model was fitted by minimizing the sum of squares of the differences between the logarithm of the mean viral load (which is a function of β, δ, ρ, and γ), and the logarithm of the corresponding measured load: the unit of measurement is TCID$_{50}$, that is 50 % Tissue Culture Infective Dose per milliliter of nasal wash.

Consider predicting the post-infection time $\operatorname{argmax}_t V_t$ when the viral load peaks, and estimating the infection's *Basic Reproductive Number* $R_0 = \rho\beta T_0/(\gamma\delta)$, which is the average number of second-generation infections produced by a single

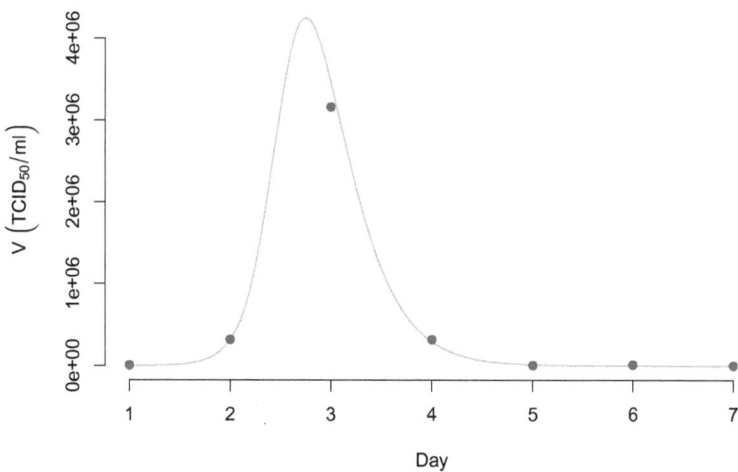

Fig. 6. Observed viral load of patient 4 [1, Table 1] at each of seven days (red dots), and fitted model (blue curve)

infected cell placed among susceptible cells: if $R_0 > 1$ the infection progresses to full course, and if $R_0 < 1$ the infection dies out prematurely.

Predicting the post-infection time $\text{argmax}_t V_t = \psi(\beta, \delta, \rho, \gamma)$ involves solving the foregoing system of differential equations and assessing the uncertainty of the result. Estimating R_0 involves estimating ρ, β, T_0, γ and δ, and propagating their associated uncertainties to obtain $u(R_0)$, the uncertainty associated with the estimate of R_0.

Both tasks can be accomplished using the parametric statistical bootstrap [7] (cf. [14]), by drawing samples from suitable, joint probability distributions for the participating quantities. The key ingredient for this is a numerical approximation to the Hessian $H(\beta, \delta, \rho, \gamma)$ of the negative log-likelihood used to fit the kinetic model to the data for Patient 4, which was obtained using function hessian of the R package numDeriv [10].

The parametric statistical bootstrap then amounts to repeating the following steps for $k = 1, 2, \ldots, K$, where K is a suitably large integer: (i) draw a sample $(\beta_k, \delta_k, \rho_k, \gamma_k)$ from a multivariate Gaussian distribution with mean $(\widehat{\beta}, \widehat{\delta}, \widehat{\rho}, \widehat{\gamma})$ and covariance matrix $H^{-1}(\widehat{\beta}, \widehat{\delta}, \widehat{\rho}, \widehat{\gamma})$, where $\widehat{\beta}$ denotes the estimate of β, and similarly for the other parameters; (ii) draw one sample from a uniform distribution for each initial condition $T_0 \pm 0.1 T_0$, $I_0 \pm 0.1 I_0$, and $V_0 \pm 0.1 V_0$ (the 10 % relative uncertainty assumed here is merely for purposes of illustration of the general idea that the uncertainty in the initial conditions also ought to be propagated); (iii) solve the kinetic model with perturbed parameters and compute $\psi(\beta_k, \delta_k, \rho_k, \gamma_k)$.

These steps produce K replicates of $\text{argmax}_t V_t$ whose spread is indicative of the corresponding measurement uncertainty. Figure 7 summarizes the results for as an estimate of the probability density that encapsulates the state of knowledge about $\text{argmax}_t V_t$.

These same K samples that were drawn from the joint probability distribution of β, δ, ρ, γ, and T_0 also yield K replicates of the Basic Reproductive Number R_0, similarly depicted in Figure 8.

4   Approximation

Suppose that we observe values corrupted by non-observable errors, $(x_1, \psi(x_1) + \epsilon_1), \ldots, (x_m, \psi(x_m) + \epsilon_m)$, of an unknown function $\psi : \mathcal{X} \mapsto \mathbb{R}$ that is "expensive" to evaluate, and that, based on such data, we wish to develop an approximant φ of ψ, and to assess its quality.

For the influenza example described in 3, the unknown function is $\psi(\beta, \delta, \rho, \gamma) = \text{argmax}_t V_t$, and we wish to build φ such that $\varphi(\beta, \delta, \rho, \gamma) \approx \psi(\beta, \delta, \rho, \gamma)$, and such that the approximation is reasonably accurate, and less "expensive" to evaluate than ψ (whose evaluations involve solving a system of differential equations).

From among the several different methods available to build such approximant, we illustrate projection pursuit regression (PPR) [8,11], which builds on the idea of projection pursuit, an algorithm that "seeks to find one- and

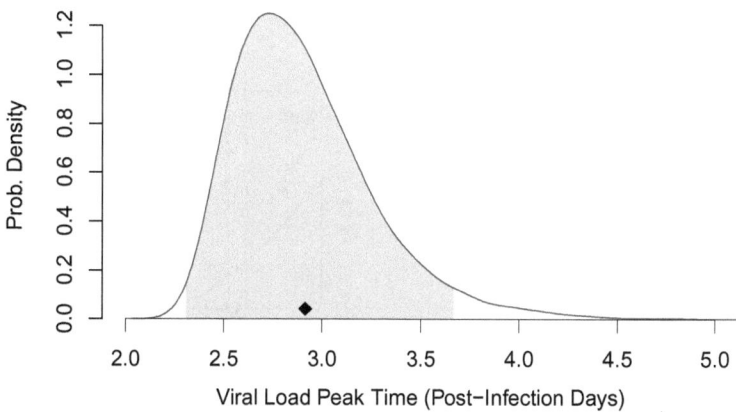

Fig. 7. Kernel probability density estimate [25] based on $K = 10\,000$ replicates of the values of $\text{argmax}_t V_t$ obtained by application of the parametric statistical bootstrap. The average of these replicates, 2.9 post-infection days (marked with a diamond), is an estimate of $\text{argmax}_t V_t$, and the associated uncertainty is the corresponding standard deviation, 0.4 post-infection days (PID). The shortest 95 % probability interval (the footprint of the shaded area under the curve) ranges from 2.3 to 3.7 PID.

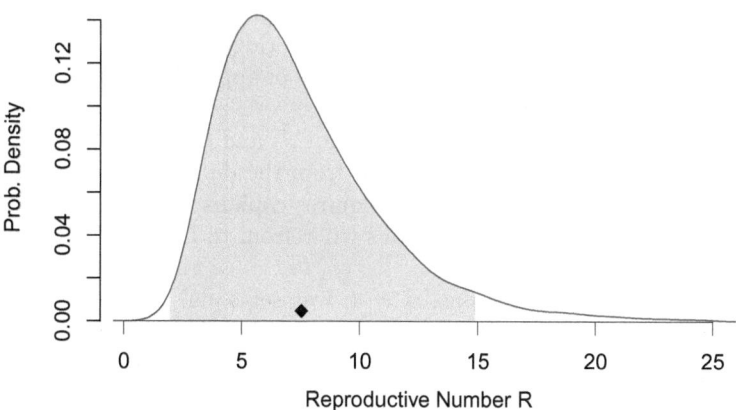

Fig. 8. Kernel probability density estimate [25] based on $K = 10\,000$ replicates of the values of R_0 obtained by application of the parametric statistical bootstrap. The average of these replicates is $\widehat{R}_0 = 7.5$ with associated standard uncertainty $u(R_0) = 3.5$, and the shortest 95 % probability interval is $(2, 15)$, thus suggesting that the basic reproductive number is greater than 1 with very high probability: in fact, the same replicates allow estimating $\Pr(R_0 > 5) = 0.76$.

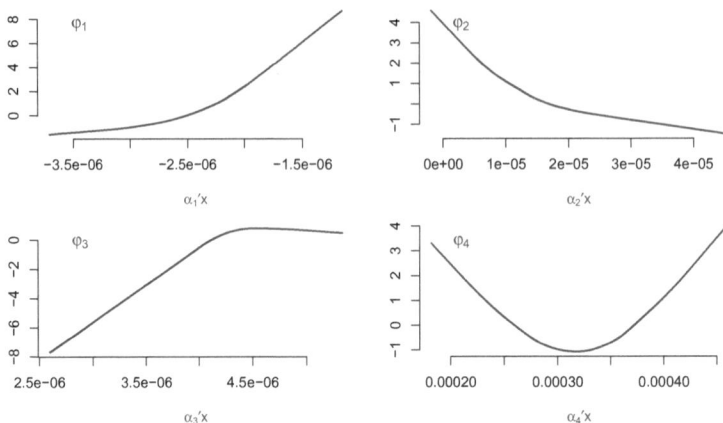

Fig. 9. Ridge functions of the projection pursuit approximant to $\psi(\beta, \delta, \rho, \gamma) =$ argmax$_t V_t$

two-dimensional linear projections of multivariate data that are relatively highly revealing" [9]. PPR builds predictors out of these "highly revealing" projections, automatically sets aside variables with little predictive power, and bypasses the *curse of dimensionality* by focussing on functions of linear combinations of the original variables, at the price of intensive computation.

The approximant is of the form $\varphi(x) = \sum_{j=1}^{J} \varphi_j(\alpha_j \cdot x)$, where $\alpha_1, \ldots, \alpha_J$ are vectors defining the projection directions, $\alpha_j \cdot x$ denotes the inner product of α_j and x, and $\varphi_1, \ldots, \varphi_J$ are the so-called *ridge functions*. Considered for all values of J, and for appropriate choices of the ridge functions, these functions form a class of universal approximants capable of approximating any continuous function [5].

In practice one needs to choose a value for J, and both the projection directions and the ridge functions are derived from the data by means of a suitable smoothing procedure, which typically requires copious amounts of data. In this case we have 10 000 sets of observations wherefrom to build a PPR approximation to $\psi(\beta, \delta, \rho, \gamma) = $ argmax$_t V_t$.

The number of ridge functions, $J = 4$, was set equal to the number of predictors: in general, one may like to use cross-validation to select the optimal number. The R function `ppr` produces the ridge functions depicted in Figure 9, when instructed to select the best model with four terms among all such models with up to eighteen terms.

The cross-validated, relative approximation error is 3 % (this is the ratio between the standard deviation of the cross-validated approximation errors and the median of the values of the function ψ that is the target of the approximation). The component of uncertainty attributable to model choice amounts to 0.7 %, which is the mean value of the relative standard deviations of the predictions made by models with $J = 2, \ldots, 6$, at each element of the testing subset: this is considerably smaller than the relative approximation error given above.

5   Conclusions

Models are necessary for measurement. And since a measurement result must comprise both an estimate of the value of the measurand and an assessment of the associated measurement uncertainty [15, 2.9], the models that are used to describe the interplay between all the quantities involved in measurement should facilitate such assessment. Since, furthermore, probability distributions are generally regarded as the best means to characterize measurement uncertainty [13, 3.3.4], measurement models ought to be statistical models.

In many measurement situations, however, more than one model may reasonably be entertained. The three examples given above all should have sufficed to make this abundantly clear, each in a particular way that may be common to many other, similar applications. The dispersion of values that such multiplicity of choices entails ought to be evaluated and expressed in the assessment of measurement uncertainty, similarly to how the contributions from all the other recognized sources of uncertainty are so evaluated and expressed.

The assessment of the component of measurement uncertainty related to model choice generally involves inter-comparing the results corresponding to alternative, comparably tenable models, done in ways that insulate the results (of such inter-comparison) from the perils of over-fitting models to data. For this reason, cross-validation, judiciously employed, presents itself as a valuable tool, not only for model selection, but also to gauge the impact that such selection has upon the uncertainty of the results.

Acknowledgements.   The dataset used in Section 2 comprises aircraft measurements of CO_2 concentrations made by Paul B. Shepson and M. Obiminda Cambaliza (Department of Chemistry, Purdue University), during a curtain flight over Indianapolis, Indiana, on June 1st, 2011, within the scope of the INFLUX project [24].

Jolene Splett (Statistical Engineering Division, Information Technology Laboratory, NIST) provided many useful suggestions that stimulated much improvement of an early draft of this paper. The author is particularly grateful to Andrew Dienstfrey and Ronald Boisvert (Applied and Computational Mathematics Division, Information Technology Laboratory, NIST) for the opportunity of presenting this material at the International Federation for Information Processing (IFIP) Working Conference on Uncertainty Quantification in Scientific Computing (August 1-4, 2011, Boulder, Colorado).

References

1. Baccam, P., Beauchemin, C., Macken, C.A., Hayden, F.G., Perelson, A.S.: Kinetics of influenza a virus infection in humans. Journal of Virology 80(15), 7590–7599 (2006)
2. Cleveland, W.S.: Robust locally weighted regression and smoothing scatterplots. Journal of the American Statistical Association 74, 829–836 (1979)

3. Cleveland, W.S., Devlin, S.J.: Locally-weighted regression: an approach to regression analysis by local fitting. Journal of the American Statistical Association 83, 596–610 (1988)
4. Cressie, N., Wikle, C.K.: Statistics for Spatio-Temporal Data. John Wiley & Sons, Hoboken (2011)
5. Diaconis, P., Shahshahani, M.: On nonlinear functions of linear combinations. SIAM Journal on Scientific and Statistical Computing 5(1) (March 1984)
6. Diggle, P.J., Ribeiro, P.J.: Model-based Geostatistics. Springer, New York (2010)
7. Efron, B., Tibshirani, R.J.: An Introduction to the Bootstrap. Chapman & Hall, London (1993)
8. Friedman, J.H., Stuetzle, W.: Projection pursuit regression. Journal of the American Statistical Association 73(376), 817–823 (1981)
9. Friedman, J.H., Tukey, J.W.: A projection pursuit algorithm for exploratory data analysis. IEEE Transactions on Computers C-23(9), 881–890 (1974)
10. Gilbert, P.: numDeriv: Accurate Numerical Derivatives (2011), http://CRAN.R-project.org/package=numDeriv, r package version 2010.11-1
11. Hastie, T., Tibshirani, R., Friedman, J.: The Elements of Statistical Learning: Data Mining, Inference, and Prediction, 2nd edn. Springer, New York (2009)
12. Hindmarsh, A.: ODEPACK, a systematized collection of ODE solvers. In: Stepleman, R.S., Carver, M., Peskin, R., Ames, W.F., Vichnevetsky, R. (eds.) Scientific Computing: Applications of Mathematics and Computing to the Physical Sciences, IMACS Transactions on Scientific Computation, vol. 1, pp. 55–64. North-Holland, Amsterdam (1983)
13. Joint Committee for Guides in Metrology: Evaluation of measurement data — Guide to the expression of uncertainty in measurement. International Bureau of Weights and Measures (BIPM), Sèvres, France (September 2008), http://www.bipm.org/en/publications/guides/gum.html, BIPM, IEC, IFCC, ILAC, ISO, IUPAC, IUPAP and OIML, JCGM 100:2008, GUM 1995 with minor corrections
14. Joint Committee for Guides in Metrology: Evaluation of measurement data — Supplement 1 to the "Guide to the expression of uncertainty in measurement" — Propagation of distributions using a Monte Carlo method. International Bureau of Weights and Measures (BIPM), Sèvres, France (2008), http://www.bipm.org/en/publications/guides/gum.html, BIPM, IEC, IFCC, ILAC, ISO, IUPAC, IUPAP and OIML, JCGM 101:2008
15. Joint Committee for Guides in Metrology: International vocabulary of metrology — Basic and general concepts and associated terms (VIM). International Bureau of Weights and Measures (BIPM), Sèvres, France (2008), http://www.bipm.org/en/publications/guides/vim.html, BIPM, IEC, IFCC, ILAC, ISO, IUPAC, IUPAP and OIML, JCGM 200:2008
16. Krige, D.G.: A statistical approach to some basic mine valuation problems on the witwatersrand. Journal of the Chemical, Metallurgical and Mining Society of South Africa 52, 119–139 (1951)
17. Loader, C.: Local Regression and Likelihood. Springer, New York (1999)
18. Loader, C.: locfit: Local Regression, Likelihood and Density Estimation (2010), http://CRAN.R-project.org/package=locfit, r package version 1.5-6
19. Matérn, B.: Spatial Variation, 2nd edn. Lecture Notes in Statistics. Springer, New York (1987)
20. Mosteller, F., Tukey, J.W.: Data Analysis and Regression. Addison-Wesley Publishing Company, Reading (1977)

21. Pebesma, E., Cornford, D., Dubois, G., Heuvelink, G., Hristopoulos, D., Pilz, J., Stoehlker, U., Morin, G., Skoien, J.: INTAMAP: the design and implementation of an interoperable automated interpolation web service. Computers & Geosciences 37, 343–352 (2011), http://dx.doi.org/10.1016/j.cageo.2010.03.019

22. R Development Core Team: R: A Language and Environment for Statistical Computing. R Foundation for Statistical Computing, Vienna, Austria (2011), http://www.R-project.org, ISBN 3-900051-07-0

23. Ribeiro, P.J., Diggle, P.J.: geoR: a package for geostatistical analysis. R-NEWS 1(2), 14–18 (2001), http://CRAN.R-project.org/doc/Rnews/ ISSN 1609-3631

24. Shepson, P., Cambaliza, M.O., Davis, K., Gurney, K., Lauvaux, T., Miles, N., Richardson, S., Sweeney, C., Turnbull, J.: The INFLUX project: Indianapolis as a case study for the accurate and high resolution determination of CO_2 and CH_4 emission fluxes from an urban center. Poster, American Geophysical Union Fall 2010 Meeting, San Francisco, CA (December 2010)

25. Silverman, B.W.: Density Estimation. Chapman and Hall, London (1986)

26. Soetaert, K., Petzoldt, T., Setzer, R.W.: Solving differential equations in R: package deSolve. Journal of Statistical Software 33(9), 1–25 (2010), http://www.jstatsoft.org/v33/i09

27. Stein, M.L.: Interpolation of Spatial Data: Some Theory for Kriging. Springer, New York (1999)

Discussion

Speaker: Antonio Possolo

Mark Campanelli: What can be said about extrapolating the influenza model to make predictions about higher risk populations (such as asthmatics, children, and the elderly), when data from lower risk populations were (presumably) used to generate the time series data from planned infections?

Antonio Possolo: The model reviewed in the presentation is generic in its formulation, and the simplest of those described by Prasith Baccam and collaborators in the 2006 article quoted in my presentation and in my corresponding contribution to these proceedings. Its purpose is to describe the progression of the disease in a particular patient, once its adjustable parameters have been estimated based on observations made of the viral load carried by the patient on different days post-infection.

Van Snyder: Are the measurements of CO_2 in Indianapolis operational *in situ* measurements, or for validation of a remote sensing system?

Antonio Possolo: The measurements of the concentration of CO_2 made in the Indianapolis area in the context of the INFLUX experiment, serve to demonstrate the operation of a multimodal, pilot network to monitor abundances of greenhouse gases, not to validate a remote sensing system.

Ruth Jacobsen: For the measurements of greenhouse gas emissions: what were the detection limits of the instrument?

Antonio Possolo: For the measurements of concentration of CO_2 that I mentioned in my presentation, my colleagues from Purdue University that participate in the INFLUX experiment, Paul Shepson and Obie Cambaliza, use a Picarro gas analyzer. The performance specifications of Picarro's analyzers are listed in data sheets published at www.picarro.com. Please note that the mention of specific products, trademarks, or brand names in the answer to this question is for purposes of identification only, and ought not to be interpreted in any way as an endorsement or certification of such products or brands by the National Institute of Standards and Technology.

Mladen Vouk: This presentation emphasizes R. Is there an issue with SAS, or Genstat, or SPSS (or other packages) regarding functions you used or customization?

Antonio Possolo: I have little first-hand knowledge of SAS and SPSS, and none of Genstat. My understanding, based on the opinions of colleagues who use them, is that all these products include reliable implementations of many statistical procedures. I find R very well-suited to my needs as a statistician, not only for the functionality that it offers, or for how well it facilitates prototyping new ideas or customizing existing procedures, but also for the scrutiny that a very large base of sophisticated, inquisitive users, constantly subject R to (including its source code), and, in the process, ensure its quality. Please note that the mention of specific products, trademarks, or brand names in the answer to this question is for purposes of identification only, and ought not to be interpreted in any way as an endorsement or certification of such products or brands by the National Institute of Standards and Technology.

Maurice Cox: On interpolation, you have measurement variances associated with the data points. You also apply a weight function that decays away from the point of interest. How can you justify that metrologically/statistically?

Antonio Possolo: Local regression is a statistical procedure with a long history and commendable performance characteristics. The work of Bill Cleveland, Susan Devlin, and of Catherine Loader, that I reference in my contribution to these proceedings, among a large body of related work, include several results on the optimality of this class of procedures under fairly general conditions.

Jon Helton: An observation: additional examples of the use of nonparametric regression techniques in uncertainty/sensitivity analysis are given in a sequence of articles by Curt Storlie *et al.* in *Reliability Engineering and System Safety*.

Uncertainty Quantification for Turbulent Mixing Flows: Rayleigh-Taylor Instability

T. Kaman[1], R. Kaufman[1], J. Glimm[1], and D.H. Sharp[2]

[1] Department of Applied Mathematics and Statistics
Stony Brook University
Stony Brook, NY USA
[2] Los Alamos National Laboratory
Los Alamos, NM, USA

Abstract. Uncertainty Quantification (UQ) for fluid mixing depends on the length scales for observation: macro, meso and micro, each with its own UQ requirements. New results are presented here for macro and micro observables. For the micro observables, recent theories argue that convergence of numerical simulations in Large Eddy Simulations (LES) should be governed by space-time dependent probability distribution functions (PDFs, in the present context, Young measures) which satisfy the Euler equation. From a single deterministic simulation in the LES, or inertial regime, we extract a PDF by binning results from a space time neighborhood of the convergence point. The binned state values constitute a discrete set of solution values which define an approximate PDF. The convergence of the associated cumulative distribution functions (CDFs) are assessed by standard function space metrics.

1 Introduction

LES convergence is an asymptotic description of numerical simulations of the inertial, or self similar, scaling range of a turbulent flow. In the LES regime we are not concerned with convergence in a conventional sense. Such mathematical convergence to a classical or weak solution, as $\Delta x \to 0$, is a property of direct numerical simulations (DNS), i.e., simulations with all length scales resolved. For practical problems of turbulence this goal may be unrealistic. By contrast, in the following we investigate LES convergence, defined as the behavior of numerical solutions in the LES (inertial) regime under mesh refinement. In this regime there is still a type of convergence but it may be weaker than that considered by traditional DNS analysis. For example, rather than convergence to a weak solution, it may be useful or even necessary to consider convergence of probability distribution functions (PDFs) to a measure valued solution (Young measure). The PDFs capture the local fluctuations of the solution, which are an important aspect of the solution in the inertial regime.

In this article, we present such a picture, still incomplete, from perspectives of mathematical theory, simulation and physical reasoning. It allows these two notions of convergence (DNS and LES; classical and w* Young measure limit) to coexist.

A. Dienstfrey and R.F. Boisvert (Eds.): WoCoUQ 2011, IFIP AICT 377, pp. 212–225, 2012.

The main result is a convergence study for PDFs and CDFs for a numerical mesh refinement study of a Rayleigh-Taylor problem. At present meshes, we find CDF but not yet PDF convergence, a minimum sampling size (supercell size) for the stochastic convergence, and suitable norms for the measurement of convergence.

2   Verification, Validation and Uncertainty Quantification for RT Mixing

2.1   The RT Mixing Rate α

The Rayleigh-Taylor (RT) instability is a classical hydrodynamical instability driven by an acceleration force applied across a density discontinuity. The result is a mixing layer, growing in time with a penetration thickness (of the bubbles, i.e. the light fluid)

$$h_b = \alpha_b A g t^2 , \qquad (1)$$

where A is the Atwood number, α_b is a dimensionless buoyancy correction factor, and g is the acceleration force. We have achieved excellent agreement with experiment in our RT simulations; see Table 1. The results of Table 1, being stronger than LES simulations of others, require detailed examination. A distinctive algorithmic feature of our simulations is the combined use of front tracking and subgrid scale models for LES, or FT/LES/SGS in brief. A second feature of our work has been careful modeling of experimental detail. We summarize here two issues important to this examination: initial conditions and mesh resolution.

2.2   Uncertainty Quantification for Initial Conditions and Mesh Convergence

For most experiments, the initial conditions were not recorded, and the possibility of influence of long wave length initial perturbations has been a subject of speculation. We have quantified the allowed long wave length perturbation amplitudes, by an analysis of the recorded early time data [5,6,4]. Including an

Table 1. Comparison of FT/LES/SGS simulation to experiment. Simulation and experimental results reported with two significant digits. Discrepancy refers to the comparison of results outside of uncertainty intervals, if any, as reported.

Ref.	Exp.	Sim. Ref.	α_{exp}	α_{sim}	Discrepancy
[21]	#112	[8]	0.052	0.055	6%
[21]	#105	[4]	0.072	0.076 ± 0.004	0%
[21,20]	10 exp.	[3]	0.055-0.077	0.066	0%
[19]	air-He	[13]	0.065-0.07	0.069	0%
[17]	Hot-cold	[8,4]	0.070 ± 0.011	0.075	0%
[17]	Salt-fresh	[4]	0.085 ± 0.005	0.084	0%

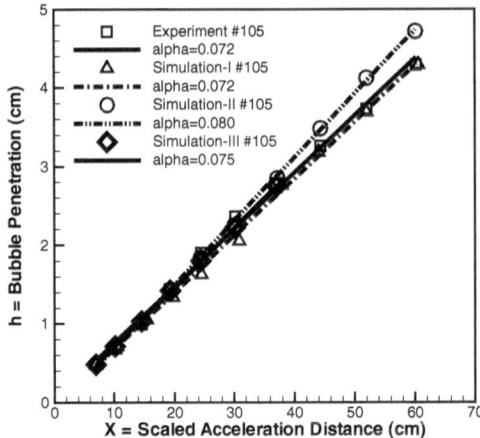

Fig. 1. Plot of the bubble penetration distance h_b vs. a scaled acceleration distance Agt^2. The slope is the mixing growth rate α_b. We plot the experimental data points and three simulation results, which have (I) $0\times$ and (II) $2\times$ our best reconstruction of the initial long wave length perturbations, as extrapolated by backward propagation in time from the early time experimental plates. (III) Inferred initial conditions for long wave length perturbations fully resolved, with a mesh $\Delta x = 111\mu m < l_{We} = 780\mu m$ where l_{We} is the critical bubble size (predicted by Weber number theory). The simulation III is still in progress.

estimate of the uncertainty of this backward extrapolation of data propagated backward to $t = 0$, we estimate the uncertainty in α_b to be 10% or less, based on simulations which included (I) no initial long wave length perturbation and (II) double the reconstructed long wave length perturbation amplitudes. This range of initial conditions encompasses our estimates in the uncertainty of the reconstruction. See Figure 1.

3   Young Measures

We explain the concept of a Young measure. For a turbulent flow, in the inertial regime, i.e, for LES simulations of turbulence, the Young measure description of the flow is a much deeper and more useful notion than is a classical weak solution or its numerical approximation. We generalize the notion of test function and of observation, using expectation values $\langle \cdots \rangle$ defined for the integration over the (state) random variables. See also [9].

To start, we suppress the spatial dependence. Thus we have a random system, whose state ξ takes on random values. We introduce a measure space Ω with $\xi \in \Omega$ and a probability measure (unit total measure) $d\nu(\xi)$ on Ω. We denote the result of integrating with respect to ν as $\langle \cdots \rangle$. Then $\langle 1 \rangle = \int_\Omega d\nu(\xi) = 1$.

A measurement is defined by a continuous function f of ξ, and defines a mean or expected value of repeated measurements of f in the random system state

$d\nu$, given by the integral

$$\langle f \rangle = \int_\Omega f(\xi) d\nu(\xi) \ . \tag{2}$$

If the expectation yields the value $1/2$, we may conclude that repeated measurements will give a 50% measurement for f, on the average. But we do not know whether the value $1/2$ occurs with each measurement (probability 1), i.e., perfect mixing with no fluctuations, or whether, at the other extreme, the value 1 occurs with probability $1/2$, that is, no mixing at all and total fluctuations. For further information, we look at moments. The second moment of the concentration ($f(\xi) = \xi$), useful for chemical reaction kinetics, is

$$\langle f(1-f) \rangle = \int_\Omega f(\xi)(1 - f(\xi)) d\nu(\xi) \ . \tag{3}$$

Eq. (3) gives information regarding the spread, or dispersion, of the measure ν. A common normalization of (3), is the coefficient of variation for f,

$$\theta = \frac{\langle f(1-f) \rangle}{\langle f \rangle \langle (1-f) \rangle} \ . \tag{4}$$

Now we add a spatial and temporal variability to all of the above. The measure $d\nu_{x,t}(\xi)$ now depends on x, t. The added value in allowing such a Young measure as a solution is that the local fluctuations are intrinsically associated with the space time point x, t.

The measurement defined by the stochastic observable $g(x, t, \xi)$ yields the expected value $\langle g(x, t, \cdot) \rangle$ at the space time point x, t. We expect this function of x, t to be a distribution, and so assuming that g is smooth (a test function) in its dependence on x, t, the outcome of the measurement is $\int \langle g \rangle dx dt$. Through this formalism, we can apply differential operators to the state $d\nu$, and as we have a governing PDE, we require $d\nu$ to be a solution of this PDE.

In contrast to multiplication by a test function for a weak solution, the values of the w* limit test function g multiply probabilities, while the state variable values (density, momentum, concentration), etc., the usual units for the values of the test function, now show up as an argument ξ of g. See Table 2.

A natural role for Young measures in a mathematical theory of the Euler equation and their relation to the Kolmogorov turbulence theory is discussed in [1]

Table 2. Comparison of weak solutions and Young measures in terms of test functions

	weak solutions	Young measures
g values multiply	state variables	probabilities
g arguments	space, time x, t	space, time, x, t; state values ξ
integration domain	space-time	space-time; state values
example	$g(x,t)$ multiplies	$g(x,t,\xi)$ multiplies
	momentum, energy, concentration	probability

and references cited there. In this reference we assume bounds from Kolmogorov theory, which serve as a type of Sobelov inequality for the approximations, and derive strong convergence for solutions of the incompressible Navier-Stokes equations (after passage to a subsequence) to weak solutions of the Euler equation limit, and w^* convergence for passive scalars coupled to the Navier-Stokes velocity field, to an Euler equation Young measure limit.

4   Verification for Stochastic (Young Measure) Convergence

The point of view presented here – w^* convergence to a Young measure solution and the coarse grain and sample algorithm to support this type of convergence numerically – needs verification and validation. Preliminary results in this direction have been established [11,10,12,7]. To discuss convergence of measures, we need to introduce function spaces for convergence. The PDFs themselves are noisy, and convergence of the PDFs directly appear to require difficult levels of mesh resolution. We introduced [10] for this purpose the indefinite integral of the PDFs, namely the probability distribution functions, i.e., the cumulative distribution functions (CDFs). These are better behaved and easier to analyze. Standard function space norms on the CDFs can be used, such as L_1 or the Kolmogorov-Smirnov norm L_∞.

We study nonlinear functions of the solution through analysis of second moments. The convergence properties of the second moments depend on the specific variables which enter into the second moment; some converge nicely while others would benefit from a larger statistical ensemble and/or further mesh refinement.

W^* convergence assumes an integration both over the solution state variables and over space and time. It applies to nonlinear functions of the solution. The idea of stochastic convergence is naturally appealing to workers versed in turbulence modeling. It is, however, a point of view which has not had extensive study in the numerical analysis literature, probably due to the requirements or perceived requirements for mesh resolution and the known limits of practicality for DNS simulations of many realistic problems. For this reason, it is of considerable interest to document exactly what is needed to achieve exactly which levels of convergence in exactly which topology.

Here we investigate multiple realizations of these ideas, in that the tradeoffs and issues related to stochastic convergence appear not to be well documented in the numerical analysis literature. We study integrated convergence through an L_1 norm (relative to integration both in solution state variables and over space-time) for the CDFs. We see that the L_1 norm for spatial integration is preferred to an L_∞ norm, and that this choice for the CDFs appears to be showing convergence. Additional mesh refinement, which we anticipate in the future as a result of increased computing power, will clarify this property.

We also explore the size of the supercell used to define the PDFs and CDFs. This size defines a tradeoff between enhanced statistical convergence and the quality of the mesh (supercell mesh) resolution. The L_1 norm convergence is enhanced with larger supercells. We study convergence of the PDFs directly. The PDFs do not show convergence in the L_1 norm with present levels of numerical and statistical resolution, but the trend of results suggests that convergence is possible with further mesh refinement.

4.1   Convergence of Second Moments

Here we show the convergence under mesh refinement of the second moments for species concentration and velocities, two quantities of interest in a miscible Rayleigh-Taylor experiment [21]. Since the quantities we report were not measured experimentally, this study is verification only, not validation. A related simulation study [9] includes comparison to the water channel experiments [15,16], in which the second moments were measured, and thus for which validation was studied.

It is commonly believed (and observed in numerical studies) that fluctuating quantities obey a type of Kolmogorov scaling law. This property, if correct, implies that the fluctuations are represented by a convergent integral, and should exhibit convergence under mesh refinement. Thus the convergence we report here should not be a surprise. Still, our results provide new information with respect to the level of refinement needed to observe convergent behavior. We generally observe satisfactory convergence through comparison between the medium and finest of the three spatial grids considered here, and unsatisfactory (poor agreement with the refined grid) properties for the coarsest grid. The limits at late time encounter a varying loss of statistical resolution due to the diminished number of statistically independent degrees of freedom at late time. The three grids have a size 520 to 130 microns (4 to 8 to 16 cells per elementary initial wave length). Of these, we have generally used the medium grid in our previous simulations, while the coarse grid is commonly favored in RT studies [2]. All second moments reported here represent mid plane values, i.e. a slice $z = $ const from the center of the mixing zone with t fixed, and are averaged over all x, y values.

The second moments of concentration, normalized to define the molecular mixing correlation $\theta = \langle f(1-f) \rangle / \langle f \rangle \langle 1-f \rangle$, were studied experimentally (distinct experiments, not reviewed here). Our value for $\theta \approx 0.8$ is consistent with values obtained numerically in related problems by others. However, significantly smaller θ values were observed in the similar fresh-salt water miscible experiments [15,16]. Since these differences are observed even at very early times, we can attribute the differences to initial conditions, specifically to the thickness of the initial diffusion layer. Fig. 2 displays numerical results for convergence of θ which model experiment [21], #112, with the three grids.

We study the turbulent correlations of density with the z component of the velocity, u_z, in Fig. 3. This correlation is related to gradient diffusion models for subscale turbulence models.

Conventionally, velocity fluctuations are studied using mass weighted averages, $\widetilde{v} = \langle \rho v \rangle / \overline{\rho}$, and as such serve to define the Reynolds stress

$$R = \langle \widetilde{v}\widetilde{v} \rangle - \frac{\langle \widetilde{v}\widetilde{v} \rangle}{\overline{\rho}} .$$
(5)

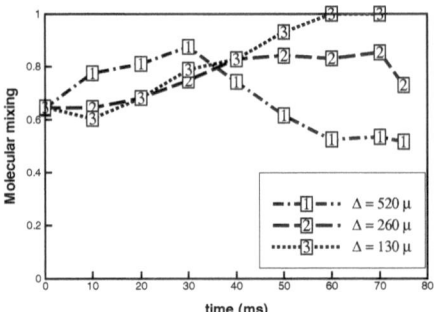

Fig. 2. Plot of the molecular mixing correlation, θ, vs. time for a numerical simulation of the experiment [21] #112. Three levels of grid refinement are shown: $\Delta x = 520, 260, 130\mu$. θ is evaluated at the mid plane value of z, as an average over all of x, y.

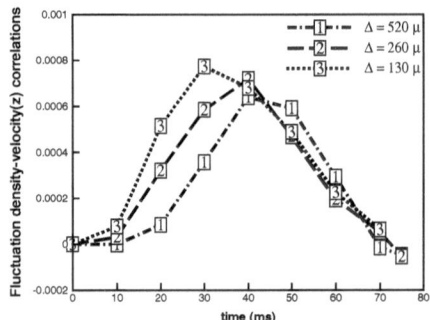

Fig. 3. Plot of $\overline{\rho' u_z'}$ vs. time. Data from the Z midplane, averaged over all of x, y.

In Fig. 4 we display the simulated Reynolds stress values for [21] experiment #112. The convergence properties for R_{zz} appear to be satisfactory (Fig. 4, left). The medium and fine grid display a reasonable level of agreement, while the coarse grid shows a significant discrepancy to the fine grid.

A sensitive comparison is that of R_{xx} to R_{yy}, see Fig. 4, right frame. These quantities should be (statistically) identical, so that the solid and dashed curves of the same mesh level family should coincide. This property holds at early but not late time, with the period of agreement increasing under mesh resolution. Moreover, the three curve families should show convergence under mesh refinement, a property which is observed at least up to the time for coincidence of R_{xx} and R_{yy}. The difficulty in the convergence of these quantities appears to be related to the inherently small size of the correlations relative to the statistical noise present in their evaluation and to the loss of statistical significance at late time. As the solution progresses, the correlation length increases, an inherent

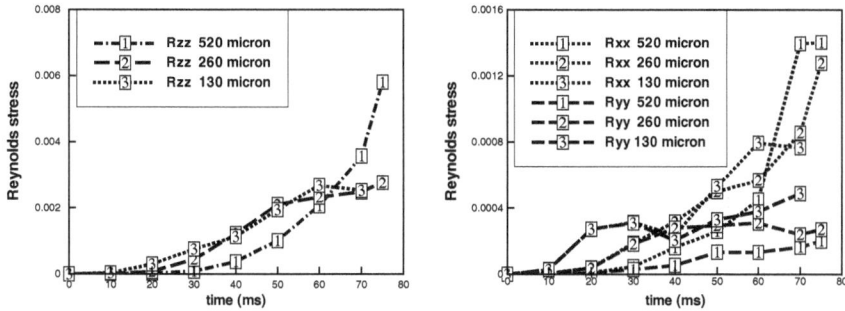

Fig. 4. Plot of Reynolds stress R_{zz} (left) and R_{xx}, R_{yy} (right) vs. time

feature of RT mixing. See the v_z gray scale plot at $t = 50$ in Fig. 5, right frame. Thus at late time, the statistical averaging to define R is drawn from a reduced number of independent degrees of freedom, introducing small sample effects into these components of R at late time.

Similar behavior is observed for R_{xz} and R_{yz}, see Fig. 5. Due to the rotational symmetry of the statistical formulation, in the case of an infinite x, y domain, these components should be zero, and any non-zero value is a finite size effect in the statistical sampling. There is satisfactory agreement with these two quantities between each other and with zero, up to a time which depends on the mesh. Because the quantities are sensitive to the sign of v_z statistically, they have enhanced randomness and decreased convergence properties relative to R_{xx} and R_{yy}; they possibly also show small sample size effects at late time.

4.2   Convergence of PDFs and CDFs

To define w^* convergence, we need to partition the simulation resolution into resources assigned to the conflicting objectives of spatial resolution and statistical resolution. We consider again the midplane $z = $ const and $t = $ const, and partition the x, y plane into supercells. We consider several values for the supercell grid, but show detailed results for an 8×2 supercell grid. Here the coarsest grid has for each supercell a resolution 9×6 with a z resolution of a single cell. For the medium and fine grids, the supercell partition is unchanged, but the number of cells in each direction increases by factors of 2 and 4.

For each supercell, we bin the concentration values into 5 bins, and count the number of values lying in each bin, to obtain a probability. In principle, the number of bins is another parameter in the analysis, variations in which are not explored here. The result of this exercise is an 8×2 array of PDFs, each represented in the form of a bar graph. The array is a graphical presentation of the Young measure at the fixed z, t value. See Fig. 6. From this array of PDFs, we can observe some level of coherence or continuity in the spatial arrangement of the PDFs, in that the central supercells have a strong heavy fluid concentration, while near the top and bottom, there is more of a mixed cell concentration.

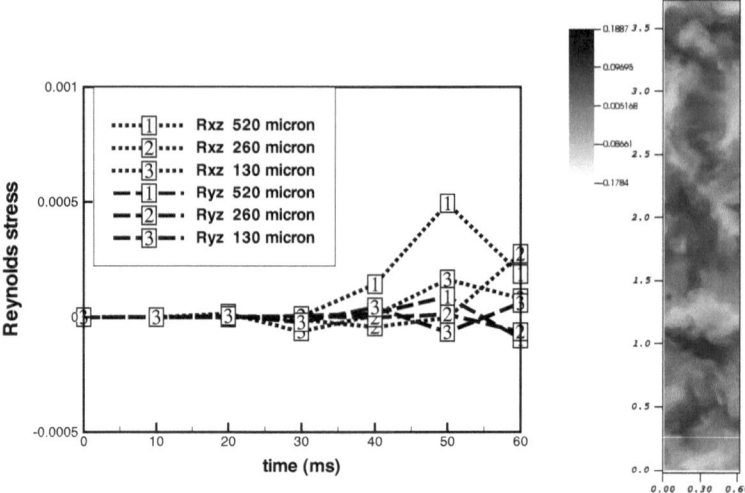

Fig. 5. Left: Plot of Reynolds stress R_{xz}, R_{yz} vs. time. Plotting time is restricted to a maximum of $t = 60$ as discussed in the text. Right: Plot of v_z (fine grid, $t = 50$) in the midplane.

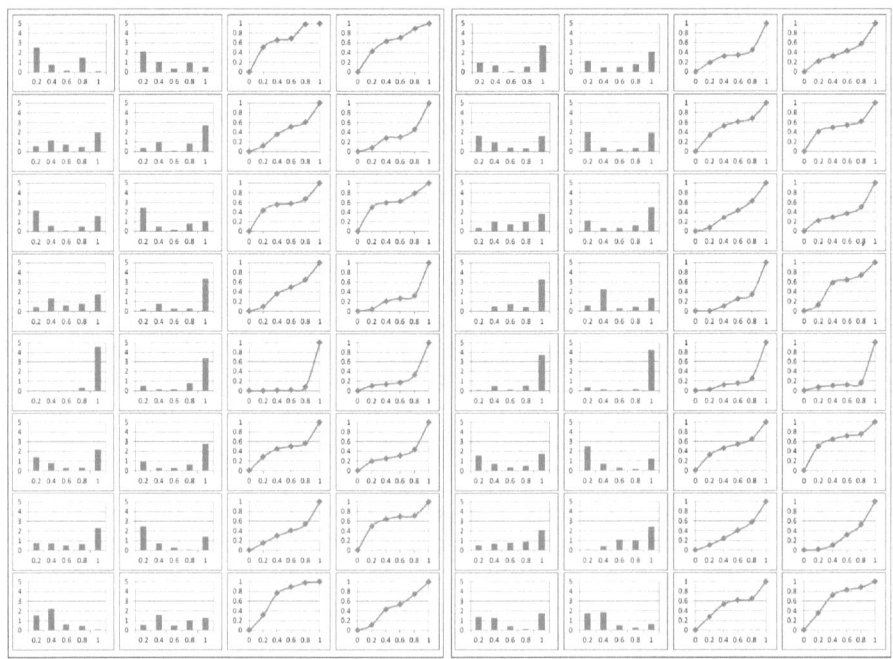

Fig. 6. Spatial array of heavy fluid concentrations at $t = 50$, for z in the midplane, as PDFs (bar graphs) and as CDFs (line graphs), Left: Medium grid. Right: Fine Grid.

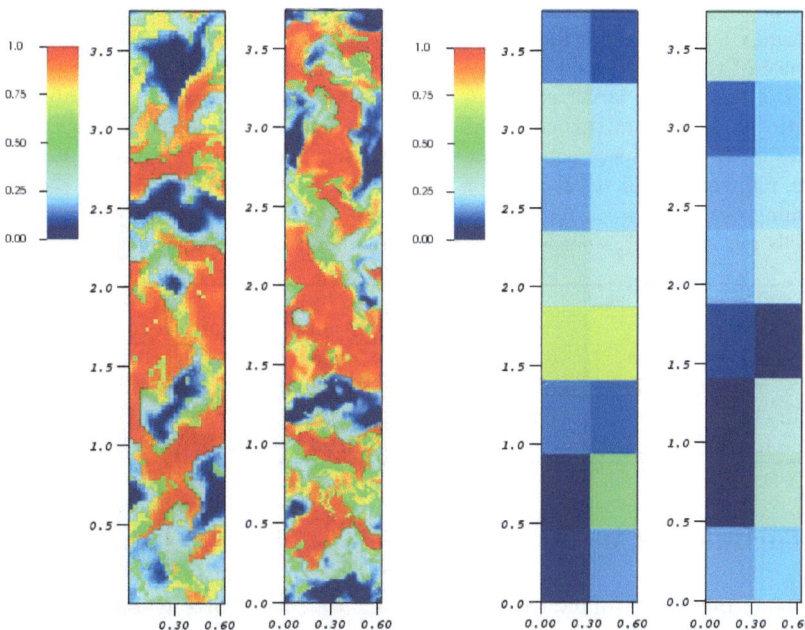

Fig. 7. Left: Plot of heavy fluid concentration at the midplane, $t = 50$. Medium grid (left). Fine grid (right). Right: Spatial array of L_1 norms of CDF mesh differences for heavy fluid concentrations at the midplane. Coarse to fine (left). Medium to fine (right).

Next we study mesh convergence of this 8×2 array of PDFs and CDFs. At the latest time completed for the fine grid, we compare the PDFs and CDFs on the coarse to fine and medium to fine grids. The comparison is to compute the L_1 norm of the pairwise differences for each of the 8×2 PDFs or CDFs. These differences yield an 8×2 array of norms, i.e. numbers, which is plotted in gray scale in Fig. 7.

The main results of this paper, namely the PDF and CDF convergence properties, are presented in Fig. 7. This data is further simplified by use of global norms. With an L_1 norm of the differences of the PDFs or CDFs for concentrations in each supercell, we consider both the L_1 and L_∞ norms relative to x, y variables. With the convergence properties thus reduced to a single number, we next explore the consequence of varying the definitions used for convergence. These are (a) the mesh, (b) PDF vs. CDF, (c) L_1 vs. L_∞ for a spatial norm and (d) the size of the supercell used to define the statistical PDF. See Table 3.

We see a convergence trend in all cases under mesh refinement, but useful results for current meshes are limited to CDF convergence. Generally L_1 norms show better convergence, and generally there is a minimum size for the supercell to obtain useful convergence. Since our convergence properties are documented for the medium grid (through comparison to the fine grid), we can speculate that the errors at the fine grid level would be smaller and that some of the above restrictions might be relaxed in this case.

Table 3. Summary norm comparison of convergence for heavy fluid concentration PDFs and CDFs at fixed values of z, t. In each supercell, an L_1 norm is applied to the difference of the PDFs or CDFs; this x, y dependent set of norms is measured by an L_1 or L_∞ norm. The larger supercell sizes, the last four columns of the table, cover the entire y domain. In this case, the space-time localization of the PDFs/CDFs are in x, z, t only. We observe convergence for CDFs; while the PDF error is decreasing, further refinement will be needed for usefully converged PDF errors. We see that a coarsening of the supercell resolution (increase of the supercell size) to 18×12 coarse grid cells per supercell is needed to obtain single digit convergence errors.

coarse grid supercell size	$9 \times 6 \times 1$		$18 \times 12 \times 1$		$36 \times 12 \times 1$	
mesh comparison	L_1 norm	L_∞ norm	L_1 norm	L_∞ norm	L_1 norm	L_∞ norm
CDFs: coarse to fine	0.26	0.98	0.16	0.48	0.15	0.39
CDFs: medium to fine	0.18	0.54	0.08	0.16	0.03	0.10
PDFs: coarse to fine	0.93	4.89	0.59	2.40	0.54	1.98
PDFs: medium to fine	0.64	2.66	0.30	0.82	0.15	0.52

5   Turbulent Combustion

Here we explain a primary rationale for our approach to convergence based on fluctuations, PDFs and Young measures. The stochastic convergence to a Young measure is certainly an increase in the complexity of the intellectual formalism in contrast to a more conventional view of convergence to weak solutions.

A simple rationale for the more complicated approach is that pointwise convergence to a weak solution generally fails in turbulent flows. New structures emerge with each new level of mesh refinement and the detailed (pointwise) flow properties are statistically unstable and in fact not observed to converge. Rather, statistical measures of the solutions, of a nature that an experimentalist would call reproducible, are used for convergence studies and these do generally display convergence. Thus we believe that our point of view finds roots in common practices for turbulent study.

In the case of reactive flow (or more generally of a nonlinear process applied to the flow), the stochastic convergence displays its power. Convergence of averages is not usable in a study of nonlinear functions, which require an independent convergence treatment. The LES formulation, moreover, is based on (grid cell or filter) averages. Thus the primitive quantities of an LES simulation cannot be used reliably if a nonlinear process (such as combustion) occurs in the fluid.

The conventional cure for LES turbulent combustion is a model of the flame structure and an assumption that the flame follows a steady state path in concentration-temperature space, with the partially burned state parametrized through a reaction progress variable [18]. This assumption leads to a model, called a flamelet model, imposed on the normal turbulent and mixing models. The approach adopted here, in contrast, allows direct computation of the chemistry of a turbulent flame in an LES framework, without the use of (flame structure) models. This approach is called finite rate chemistry. Conventionally, finite rate chemistry is possible for DNS only and the extension to LES is a major benefit derived from the stochastic convergence ideas advanced here.

Preliminary results are presented in [9] and will not be reviewed here.

Acknowledgements.   This work is supported in part by the Nuclear Energy University Program of the Department of Energy, project NEUP-09-349, Battelle Energy Alliance LLC 00088495 (subaward with DOE as prime sponsor), Leland Stanford Junior University 2175022040367A (subaward with DOE as prime sponsor), Army Research Office W911NF0910306. Computational resources were provided by the Stony Brook Galaxy cluster and the Stony Brook/BNL New York Blue Gene/L IBM machine. This research used resources of the Argonne Leadership Computing Facility at Argonne National Laboratory, which is supported by the Office of Science of the U.S. Department of Energy under contract DE-AC02-06CH11357.

References

1. Chen, G.Q., Glimm, J.: Kolmogorov's theory of turbulence and inviscid limit of the Navier-Stokes equations in R^3. Commun. Math. Phys. (2010) (in press)
2. Dimonte, G., Youngs, D.L., Dimits, A., Weber, S., Marinak, M., Wunsch, S., Garsi, C., Robinson, A., Andrews, M., Ramaprabhu, P., Calder, A.C., Fryxell, B., Bielle, J., Dursi, L., MacNiece, P., Olson, K., Ricker, P., Rosner, R., Timmes, F., Tubo, H., Young, Y.N., Zingale, M.: A comparative study of the turbulent Rayleigh-Taylor instability using high-resolution three-dimensional numerical simulations: The alpha-group collaboration. Phys. Fluids 16, 1668–1693 (2004)
3. George, E., Glimm, J., Li, X.L., Li, Y.H., Liu, X.F.: The influence of scale-breaking phenomena on turbulent mixing rates. Phys. Rev. E 73, 016304 (2006)
4. Glimm, J., Sharp, D.H., Kaman, T., Lim, H.: New directions for Rayleigh-Taylor mixing. Philosophical Transactions of The Royal Society A: Turbulent Mixing and Beyond (2011), submitted for publication; Los Alamos National Laboratory National Laboratory preprint LA UR 11-00423. Stony Brook University preprint number SUNYSB-AMS-11-01
5. Kaman, T., Glimm, J., Sharp, D.H.: Initial conditions for turbulent mixing simulations. Condensed Matter Physics 13, 43401 (2010), Stony Brook University Preprint number SUNYSB-AMS-10-03 and Los Alamos National Laboratory Preprint number LA-UR 10-03424
6. Kaman, T., Glimm, J., Sharp, D.H.: Uncertainty quantification for turbulent mixing simulation. In: ASTRONUM, Astronomical Society of the Pacific Conference Series (2010), Stony Brook University Preprint number SUNYSB-AMS-10-04. Los Alamos National Laboratory preprint LA-UR 11-00422 (submitted)
7. Kaman, T., Lim, H., Yu, Y., Wang, D., Hu, Y., Kim, J.D., Li, Y., Wu, L., Glimm, J., Jiao, X., Li, X.L., Samulyak, R.: A numerical method for the simulation of turbulent mixing and its basis in mathematical theory. In: Lecture Notes on Numerical Methods for Hyperbolic Equations: Theory and Applications: Short Course Book, pp. 105–129. CRC/Balkema, London (2011), Stony Brook University Preprint number SUNYSB-AMS-11-02

8. Lim, H., Iwerks, J., Glimm, J., Sharp, D.H.: Nonideal Rayleigh-Taylor mixing. PNAS 107(29), 12786–12792 (2010), Stony Brook Preprint SUNYSB-AMS-09-05 and Los Alamos National Laboratory preprint number LA-UR 09-06333

9. Lim, H., Kaman, T., Yu, Y., Mahadeo, V., Xu, Y., Zhang, H., Glimm, J., Dutta, S., Sharp, D.H., Plohr, B.: A mathematical theory for LES convergence. Acta Mathematica Scientia (2011) (submitted for publication), Stony Brook Preprint SUNYSB-AMS-11-07 and Los Alamos National Laboratory preprint number LA-UR 11-05862

10. Lim, H., Yu, Y., Glimm, J., Li, X.L., Sharp, D.H.: Subgrid models in turbulent mixing. In: Astronomical Society of the Pacific Conference Series, vol. 406, p. 42 (2008), Stony Brook Preprint SUNYSB-AMS-09-01 and Los Alamos National Laboratory Preprint LA-UR 08-05999

11. Lim, H., Yu, Y., Glimm, J., Li, X.L., Sharp, D.H.: Subgrid models for mass and thermal diffusion in turbulent mixing. Physica Scripta T142, 014062 (2010), Stony Brook Preprint SUNYSB-AMS-08-07 and Los Alamos National Laboratory Preprint LA-UR 08-07725

12. Lim, H., Yu, Y., Glimm, J., Sharp, D.H.: Nearly discontinuous chaotic mixing. High Energy Density Physics 6, 223–226 (2010), Stony Brook University Preprint SUNYSB-AMS-09-02 and Los Alamos National Laboratory preprint number LA-UR-09-01364

13. Liu, X.F., George, E., Bo, W., Glimm, J.: Turbulent mixing with physical mass diffusion. Phys. Rev. E 73, 056301 (2006)

14. Moin, P., Squires, K., Cabot, W., Lee, S.: A dynamic subgrid-scale model for compressible turbulence and scalar transport. Phys. Fluids A3, 2746–2757 (1991)

15. Mueschke, N., Schilling, O.: Investigation of Rayleigh-Taylor turbulence and mixing using direct numerical simulation with experimentally measured initial conditions. i. Comparison to experimental data. Physics of Fluids 21, 014106-1–014106-19 (2009)

16. Mueschke, N., Schilling, O.: Investigation of Rayleigh-Taylor turbulence and mixing using direct numerical simulation with experimentally measured initial conditions. ii. Dynamics of transitional flow and mixing statistics. Physics of Fluids 21, 014107-1–014107-16 (2009)

17. Mueschke, N.J.: Experimental and numerical study of molecular mixing dynamics in Rayleigh-Taylor unstable flows. Ph.D. thesis, Texas A and M University (2008)

18. Pitsch, H.: Large-eddy simulation of turbulent combustion. Annual Rev. Fluid Mech. 38, 453–482 (2006)

19. Ramaprabhu, P., Andrews, M.: Experimental investigation of Rayleigh-Taylor mixing at small atwood numbers. J. Fluid Mech. 502, 233–271 (2004)

20. Read, K.I.: Experimental investigation of turbulent mixing by Rayleigh-Taylor instability. Physica D 12, 45–58 (1984)

21. Smeeton, V.S., Youngs, D.L.: Experimental investigation of turbulent mixing by Rayleigh-Taylor instability (part 3). AWE Report Number 0 35/87 (1987)

Discussion

Speaker: James Glimm

Bill Oberkampf: You have presented several new ideas in both V&V as well as UQ that are very innovative. I have two questions. What advantages do you see in considering temporal and statistical convergence of PDF's of quantities of interest, as opposed to convergence of time averaged quantities at a point?

James Glimm: See Sec. 5.

Bill Oberkampf: When you examine mesh and temporal convergence in LES simulations you are merging changing sub-grid scales (resulting in changes in the math model) and changing numerical solution error. Since these are very different sources of uncertainty, what are your ideas for separating these uncertainties?

James Glimm: This is an excellent and deep question, whose answer is context dependent. For the case of turbulence models, we will address this issue in a separate publication (manuscript in preparation). Briefly, and for turbulent mixing, the SGS turbulent terms to be added to the Navier Stokes equations have a formulation in terms of gradients of primitive solution variables, if the dynamic subgrid models [14] are used. For this closure, convergence of the turbulence SGS model terms is a numerical analysis issue, as is already defined from the perspective of physics and modeling. Verification and validation of these models (a much studied topic) should be addressed in each separate simulation or flow regime. See for example, Sec. 4.

From Quantification to Visualization: A Taxonomy of Uncertainty Visualization Approaches

Kristin Potter, Paul Rosen, and Chris R. Johnson

Scientific Computing and Imaging Institute, University of Utah
72 S. Central Campus Dr., RM 3750, Salt Lake City, UT 84112, USA
http://www.sci.utah.edu

Abstract. Quantifying uncertainty is an increasingly important topic across many domains. The uncertainties present in data come with many diverse representations having originated from a wide variety of disciplines. Communicating these uncertainties is a task often left to visualization without clear connection between the quantification and visualization. In this paper, we first identify frequently occurring types of uncertainty. Second, we connect those uncertainty representations to ones commonly used in visualization. We then look at various approaches to visualizing this uncertainty by partitioning the work based on the dimensionality of the data and the dimensionality of the uncertainty. We also discuss noteworthy exceptions to our taxonomy along with future research directions for the uncertainty visualization community.

Keywords: uncertainty visualization.

1 Introduction

In the past few years, quantifying uncertainty has become an increasingly important research area, especially in regard to computational science and engineering applications. Just as we need to quantify simulation accuracy and uncertainty, we must also convey uncertainty information, often through visualization. As the number of techniques for visualizing uncertainty grows, the broadening scope of uses and applications can make classifying uncertainty visualization techniques difficult. Uncertainty is often defined, quantified, and expressed using models specific to individual application domains. In visualization however, we are limited in the number of visual channels (3D position, color, texture, opacity, etc.) available for representing the data. Thus, when moving from quantified uncertainty to visualized uncertainty, we often simplify the uncertainty to make it fit into the available visual representations. In this paper, we identify traditional types of uncertainty quantification and reduce their representations to those that are familiar to uncertainty visualization researchers. We then give an overview of different uncertainty visualization approaches targeted at these uncertainty representations. We then further differentiate uncertainty visualization approaches

A. Dienstfrey and R.F. Boisvert (Eds.): WoCoUQ 2011, IFIP AICT 377, pp. 226–249, 2012.

based on the dimensionality of the data and the dimensionality of the uncertainty. We conclude with a discussion of a few noteworthy exceptions to our taxonomy. Our main goal throughout is to position previous work on uncertainty visualization within the scope of uncertainty quantification in order to better connect the two.

2   Quantifying Uncertainty

To begin a discussion of uncertainty quantification, we must first define uncertainty into two overall, broad types: epistemic and aleatoric. *Epistemic* uncertainty describes uncertainties due to lack of knowledge and limited data which could, in principle, be known, but in practice are not. Such uncertainties are introduced through deficient measurements, poor models, or missing data. Quantification and characterization of epistemic uncertainty aims to better understand the underlying processes of the system and use methods such as fuzzy logic. *Aleatoric* uncertainty is defined as uncertainties that arise from, for example, running an experiment and getting slightly different results each time. This type of uncertainty is the random uncertainty inherent to the problem and cannot be reduced or removed by things such as model improvements or increases in measurement accuracy. Aleatory uncertainty can be characterized statistically and is often represented as a probability density function (PDF). The visualization of uncertainty focuses enhancing data understanding by unlocking and communicating the known aleatoric uncertainties present within data.

According to the NIST report on evaluating and expressing uncertainty [88], aleatoric uncertainty can be classified into two groups: type A and type B. While the distinction between the two classes may not always be apparent, they can be described as type A uncertainties arising from a "random" effect, whereas type B uncertainties arise from a "systematic" effect, where the former can give rise to a possible random error in the current measurement process and the latter gives rise to a possible systematic error in the current measurement process. The main difference between these two is in the evaluation of the uncertainties. Type A evaluation may be based on any valid statistical measure. However, the evaluation of type B is based on scientific judgment that will use all relevant information available, which often can include statistical reasoning.

While these classifications are important to note, and often have great impact on the quantification of uncertainty, their impact lessens when moving from quantification to visualization. The most straightforward understanding of uncertainty is often the easiest to expose visually, and thus uncertainty within the field of visualization is often thought of as type A - that is entirely statistically defined. Thus, unless otherwise noted, all of the papers in this taxonomy deal with statistically quantifiable uncertainty.

3   From Quantification to Visualization

The growing need to understand the effects of errors, randomness, and other unknowns within systems has lead to the recent upswing in research on uncertainty quantification. This growing body of work is creating an array of definitions of uncertainty differing in not only the mathematical measures defining uncertainty, but also in the way the uncertainty is expressed and used. These differences are often most apparent when crossing boundaries between scientific fields, but can also arise within the same field through various sources including data acquisition, transformation, sampling, interpolation, quantization, and visualization [70]. While understanding the measurement and propagation of uncertainty throughout a workflow pipeline is very challenging for quantifying the overall uncertainty of a system, this complexity can be prohibitive for visualization.

Using visualization as a tool for understanding leverages the high bandwidth of the human visual system, allowing for the fast understanding of large amounts of data. However our visual channels can be overwhelmed when increasing the amount and dimensionality of the data. For computational science applications, visualizing time-dependent, three-dimensional scalar, vector, or tensor field data is often the goal. However, depending on the complexity of the underlying geometry, such visual representations can suffer from problems such as occlusion, which may require user interaction to relieve. Even two-dimensional displays can suffer from visual clutter and overload leading to ineffective visualizations. Thus, regardless of additional uncertainty information, the visualization of data alone can be difficult to visually display in an effective way.

Adding uncertainty information is not only challenging in the design of the visual abstraction, it is also difficult to fully express the complexity of the uncertainty itself. While typically expressed as a PDF, very few visualization approaches can directly display this function, and those that can are restricted to 1D or limited 2D. Thus, to visualize the uncertainty, some type of assumptions are typically imposed on the data in order to reduce it to a manageable size or dimension. This is most often done by aggregating the uncertainty into a single value, such as standard deviation or defining an interval along which the value could possibly lie. This reduces the uncertainty to one or two values, which considerably eases its visual expression. However, this can often misrepresent characteristics of the actual data as mean and standard deviation often imply a normal distribution whereas an interval can be interpreted as uniform.

For visualization, these types of assumptions are often accepted since there are not yet readily available visual abstracts to address non normal distributions nor visual representations of high dimensional PDFs. It is very important to understand that these problems exist and that beyond the uncertainties associated with the data, there also exist uncertainties in the visualization - both in the technical mechanisms used to create the visual presentation, but also in the perception of the visualization itself. A handful of approaches have looked at exposing these assumptions by presenting information on the underlying PDF, however this greatly increases the complexity of the visualization, and most work to date uses a simplified view of uncertainty.

Table 1. Our taxonomy of uncertainty visualization approaches. Cells in light yellow represent categories with no known work. Citations in red refers to work with an emphasis on evaluation.

Data Dim.	Uncertainty Dimensionality		
	Scalar	Vector	Tensor
1D	[62] [77] [85] [82]		
2D	[7] [13] [14] [22] [27] [30] [31] [34] [43] [45] [44] [49] [53] [51] [56] [60] [64] [69] [72] [78] [79] [77] [76] [83] [91] [95] [16] [17] [28] [82]	[8] [9] [33] [53] [56] [63] [67] [66] [92] [97]	
3D	[12] [20] [19] [18] [42] [46] [47] [50] [54] [55] [59] [58] [71] [72] [73] [75] [80] [81] [86] [87] [93] [96] [15] [82] [61]	[5] [50] [53] [52] [68] [92]	[11] [35] [37] [41]
ND	[2] [23] [26] [32] [90]		

4 Taxonomy for Visualization Approaches

A wide array of taxonomies and typologies exist to help understand the field of uncertainty visualization. One of the first taxonomies to address uncertainty visualization separates methods by data type (scalar, multivariate, vector, and tensor) and visualization form (discrete and continuous) and proposes appropriate visual representations for each combination [70]. Skeels et al. [84] create a classification for information visualization which organizes the type of uncertainty by what it is trying to describe as well as commonalities between types and discusses exemplary visualizations for each type. Uncertainty has been a major theme in the area of geographic and information systems (GIS) and typologies have been created to focus on geospatial information visualization in the context of intelligence analysis [57, 89]. In contrast to these previous works, we differentiate our taxonomy by focusing on presenting the to-date uncertainty visualization approaches in as simple of a form as possible. We categorize approaches by two qualities: data dimension and data uncertainty dimension, and discuss the various visualization approaches based on these two categories.

4.1 Data Dimension

The data dimension is the most obvious of the categorization attributes. This is the dimension that the data lives in and may, or may not be the dimension that the visualization exists, or that the uncertainty is quantified. From a mathematical standpoint, this typically refers to the range of the function. For example, we have a computational science simulation that uses a model characterized by input parameters. The range refers to the output space of the simulation, which in many instances is spatial. This is the typical viewpoint for 1D-3D spatial data dimensions, however when moving to ND the interest may move to understanding the relationship between the parameter space, or domain, or of the function and the output. These types of questions are often answered by parameter-space studies and are treated in this work as ND.

1D. A one dimensional data dimension can be thought of as a single variable at a single point such that the uncertainty describes the variation or possible values of that single data value. This type of data is rarely found alone, we typically see it expressed as bar charts where each bar expresses a single independent variable, however the collection of bars may have some relationship - for example populations of countries. Here, a bar chart will have a bar for each country, however there is no intrinsic relationship between country population values. Thus, the data can be represented by a single 1D PDF and any higher-dimensional aspect of the data is implied, rather than intrinsic to the data.

2D. In contrast to 1D data, 2D can have a number of possibilities when it comes to the interpretation of the data. The data may be a 2D PDF, meaning a PDF defined over two variables, in which case the data is truly multivariate and can often be simplified to two distinct 1D PDFs. Alternatively, we can think of the data as having a 2D spatial domain, in which every location across the domain has a 1D PDF. This can be interpreted as a collection of 1D PDFs in space, or alternatively as a series of realizations across the 2D space where a single surface is made up of a sample from each of the PDFs. The term "ensemble" often comes up in this context and refers to the collection of output realizations, but may also include the particular parameter set associated with each ensemble member.

3D. Similarly to 2D, 3D data in general refers to a variable defined across a spatial volume where a single PDF exists at each position within the volume. In contrast to much of the work in 2D, 3D often deals with spatial positioning and boundaries rather than variable value across the space.

ND. Non-spatial, multivariate, and time-varying data is the final category of data dimension. The most often seen example of ND data is the addition of time, which can be added to 1,2, or 3D. Alternatively, ND data can refer to high-dimensional data often seen in parameter space explorations. In this case, there exists a parameter-space in R^M which maps to the target space in R^N. The M parameters can be modeled in some way such that an understanding of the relationship between the parameters and target is gained. While this relationship is often quite complex, the resulting data set is simply a set of realizations of the target space, and thus a collection of 1D PDFs across N and measures of uncertainty can be imposed on those 1D PDFs. While it can be the case that the dimension of the target space N is limited to 1, 2, or 3D, we are distinguishing work focusing on parameter-space uncertainties because the uncertainty in these works are more often focused on simulations that do not necessarily have a constrained dimension - they assume the target dimension as N and thus the actual value of N is irrelevant. Finally, multivariate data considers many variables simultaneously. This type of data may often be viewed as a collection of 1D data sets, unless there exists an inherent relationship between the variables. For our categorization, we reserve the ND classification to work which deals specifically with high-dimensional studies, for multivariate data which cannot be reduced to a collection of 1D variables, or for time-varying work which is separated from the lower-dimensional work by more than just animation.

4.2   Uncertainty Dimension

The uncertainty dimension refers to the dimensionality across which the uncertainty is quantified. This can often be a different dimension than the data. For example, many data sets attach a single value, i.e. a scalar, to points in 1D, 2D, or beyond. The uncertainty represented by these scalar values is still a 1D PDF. The data uncertainty dimension includes the categories of scalar, vector, and tensor representations.

Scalar. The term *scalar* typically refers to a single data value. For uncertainty, we can think of the term *scalar uncertainty* as the uncertainty associated with a scalar variable. For a scalar variable, we define the scalar uncertainty as a 1D PDF.

Vector. A vector is usually thought of as consisting of two quantities, such as a magnitude and direction, defined over a grid and often changing with time. The uncertainty typically investigated using vectors looks at the quantities not as precise values, but rather random variables, which can be characterized as PDFs. These PDFs are influenced locally by noise, measurement and simulation errors, uncertain parameters, boundary and initial conditions, and inherent randomness due to turbulence.

Tensor. Tensors are data types that define linear relationships between values for any dimensionality. While scalars and vectors are both technically low-order tensor data, we differentiate our discussion of tensors to be higher-order tensors only. These approaches do not visualize the tensors directly, but instead visualize some derived representation. For example, in [36], the authors visualize uncertainty in white matter tract reconstruction based upon ensembles of orientation distribution functions from diffusion tensor images.

5   Scalar Field Uncertainty

5.1   1D Data

As mentioned in Section 4.1, we typically see uncertainty in 1D scalar field data expressed as error bars or boxplots [74] in charts and graphs. These mechanisms typically show the expected value along with a range of possibilities.

While this often may be enough to express the "unknownness" of the value, both of these techniques can be misleading by implying a normal or Gaussian distribution. Most work in 1D scalar fields have been on trying to express the actual distribution of the variable, in order to get away from the assumption of a specific distribution and more accurately express the uncertainty. An example of visualizing non-Gaussian distributions comes from work on dealing with bounded uncertainty. This type of uncertainty is defined as an interval in which the actual data value lies. To express this visually, rather than having a line for the expected value and the range with error bars, the entire interval is depicted as fuzzy [62]. Thus, there is no line for expected value and the user can clearly see the location of where the data may lie. This "ambiguation" can be used for graphs and charts with an absolute scale, such as bar charts, and can also be applied to absolute scale charts such as pie charts. The expression of bounded uncertainty can be thought of as displaying a uniform distribution within the range, as each value within the range is equally likely, and values outside of the range are not possible. A similar idea is to express more characteristics of the data in order to fully express the distribution. Potter et al. [77] use this idea by expressing higher-order statistics of the distribution. As seen in Figure 1, the *summary plot* shows not only the traditional box plot (abbreviated to reduce visual clutter) but also a histogram which shows an approximation of the probability distribution function, and a glyph-based moment "signature" which shows the mean, standard deviations, skew, kurtosis, and tail. This hybrid plot allows for a better understanding of the distribution underlying the uncertainty

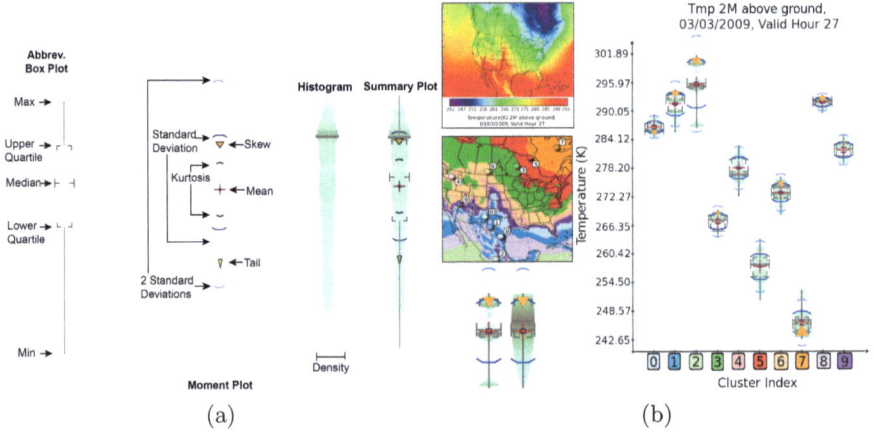

Fig. 1. Construction (a) and application (b) of the Summary Plot used by Potter et al. [77]. The Summary Plot highlights the variation of a distribution from normal by combining three glyph-based plots of statistical characteristics of the data. Similar to error bars, the application of the Summary Plot must be constrained to individual 1D points to avoid overwhelming visual clutter. Here (b), a clustering technique is used to select regions of interest for further exploration using the Summary Plot.

and quickly shows the non-normal behaviour of the data. These characteristics of 1D uncertainty are also present in tabular data where the cell value may be interpreted as average, estimated, possible, or likely. These terms express different understandings of the value, and may or may not be statistically grounded. To show the difference between these meanings, different line types are used to plot the value. For example, a dashed line is used for *estimated* and *possible* is expressed by widening the line to cover all valid values [85]. A few visualization methods have been developed to explore the characteristics of the PDFs underlying the uncertainty in 2D scalar fields. Most of these techniques employ some sort of dimensionality reduction or abstraction because, even as a low resolution grid, having a PDF at every point leads to too much visual complexity. Clustering is a common technique for grouping similar things. Bordoli et al. [7] use clustering techniques to group similar PDFs across the 2D spatial domain or to group 2D realizations. In a similar manner, Kao et al. [43, 44] use pixel-wise or feature-wise summaries to reduce the data to groups. Difference measures have been developed to compare a collection of PDFs against each other [76] to show the differences or similarities between them (shown in Figure 2b), and a defined set of operators has been used to reduce the distributions down to scalars [56]. It should be noted that the interpretation of the data as a set of 1D PDFs is an approximation and that linear interpolation between the points across the surface may not always be the most accurate or correct representation. Gerharz et al. [27] advocating looking at full joint PDFs and compare statistical methods for both marginal and joint PDFs defined across the spatio-temporal domain. All of the above techniques allow for the application of traditional 2D visualization techniques such as color mapping, however this leaves the third dimension free to be leveraged for the exploration of the PDFs. A density estimate volume can be computed [45] that creates a comparison volume across all PDFs allowing for the interrogation with cutting planes, local surface graphs, PDF isosurfaces, and glyphs. Thinking of the data as a set of realization surfaces allows for the creation of a volume which can be visualized using volume rendering and streamlines [78]; however this type of interpretation of the data imposes some sort of ordering on the realization surfaces which is not actually existent in the data. While the above techniques attempt to maintain the presence of the PDF in the visualization, it is often easier to reduce the understanding of uncertainty down to mean and standard deviation, a range of uncertainty, or a single scalar value depicting the magnitude of uncertainty. While this type of interpretation may impose assumptions on characteristics of the uncertainty quantification, it greatly reduces the difficulty in visualization.

5.2   2D Data

Methods that use this type of approach often employ color maps [13, 51, 60, 79, 83, 91] such as those in Figure 2, texture irregularities, opacity [34, 69], surface displacement [31], animation [22, 30] and glyphs [13, 51, 60, 83, 91, 95] to show uncertainty. Modifying contour color, thickness, and opacity [64, 72, 83] can show regions of uncertainty across the spatial domain. These types of displays can

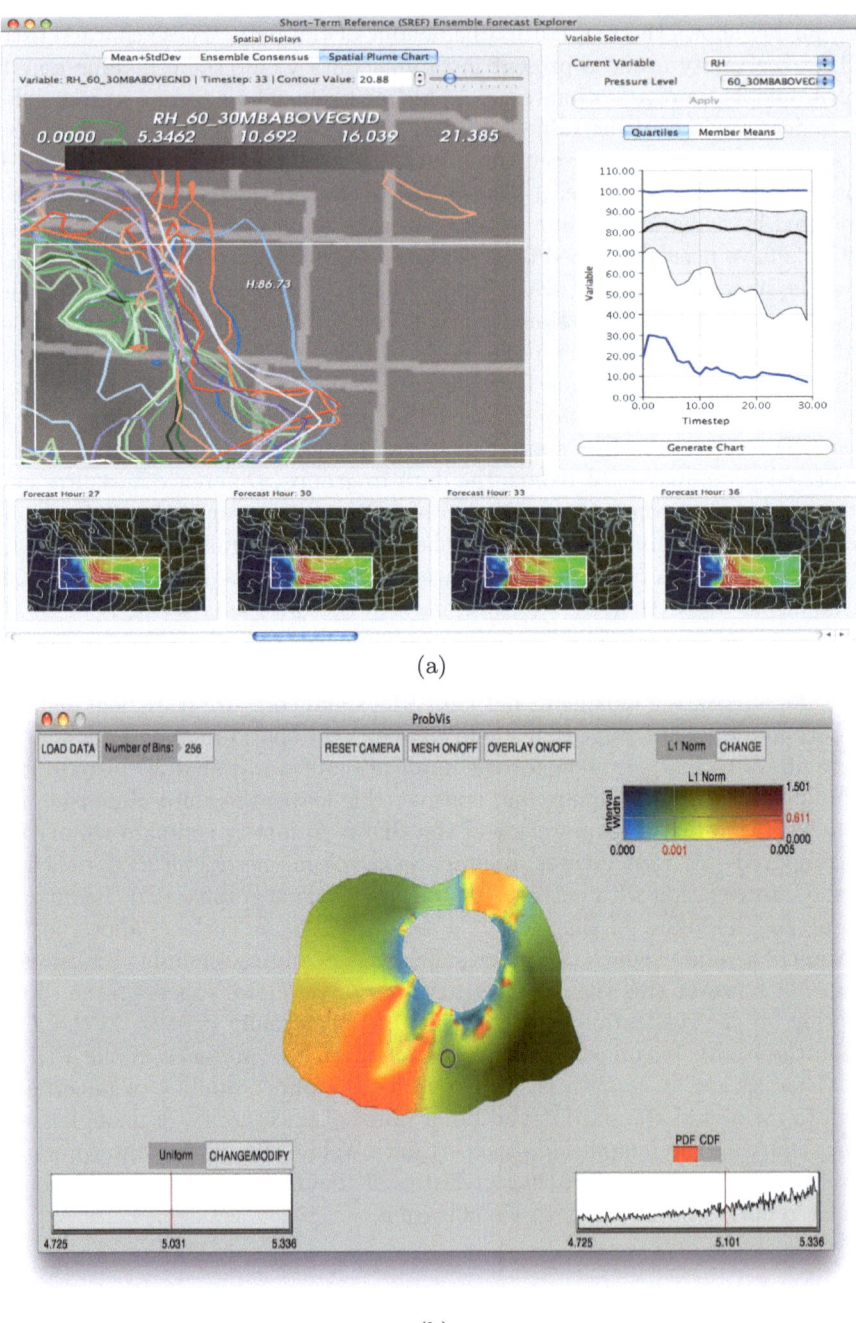

Fig. 2. Two examples of the visualization of 2D scalar data. (a) EnsembleVis [79] uses multiple windows to show various characteristics of the uncertainty and provides linking and brushing through a gui. (b) PDFVis [76] uses a color map to compare all PDFs across the 2D spatial domain.

be augmented with uncertainty annotations [13] which modulate properties of information external to the data display, such as longitude and latitude lines, in order to show uncertainty in a way that does not interfere with the data display. Multiwindow methods can help expose underlying information of the PDF that these types of approaches hide   [79, 83], as shown in Figure 2a, and can also provide for application specific types of visualizations. Finally, in contrast to 2D spatial domains, uncertainty can exist in 2D lattice and tree structures. This type of uncertainty arises as data structures for many statistical processing systems where the structure usually represents the "best guess" and alternate branches or leaves are shown with reduced opacity, or variations in positioning, color, or size [14, 49].

5.3   3D Data

Moving into 3D data, the number of visualization channels available has significantly diminished which limits the amount of information that can be readily displayed on the screen. The direct display of each PDF contributing the data set possible in 1D and less so in 2D, is now greatly diminished. Rather than expressing these full PDFs in this context, it is necessary to reduce the uncertainty information into an aggregated form, such a summarizing through a small set of numbers, or as an interval. The emphasis of 3D techniques is more often on displaying the location and relative size, rather than the exact quantification of the uncertainty.

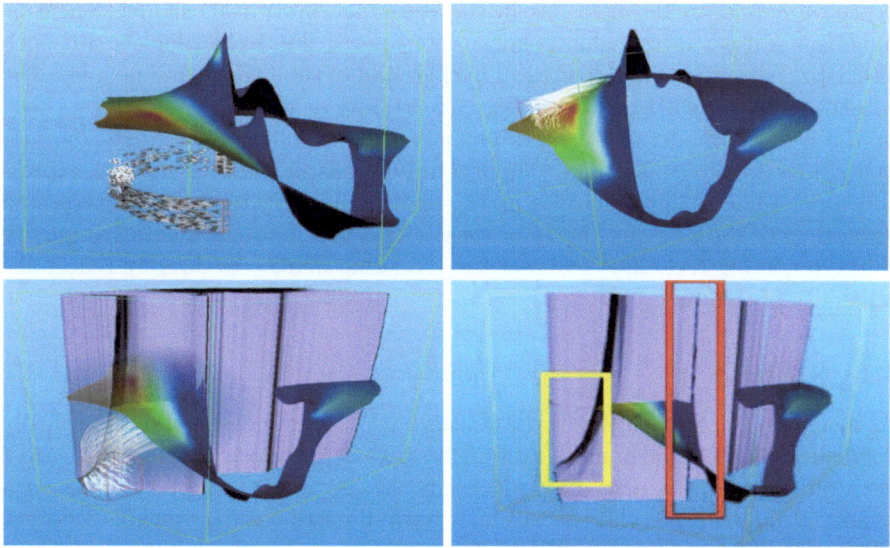

Fig. 3. 3D scalar field uncertainty visualization using glyphs, color maps, isosurfacing, and volume rendering [78]

The most commonly found techniques for showing uncertainty in 3D include color mapping, opacity, texture, and glyphs [15,50,61,82], with Figure 3 showing some examples. This is used in volume rendering [19] where the transfer function is used to encode uncertainty with color and opacity, or as a post-processes composite with texture. This work was later extended to include depth cuing and improved transfer function selection [18]. Rather than simply encoding a single value of uncertainty, the transfer function can be used to encode different measures of uncertainty, such as risk or fuzzy classifications of tissues [47,81]. This idea can also be applied to the fuzzy classification of isosurfaces [54].

In contrast to mapping a quantity of uncertainty onto a 3D visualization, it is noteworthy to point out uncertainties created by the visualization itself. Probabilistic marching cubes and uncertain isocontours [72,73] are techniques which investigate the uncertainties in calculating underlying 2D and 3D visual representations. Djurcilov et al. [20] construct visualization geometries (point-clouds, contours, isosurfaces & volume rendering) with missing data, and Pauly et al. [71] investigate surfaces generated from 3D data acquired from scanners. Overlay, pseudo-coloring, transparency, and glyphs can be used to compare differences in isosurface generation algorithms and volumetric interpolation techniques [42,80], and color mapping using flowline curvature [46] can be used to gain knowledge on the quality of an isosurface for representing the underlying data. Animation is often used to highlight these discrepancies by vibrating through possible isosurface positions [12], or looping through visualization parameter settings [55].

The last body of work on uncertainty in 3D deals with data reliability data and is most often found in fields such as archeology and virtual architectural reconstruction [58,86]. In these works, the uncertainty is defined as the confidence an expert in the field has in the construction of a 3D model. Scientific judgment is used to fuse what is know about particular archaeological site, such as existing structures, and historical background of the regions and peoples. The construction of the 3D models reflects uncertainty or points of contention based on the way the model is rendered. Opacity [87], sketch-based texture [75], animated line drawings [59], and temporal animations [96] can be used to express this type of uncertainty. Because highly-realistic imagery tend to be interpreted as truth [21], the unifying theme of these works is to add an illustrative quality to the rendering technique to lower the rendering quality to directly reflect the reliability of the data [93].

5.4 ND Data

As mentioned in Section 4, ND data deals with high-dimensional data typically defined as time-varying, multivariate, or parameter space explorations. Most work on time-varying data simply extends the 2D or 3D using techniques such as animation, and thus these works have been discussed in the previous sections. Here, we will focus our discussion on multivariate and parameter space data.

Multivariate data involves many related variables. Simple visualization of multivariate data is, in itself, a challenge and much work towards visualizing this type of data has been done [24]. A common approach for this type of data is parallel coordinates, which creates a coordinate system and plots the location of points across all axes. Adding uncertainty to parallel coordinates can be done through blurring, opacity, and color [23, 25]. While parallel coordinates do indeed display many dimensions within the same window, they are often hard to understand. As an alternative to parallel coordinates, multiple visualization windows can be used to expose uncertainties in relationships between spatial, temporal, and other dimensionalities [32]. This type of approach, however, reduces the multivariate aspect of the data to a lower dimensional representation more appropriate to visualization.

Parameter-space explorations expose the uncertainties within systems by analysing the relationships between input parameters and outcomes and are often used to better understand and improve simulations. While a full discussion of work in parameter-space analysis is outside the scope of this paper, the connection to uncertainty visualization is of interest. Here, we discuss a few notable works that relate parameter-space analysis to visualization, and we refer the reader to the papers for a treatment of the underlying mechanics.

The first exemplary work uses a combination of parallel coordinates and scatterplots to show the parameter-space sensitivity [2]. For each dimension, a PDF defines the uncertainty, which is then expressed as a histogram on top of each axes in the parallel coordinate display, or the user can select two dimensions to be displayed as a scatterplot with overlaid boxplots. An alternative to parallel coordinates, Gerber et al. [26] propose using the Morse-Smale complex to summarize the high-dimensional parameter space with a 2D representation that preserves salient features, and provides an interactive framework for a qualitative understanding of the effect of simulation parameters on simulation outcomes. The final example is World Lines which [90], using the demonstrative application of a flooding scenario, visualizes the multiple output scenarios individually, allowing the user to interactively explore the various world outcomes of the simulation.

6 Vector Fields

Vector fields are typically found as 2D and 3D with a time component. While these are different domains, both have equal treatment in the visualization space; the majority of techniques are either applied to both 2D and 3D or have been extended from 2D to handle 3D. 1D vector fields equate to a 1D scalar fields which are discusses in Section 5.1. Both 2D and 3D fields often have a time component and thus, for sake of our discussion here, we classify data with a time-component as 2D or 3D and assume any ND vector field work deals with parameter exploration. This eliminates the discussion of 1D vector fields, and postpones the consideration of ND vector fields to a later date, as no work in this area has been done at this time.

The visualization of uncertain vector fields can be classified into four types and we will assume the description of each type applies to both 2D and 3D, unless otherwise noted.

A common visualization technique for both 2D and 3D vector fields are glyphs [68]. Glyphs typically encode the two variables of the vector within their construction, such as an arrow pointing towards the direction with length scaled by magnitude. Expansions of glyphs to uncertainty information include using area, direction, length, and additional geometry to indicate uncertainty [92], line segment or barbell glyphs [52], or ellipsoidal glyphs depicting regions of possible vector positions [50]. Finally, time can be included in the glyph itself [33] or through animation of the glyphs [97].

Stream and particle lines show the path of flow from a particular seed point through time. In 2D these can be represented as lines [56] and as ribbons or tubes in 3D [52], both of which can use color, opacity, width, and animation to show uncertainties such as interpolation error in meteorological trajectory [5, 50] or differences in integration methods for particle tracing [52]. Texture-based streakline methods are more often used for 2D vector fields and use attributes such as noise, color and fog to modulate streaklines with uncertainty as they move across the domain [8, 9, 63] .

Most recently, Otto et. al. suggested that instead of creating new uncertainty representations for vector fields, we should instead recycle the approaches used in scalar field visualization. Their approach calculates multiple scalar fields with probabilities that represent topological features such as sinks, sources, and basins. Then, scalar field uncertainty approaches can be applied for 2D [66] and 3D [65] or used to analyze uncertainty of motion in video data [67]. In another topological approach, [3, 4] create a new data structure called the "edge map" to represent 2D flow and bound error. Once they have quantified the associated error, they use streamwaves to visualize the fuzzy topological constructions.

A notable break from traditional visualization techniques, sonification is the use of sound to indicate areas of uncertainty and has been included in the UFLOW system [53] to visualize flow fields using glyphs and streamlines as well as a system for visualizing the uncertainty of surface interpolants.

7   Tensor Fields

Most of the work in 3D tensors fields has focused on brain fiber tracking in diffusion tensor imaging. The first set of methods display the data as a glyph representation [35, 41], which indicates the fiber directional or fiber crossing uncertainty at a given location. The other set of approaches track fibers under uncertain conditions, giving either a color map for confidence [11] or an envelop of potential fiber routes [37]. Figure 4 shows a recent effort by [36] to use volume rendering with multi-dimensional transfer functions to visualize the uncertainty associated with high angular resolution diffusion imaging (HARDI). The authors use ensembles of orientation distribution functions (ODF) and apply volume rendering to 3D ODF glyphs, which they term SIP functions of diffusion shapes to capture their variability due to underlying uncertainty. Beyond these few approaches, very little work has been performed on visualizing uncertainty within tensor fields.

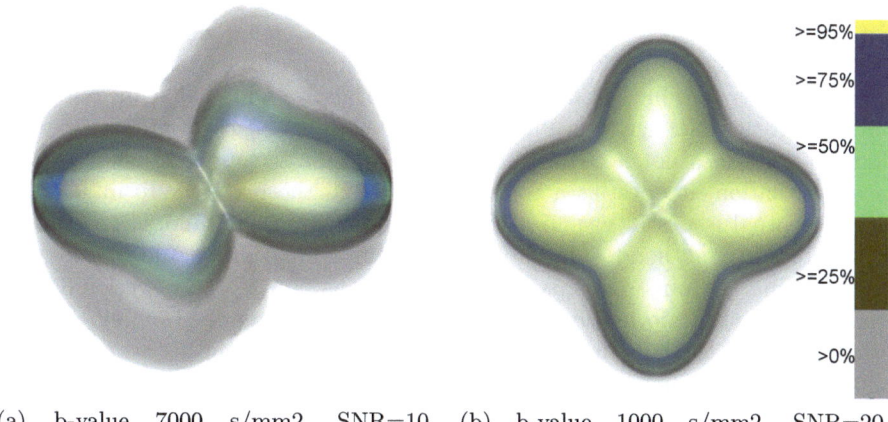

(a) b-value 7000 s/mm2, SNR=10,   (b) b-value 1000 s/mm2, SNR=20,
(0.6,0.4) 60 degrees   (0.5,0.5) 90 degrees

Fig. 4. Visualization of the uncertainty in two diffusion shapes from diffusion tensor imaging using volume rendering applied to an ensemble of 3D orientation distribution functions. (a) Two fibers crossing at 60 degrees with relative weight of 0.6:0.4 and SNR of 10. (b) Two fibers crossing at 90 degrees with equal weight and SNR of 20 (with much less uncertainty).

8   Evaluation

An often overlooked aspect in the field of visualization is evaluation. This is also the case in uncertainty visualization, which is doubly problematic in that the visualizations often represent highly complex phenomenon, and the assessment of effectiveness must take into account not only good visual design, but also appropriate understanding, transformation, and expression of the data.

A handful of papers have been dedicated to evaluating visualizations in the context of uncertainty. Most of these look at the method of visual encoding such error bars, glyph size, and colormapping in 1 and 2D [82], glyph type in 3D [61], or comparing adjacent, sequential, integrated, and static vs dynamic displays [28]. While each work identified a "better" technique for their unique study; surface and glyph color work better than size, multi-point glyphs perform better than ball, arrow, and cone glyphs, and adjacent displays with simple indications of data and uncertainty were preferred by the users, however none of the techniques performed well enough to be called the best display of uncertainty in all circumstances. A more human-centered approach evaluated the psychophysical sensitivity to contrast and identified particular noise ranges appropriate for uncertainty visualization [15]. Finally, indications of uncertainty in a visualization were found to influence confidence levels in the context of decision making [16, 17].

While each of the above works is significant in improving our understanding of the effectiveness of uncertainty visualizations, much more work must be done. The number of fields turning to visualization for understanding and decision making is growing, as well as the range of users, and this again reiterates both the great challenge in evaluation as well as the great need. While the work done to date does not necessarily point out specific techniques that will work in any situation, it does, as a collection, point to the necessity to understand the perceptual issues in visualization, as well as the needs tailored specifically to the problem at hand. Thus future work in evaluation should continue to study in what circumstances particular visual devices work, how overloading visual displays with information such as uncertainty effect understanding, and the ramifications of the human visual system. A formal treatment in this regard will allow future developers of uncertainty visualizations to position work within a tested set of constraints which, while not guaranteeing a successful visualization, may help foster good design.

9 Conclusion

Uncertainty visualization has been identified as a top visualization research problem [38,39]. The increased need for uncertainty visualization is demonstrated not only by the various taxonomies and typologies referenced in Section 4, but by the discussions found in numerous positional papers and reports [40], which motivate the need from the viewpoints of several scientific domains, including GIS and decision making [1, 10, 29, 48, 69]. This has inspired significant growth of work in the area of uncertainty visualization and with this amassing number of emerging works there has also become a need for an organization of that work. As a compendium to previous surveys on uncertainty visualization [6,57,68], this paper organizes the state of the art in uncertainty visualization based on the dimensionality of the data and of the uncertainty. Our main contribution is the classification of the work into groups, as well as identifying common visualization techniques for each group and point out specific unique techniques.

9.1 Directions for Uncertainty Visualization

Below we outline areas of uncertainty visualization we have identified as still needing further study.

Scalar Fields. The majority of work in uncertainty visualization has focused on scalar fields. These visualization methods almost always depend upon a single uncertainty value for their visualization. This limits the uncertainty information they can convey. New methods of visualizing the underlying PDF would allow visualizations to more accurately convey the possibilities for the shape of the underlying data without increasing visual clutter. We see new glyphs representations as one promising direction for solutions. Additionally, clustering of similar uncertainty might offer a possibility to reign in visual cluster.

Significant needs also remain for parameter space visualizations of uncertain data, as this type of data are becoming more widespread. The current approaches most often take standard parameter space visualizations, such as parallel coordinates, and apply standard scalar field uncertainty approaches. New abstractions and visual designs are needed to better convey the richness of this data.

Vector Fields. One of the more intriguing directions of work is that suggested by Otto et al., recycling scalar field visualization techniques by finding topological features within the vector field. Topological analysis of vector fields is a robust field of research. The uncertainty visualizing community could certainly leverage topological techniques as a way to better communicate uncertain vector field data, such as that put forth by Bhatia et al. [3].

There has been limited work performed on joint-histograms and correlations between two variables. While we do not consider this as a vector field within our taxonomy, it is in some sense related. Most techniques assume an independent 1D PDF for each variable, no matter the number of variables, with higher dimensional PDFs only available combinatorially. In fact, multiple variables can correlated in such a way that the structure of their uncertainty is not separable into multiple 1D PDFs. Instead they need higher dimensional PDFs and new visualization methods for those additional dimensions. Potter et al. [77] suggest some simple techniques for addressing these problems using 2D plots but more work is still needed.

Tensor Fields. Little work has been done for visualizing uncertainty associated with tensor field. The small body of work in visualizing uncertainty within tensors has mostly focused its efforts towards visualizing uncertainty of derived values, such as white matter fibers. Future work on this problem must focus on both the uncertainty of derived values and the uncertainty present in the tensor itself. However, visualizing a tensor directly, irregardless of uncertainty, is in itself a challenging problem, especially as the order of tensor increases. For high dimensional uncertainty tensor visualization, one avenue for future research is to combine traditional tensor field visualization techniques from information and statistical visualization techniques, such as [94], which combines 3D view of diffusion tensor fiber tracts with 2D and 3D embedded points along with multiple histograms that show derived quantities including fractional anisotropy, fiber length, and average curvature.

Acknowledgements. This work was supported in part by grants from the DOE SciDAC and DOE NETL, Award No. KUS-C1-016-04, made by King Abdullah University of Science and Technology (KAUST), and the NIH Center for Integrative Biomedical Computing, 2P41 RR0112553-12. We would also like to thank Samuel Gerber and Mike Kirby for their valuable input.

References

1. Beard, M.K., Buttenfield, B.P., Clapham, S.B.: Ncgia research initiative 7 visualization of spatial data quality, technical paper 91-26. Tech. rep., Natl Center for Geographic Information and Analysis (1991)
2. Berger, W., Piringer, H., Filzmoser, P., Gröller, E.: Uncertainty-aware exploration of continuous parameter spaces using multivariate prediction. CGF 30(3), 911–920 (2011)
3. Bhatia, H., Jadhav, S., Bremer, P.T., Chen, G., Levine, J., Nonato, L., Pascucci, V.: Edge maps: Representing flow with bounded error. In: Proceedings of IEEE Pacific Visualization Symposium, pp. 75–82 (2011)
4. Bhatia, H., Jadhav, S., Bremer, P.T., Chen, G., Levine, J., Nonato, L., Pascucci, V.: Flow visualization with quantified spatial and temporal errors using edge maps. IEEE TVCG (to appear, 2012)
5. Boller, R.A., Braun, S.A., Miles, J., Laidlaw, D.H.: Application of uncertainty visualization methods to meteorological trajectories. Earth Science Informatics 3(1-2), 119–126 (2010)
6. Bonneau, G.P., Hege, H.C., Johnson, C.R., Oliveria, M., Potter, K., Rheingans, P.: Overview and State-of-the-Art of Uncertainty Visualization (to appear, 2012)
7. Bordoloi, U.D., Kao, D.L., Shen, H.W.: Visualization techniques for spatial probability density function data. Data Science Journal 3, 153–162 (2004)
8. Botchen, R.P., Weiskop, D., Ertl, T.: Texture-based visualization of uncertainty in flow fields. In: IEEE Visualization 2005, pp. 647–654 (2005)
9. Botchen, R.P., Weiskopf, D., Ertl, T.: Interactive visualisation of uncertainty in flow fields using texture-based techniques. In: 12th Iternational Symposium on Flow Visualisation (2006)
10. Boukhelifa, N., Duke, D.J.: Uncertainty visualization - why might it fail? In: CHI 2009: Proceedings of the 27th International Conference Extended Abstracts on Human Factors in Computing Systems, pp. 4051–4056 (2009)
11. Brecheisen, R., Platel, B., Vilanova, A., ter Haar Romeny, B.: Parameter sensitivity visualization for dti fiber tracking. IEEE TVCG 15(6), 1441–1448 (2009)
12. Brown, R.: Animated visual vibrations as an uncertainty visualisation technique. In: GRAPHITE 2004: Proc. of the 2nd International Conference on Computer Graphics and Interactive Techniques, pp. 84–89 (2004)
13. Cedilnik, A., Rheingans, P.: Procedural annotation of uncertain information. In: IEEE Visualization 2000, pp. 77–84 (2000)
14. Collins, C., Carpendale, S., Penn, G.: Visualization of uncertainty in lattices to support decision-making. In: Proceedings of Eurographics/IEEE VGTC Symposium on Visualization (EuroVis 2007), pp. 51–58 (2007)
15. Coninx, A., Bonneau, G.P., Droulez, J., Thibault, G.: Visualization of uncertain scalar data fields using color scales and perceptually adapted noise. In: Applied Perception in Graphics and Visualization (2011)
16. Deitrick, S., Edsall, R.: The influence of uncertainty visualization on decision making: An empirical evaluation. In: Progress in Spatial Data Handling, pp. 719–738. Springer, Heidelberg (2006)
17. Deitrick, S.A.: Uncertainty visualization and decision making: Does visualizing uncertain information change decisions? In: Proceedings of the XXIII International Cartographic Conference (2007)

18. Djurcilov, S., Kim, K., Lermusiaux, P., Pang, A.: Visualizing scalar volumetric data with uncertainty. Computers and Graphics 26, 239–248 (2002)
19. Djurcilov, S., Kim, K., Lermusiaux, P.F.J., Pang, A.: Volume rendering data with uncertainty information. In: Data Visualization (Proceedings of the EG+IEEE VisSym), pp. 243–252 (2001)
20. Djurcilov, S., Pang, A.: Visualizing sparse gridded data sets. IEEE Computer Graphics and Applications 20(5), 52–57 (2000)
21. Dooley, D., Cohen, M.F.: Automatic illustration of 3d geometric models: Lines. In: Proceedings of the Symposium on Interactive 3D Graphics, pp. 77–82 (1990)
22. Ehlschlaeger, C.R., Shortridge, A.M., Goodchild, M.F.: Visualizing spatial data uncertainty using animation. Computers in GeoSciences 23(4), 387–395 (1997)
23. Feng, D., Kwock, L., Lee, Y., Taylor II, R.M.: Matching visual saliency to confidence in plots of uncertain data. IEEE TVCG 16(6), 980–989 (2010)
24. Fuchs, R., Hauser, H.: Visualization of multi-variate scientific data. Computer Graphics Forum 28(6), 1670–1690 (2009)
25. Ge, Y., Li, S., Lakhan, V.C., Lucieer, A.: Exploring uncertainty in remotely sensed data with parallel coordinate plots. International Journal of Applied Earth Observation and Geoinformation 11(6), 413–422 (2009)
26. Gerber, S., Bremer, P.T., Pascucci, V., Whitaker, R.: Visual exploration of high dimensional scalar functions. IEEE TVCG 16(6), 1271–1280 (2010)
27. Gerharz, L., Pebesma, E., Hecking, H.: Visualizing uncertainty in spatio-temporal data. In: Spatial Accuracy 2010, pp. 169–172 (2010)
28. Gerharz, L.E., Pebesma, E.J.: Usability of interactive and non-interactive visualisation of uncertain geospatial information. In: Reinhardt, W., Krüger, A., Ehlers, M. (eds.) Geoinformatik 2009 Konferenzband, pp. 223–230 (2009)
29. Gershon, N.: Visualization of an imperfect world. IEEE Computer Graphics and Applications 18(4), 43–45 (1998)
30. Gershon, N.D.: Visualization of fuzzy data using generalized animation. In: Proceedings of the IEEE Conference on Visualization, pp. 268–273 (1992)
31. Grigoryan, G., Rheingans, P.: Point-based probabilistic surfaces to show surface uncertainty. IEEE Transactions on Visualization and Computer Graphics 10(5), 546–573 (2004)
32. Haroz, S., Ma, K.L., Heitmann, K.: Multiple uncertainties in time-variant cosmological particle data. In: IEEE Pacific Visualization Symposium, pp. 207–214 (2008)
33. Hlawatsch, M., Leube, P., Nowak, W., Weiskopf, D.: Flow radar glyphs & static visualization of unsteady flow with uncertainty. IEEE Transactions on Visualization and Computer Graphics 17(12), 1949–1958 (2011)
34. Interrante, V.: Harnessing natural textures for multivariate visualization. IEEE Computer Graphics and Applications 20(6), 6–11 (2000)
35. Jiao, F., Gur, Y., Phillips, J., Johnson, C.: Uncertainty visualization in hardi based on ensembles of odfs. To appear 2012 IEEE Pacific Visualization Symposium (PacificVis) (2012)
36. Jiao, F., Phillips, J.M., Gur, Y., Johnson, C.R.: Uncertainty visualization in hardi based on ensembles of odfs metrics for uncertainty analysis and visualization of diffusion tensor images. In: IEEE Pacific Visualization Symposium (to appear, 2012)
37. Jiao, F., Phillips, J.M., Stinstra, J., Krger, J., Varma, R., Hsu, E., Korenberg, J., Johnson, C.R.: Metrics for Uncertainty Analysis and Visualization of Diffusion Tensor Images. In: Liao, H., "Eddie" Edwards, P.J., Pan, X., Fan, Y., Yang, G.-Z. (eds.) MIAR 2010. LNCS, vol. 6326, pp. 179–190. Springer, Heidelberg (2010)

38. Johnson, C.R.: Top scientific visualization research problems. IEEE Computer Graphics and Applications 24(4), 13–17 (2004)
39. Johnson, C.R., Sanderson, A.R.: A next step: Visualizing errors and uncertainty. IEEE Computer Graphics and Applications 23(5), 6–10 (2003)
40. Johnson, C., Moorhead, R., Munzner, T., Pfister, H., Rheingans, P., Yoo, T.: Nih/nsf visualization research challenges report summary. IEEE Computing in Science and Engineering, 66–73 (2006)
41. Jones, D.K.: Determining and visualizing uncertainty in estimates of fiber orientation from diffusion tensor mri. Magnetic Resonance in Medicine 49, 7–12 (2003)
42. Jospeh, A.J., Lodha, S.K., Renteria, J.C., Pang, A.: Uisurf: Visualizing uncertainty in isosurfaces. In: Proceedings of the Computer Graphics and Imaging, pp. 184–191 (1999)
43. Kao, D., Dungan, J.L., Pang, A.: Visualizing 2d probability distributions from eos satellite image-derived data sets: a case study. In: Proceedings Visualization, pp. 457–561 (2001)
44. Kao, D., Kramer, M., Luo, A., Dungan, J., Pang, A.: Visualizing distributions from multi-return lidar data to understand forest structure. The Cartographic Journal 42(1), 35–47 (2005)
45. Kao, D., Luo, A., Dungan, J.L., Pang, A.: Visualizing spatially varying distribution data. In: Proceedings of the Sixth International Conference on Information Visualisation 2002, pp. 219–225 (2002)
46. Kindlmann, G., Whitaker, R., Tasdizen, T., Moller, T.: Curvature-based transfer functions for direct volume rendering: Methods and applications. In: Proceedings of the 14th IEEE Visualization 2003 (VIS 2003), pp. 67–74 (2003)
47. Kniss, J.M., Uitert, R.V., Stephens, A., Li, G.S., Tasdizen, T., Hansen, C.: Statistically quantitative volume visualization. In: Proceedings of IEEE Visualization 2005, pp. 287–294 (2005)
48. Kunz, M., Hurni, L.: Hazard maps in switzerland. In: Proceedings of the 6th ICA Mountain Cartography Workshop, pp. 125–130 (2008)
49. Lee, B., Robertson, G.G., Czerwinski, M., Parr, C.S.: Candidtree: visualizing structural uncertainty in similar hierarchies. Information Visualization 6, 233–246 (2007)
50. Li, H., Fu, C.W., Li, Y., Hanson, A.J.: Visualizing large-scale uncertainty in astrophysical data. IEEE TVCG 13(6), 1640–1647 (2007)
51. Lodha, S., Sheehan, B., Pang, A., Wittenbrink, C.: Visualizing geometric uncertainty of surface interpolants. In: Proceedings of the Conference on Graphics Interface 1996, pp. 238–245 (1996)
52. Lodha, S.K., Pang, A., Sheehan, R.E., Wittenbrink, C.M.: Uflow: visualizing uncertainty in fluid flow. In: Proceedings Visualization 1996, pp. 249–254 (1996)
53. Lodha, S.K., Wilson, C.M., Sheehan, R.E.: Listen: sounding uncertainty visualization. In: Proceedings Visualization 1996, pp. 189–195 (1996)
54. Lucieer, A.: Visualization for exploration of uncertainty related to fuzzy classification. In: IEEE International Conference on Geoscience and Remote Sensing Symposium, IGARSS 2006, pp. 903–906 (2006)
55. Lundström, C., Ljung, P., Persson, A., Ynnerman, A.: Uncertainty visualization in medical volume rendering using probabilistic animation. IEEE Transactions on Visualization and Computer Graphics 13(6), 1648–1655 (2007)
56. Luo, A., Kao, D., Pang, A.: Visualizing spatial distribution data sets. In: VISSYM 2003: Proceedings of the Symposium on Data Visualisation 2003, pp. 29–38 (2003)

57. MacEachren, A.M., Robinson, A., Hopper, S., Gardner, S., Murray, R., Gahegan, M., Hetzler, E.: Visualizing geospatial information uncertainty: What we know and what we need to know. Cartography and Geographic Information Science 32(3), 139–160 (2005)

58. Masuch, M., Freudenberg, B., Ludowici, B., Kreiker, S., Strothotte, T.: Virtual reconstruction of medieval architecture. In: Proceedings of EUROGRAPHICS 1999, Short Papers, pp. 87–90 (1999)

59. Masuch, M., Strothotte, T.: Visualising ancient architecture using animated line drawings. In: Proceedings of the IEEE Conference on Information Visualization, pp. 261–266 (1998)

60. Miler, J.R., Cliburn, D.C., Feddema, J.J., Slocum, T.A.: Modeling and visualizing uncertainty in a global water balance model. In: Proceedings of the 2003 ACM Symposium on Applied Computing, pp. 972–978 (2003)

61. Newman, T.S., Lee, W.: On visualizing uncertainty in volumetric data: techniques and their evaluation. Journal of Visual Languages and Computing 15, 463–491 (2004)

62. Olston, C., Mackinlay, J.D.: Visualizing data with bounded uncertainty. In: Proceedings of IEEE InfoVis 2002, pp. 37–40 (2002)

63. Osorio, R.S.A., Brodlie, K.W.: Uncertain flow visualization using lic. In: 7th EG UK Theory and Practice of Computer Graphics, pp. 1–9 (2009)

64. Osorio, R.A., Brodlie, K.: Contouring with uncertainty. In: 6th Theory and Practice of Computer Graphics Conference, pp. 59–66 (2008)

65. Otto, M., Germer, T., Theisel, H.: Uncertain topology of 3d vector fields. In: IEEE Pacific Visualization Symposium (PacificVis), pp. 67–74 (March 2011)

66. Otto, M., Germer, T., Hege, H.C., Theisel, H.: Uncertain 2d vector field topology. Computer Graphics Forum 29(2), 347–356 (2010)

67. Otto, M., Nykolaychuk, M., Germer, T., Richter, K., Theisel, H.: Evaluation of motion data with uncertain 2d vector fields. In: Poster Abstracts at Eurographics/IEEE-VGTC Symposium on Visualization, pp. 1–2 (2010)

68. Pang, A.: Visualizing uncertainty in geo-spatial data. In: Proceedings of the Workshop on the Intersections between Geospatial Information and Information Technology (2001)

69. Pang, A., Furman, J.: Data quality issues in visualization. In: Visual Data Exploration and Analysis, SPIE, vol. 2278, pp. 12–23 (February 1994)

70. Pang, A., Wittenbrink, C., Lodha, S.: Approaches to uncertainty visualization. The Visual Computer 13(8), 370–390 (1997)

71. Pauly, M., Mitra, N.J., Guibas, L.: Uncertainty and variability in point cloud surface data. In: Symposium on Point-Based Graphics, pp. 77–84 (2004)

72. Pöthkow, K., Hege, H.C.: Positional uncertainty of isocontours: Condition analysis and probabilistic measures. IEEE TVCG PP(99), 1–15 (2010)

73. Pöthkow, K., Weber, B., Hege, H.C.: Probabilistic marching cubes. Computer Graphics Forum 30(3), 931–940 (2011)

74. Potter, K.: Methods for presenting statistical information: The box plot. In: Visualization of Large and Unstructured Data Sets, GI-Edition Lecture Notes in Informatics (LNI), vol. S-4, pp. 97–106 (2006)

75. Potter, K., Gooch, A., Gooch, B., Willemsen, P., Kniss, J., Riesenfeld, R., Shirley, P.: Resolution independent npr-style 3d line textures. Computer Graphics Forum 28(1), 52–62 (2009)

76. Potter, K., Kirby, R.M., Xiu, D., Johnson, C.R.: Interactive visualization of probability and cumulative density functions. International Journal for Uncertainty Quantification (to appear, 2012)
77. Potter, K., Kniss, J., Riesenfeld, R., Johnson, C.R.: Visualizing summary statistics and uncertainty. Computer Graphics Forum 29(3), 823–831 (2010)
78. Potter, K., Krüger, J., Johnson, C.: Towards the visualization of multi-dimensional stochastic distribution data. In: Proceedings of The International Conference on Computer Graphics and Visualization (IADIS 2008) (2008)
79. Potter, K., Wilson, A., Bremer, P.T., Williams, D., Doutriaux, C., Pascucci, V., Johhson, C.R.: Ensemble-vis: A framework for the statistical visualization of ensemble data. In: IEEE Workshop on Knowledge Discovery from Climate Data: Prediction, Extremes, pp. 233–240 (2009)
80. Rhodes, P.J., Laramee, R.S., Bergeron, R.D., Sparr, T.M.: Uncertainty visualization methods in isosurface rendering. In: EUROGRAPHICS 2003 Short Papers, pp. 83–88 (2003)
81. Saad, A., Hamarneh, G., Möller, T.: Exploration and visualization of segmentation uncertainty using shape and appearance prior information. IEEE Transactions on Visualization and Computer Graphics 16(6), 1366–1375 (2010)
82. Sanyal, J., Zhang, S., Bhattacharya, G., Amburn, P., Moorhead, R.J.: A user study to compare four uncertainty visualization methods for 1d and 2d datasets. IEEE TVCG 15(6), 1209–1218 (2009)
83. Sanyal, J., Zhang, S., Dyer, J., Mercer, A., Amburn, P., Moorhead, R.J.: Noodles: A tool for visualization of numerical weather model ensemble uncertainty. IEEE TVCG 16(6), 1421–1430 (2010)
84. Skeels, M., Lee, B., Smith, G., Robertson, G.: Revealing uncertainty for information visualization. Information Visualization 9(1), 70–81 (2010)
85. Streit, A., Pham, B., Brown, R.: A spreadsheet approach to facilitate visualization of uncertainty in information. IEEE TVCG 14(1), 61–72 (2008)
86. Strothotte, T., Masuch, M., Isenberg, T.: Visualizing knowledge about virtual reconstructions of ancient architecture. In: Proceedings of Computer Graphics International, pp. 36–43 (June 1999)
87. Strothotte, T., Puhle, M., Masuch, M., Freudenberg, B., Kreiker, S., Ludowici, B.: Visualizing uncertainty in virtual reconstructions. In: Proceedings of Electronic Imaging and the Visual Arts, EVA Europe 1999, p. 16 (1999)
88. Taylor, B.N., Kuyatt, C.E.: Guidelines for evaluating and expressing the uncertainty of nist measurement results. Tech. rep., NIST 1297 (1994)
89. Thomson, J., Hetzler, B., MacEachren, A., Gahegan, M., Pavel, M.: A typology for visualizing uncertainty. In: Proc. of SPIE, vol. SPIE-5669, pp. 146–157 (2005)
90. Waser, J., Fuchs, R., Ribičič, H., Schindler, B., Blöschl, G., Gröller, M.E.: World lines. IEEE TVCG 16(6), 1458–1467 (2010)
91. Wittenbrink, C., Pang, A., Lodha, S.: Verity visualization: Visual mappings. Tech. rep., University of California at Santa Cruz (1995)
92. Wittenbrink, C.M., Pang, A.T., Lodha, S.K.: Glyphs for visualizing uncertainty in vector fields. IEEE TVCG 2(3), 266–279 (1996)
93. Wray, K.: Using the creative design process to develop illustrative rendering techniques to represent information quality. The Journal of Young Investigators 17(2) (2007)
94. Yang, X., Wu, R., Ding, Z., Chen, W., Zhang, S.: A comparative analysis of dimension reduction techniques for representing dti fibers as 2d/3d points. In: Laidlaw, D., Bartroli, A. (eds.) New Developments in the Visualization and Processing of Tensor Fields, pp. 183–198. Springer (to appear, 2012)

95. Zehner, B., Watanabe, N., Kolditz, O.: Visualization of gridded scalar data with uncertainty in geosciences. Computers & Geosciences 36(10), 1268–1275 (2010)
96. Zuk, T., Carpendale, S., Glanzman, W.D.: Visualizing temporal uncertainty in 3d virtual reconstructions. In: Proceedings of the 6th International Symposium on Virtual Reality, Archaeology and Cultural Heritage (VAST 2005), pp. 99–106 (2005)
97. Zuk, T., Downton, J., Gray, D., Carpendale, S., Liang, J.: Exploration of uncertainty in bidirectional vector fields. In: Society of Photo-Optical Instrumentation Engineers (SPIE) Conference Series, vol. 6809 (2008) (published online)

Discussion

Speaker: Chris Johnson

Brian Ford: You talked of neurosurgeons not trusting the images you provide and so not using them in their work. Do you meet these doubts in many fields of investigation? How do you seek to overcome them, e.g. by seeking an approach through the mind set of the field of investigation or through explaining the techniques of the visualization etc.?

Chris Johnson: In the early days of scientific visualization, we could not address the tremendous complexity of many biomedical (and other science and engineering) applications. As hardware became faster and software and algorithms more sophisticated, we have been able to address more and more complexity. At the same time, we still must make simplifying assumptions in our models and visual abstractions. Anytime a new visualization technique is created, there is always a learning curve in understanding how to use it effectively and often there is initial skepticism that the new technique will prove useful.

The most successful visualization techniques and tools we have created have often been in close collaboration with scientists, engineers, and biomedical researchers and clinicians. Visualization researchers are not expert in the particular needs of the application researcher and the application research is not expert in visualization techniques. Collaborations often start with a presentation by the application researcher with an overview of the current visualization tools they use, with emphasis on what current and future needs they have that are not being satisfied by their current visualization software. We then can follow up with an overview of recent visualization tools and discuss how such tools might be modified (or new techniques and tools created) to meet the needs of the application researcher. We then proceed in an iterative way, moving towards a successful collaboration, which in my mind is when we can create new visualization techniques that can help solve or better understand the application researcher's problem. I get tremendous satisfaction when we can work together with biomedical, engineering, and science researchers to solve problems together that neither of us could have solved independently.

Pasky Pascual: Do you have plans to make your visualization tools widely available? For example, have you considered developing some of your visualization tools as R packages?

Chris Johnson: At the Scientific Computing and Imaging (SCI) Institute, we make all of our software available as open source and usually support the software on multiple platforms (OSX, Linux, Windows). You can download our visualization, image analysis, and scientific computing software from www.sci.utah.edu/software.html.

Kyle Hickman: Many of the visualizations seem that they would not lend themselves to publication in current journals. What do you think can be done? What do you think the future of visualization in scientific publishing will be?

Chris Johnson: We are making some progress in this regard. High quality color figures are becoming more the norm in journals and some journals are allowing authors to submit videos along with their papers. More recently, Adobe has partnered with Tech Soft 3D to create embedded 3D PDF and geospatial PDF. One can click on the figure and be presented with a 3D visualization that one can rotate and interact with. We are also seeing more on-line journals that can feature more visualization and media capabilities.

Michael Goldstein: I fully agree that visualization of uncertainty is an essential component of any complex analysis. There are two types of visualization that are required. Firstly, there is visualization uncertainty for the final outcome of the analysis, as was beautifully illustrated in your talk. Secondly, there is the need for uncertainty visualization for the analysis, to understand and criticize diagnostically the uncertainty flow through each of the stages of the analysis. I wonder if the speaker has any suggestions for such displays. As a comment, I have found structuring such a display through graphical overlays on Bayesian graphical models underlying the analysis to be a very useful tool, but of course I am no expert in the field of uncertainty visualization.

Chris Johnson: I agree completely. We need visualization for both display and analysis. In the early days of scientific visualization, many researchers focused on creating images for display, presentation and publication. As the field of visualization matured, more emphasis is being placed on analysis. I see visualization techniques and tools being an integral part of the scientific problem solving process, along side of, and integrated along with, other important techniques and tools, e.g. statistics, numerical modeling and simulation.

Efficient Computation of Observation Impact in 4D-Var Data Assimilation

Alexandru Cioaca[1], Adrian Sandu[1], Eric De Sturler[2],
and Emil Constantinescu[3]

[1] Department of Computer Science
[2] Department of Mathematics
Virginia Tech
Blacksburg, VA, USA
[3] Mathematics and Computer Science Division
Argonne National Laboratory Argonne, IL, USA
{alexgc,sandu,sturler,emconsta}@vt.edu

Abstract. Data assimilation combines information from an imperfect model, sparse and noisy observations, and error statistics, to produce a best estimate of the state of a physical system. Different observational data points have different contributions to reducing the uncertainty with which the state is estimated. Quantifying the observation impact is important for analyzing the effectiveness of the assimilation system, for data pruning, and for designing future sensor networks. This paper is concerned with quantifying observation impact in the context of four dimensional variational data assimilation. The main computational challenge is posed by the solution of linear systems, where the system matrix is the Hessian of the variational cost function. This work discusses iterative strategies to efficiently solve this system and compute observation impacts.

Keywords: Four dimensional variational data assimilation, observation impact, nonlinear optimization, preconditioning, adjoint model, iterative linear solvers.

1 Introduction

This paper discusses a framework for computing the impact of observations in four dimensional variational (4D-Var) data assimilation. The purpose of this framework is to reveal the degree of usefulness of the observational data in the process of correcting the *a priori* knowledge of the state of the system. From the analysis of observation impact, certain decisions can be inferred about the strategy to collect observations. The main computational challenge associated with estimating observation impact is the solution of a linear system where the matrix is the Hessian of the 4D-Var cost function. This Hessian matrix is usually very large and accessible only through matrix-vector products. A high accuracy solution is typically sought. In applications such as real-time sensor deployment, this solution needs to be obtained quickly.

A. Dienstfrey and R.F. Boisvert (Eds.): WoCoUQ 2011, IFIP AICT 377, pp. 250–264, 2012.

This paper discusses the particular characteristics of the linear system involved in observation impact calculations. We study the use of several iterative linear solvers that use only Hessian-vector products, and propose several inexpensive preconditioners to improve convergence. Numerical tests using shallow water equations illustrate the improvements in efficiency resulting from this methodology.

The paper is organized as follows. Sect. 2 introduces variational data assimilation. Sect. 3 zooms in on the topic of observation impact in the context of data assimilation. Sect. 4 presents the shallow water equations test problem and its implementation. Sect. 5 gives details of the data assimilation experiment. A discussion of the main computational issues and the proposed solutions can be found in Sect. 6.

2   Data Assimilation

Data assimilation is the process by which measurements are used to constrain model predictions [1,2]. The information from measurements can be used to obtain initial and boundary conditions that approximate better the real state of the model at a particular moment. Variational data assimilation allows the optimal combination of three sources of information: *a priori* (background) estimate of the state of the system, knowledge of the interrelationships among the variables of the model, and observations of some of the state variables. The optimal initial state $\mathbf{x}^a$ (also called the analysis) is obtained by minimizing the cost function

$$\mathcal{J}(\mathbf{x}_0) = \frac{1}{2}(\mathbf{x}_0 - \mathbf{x}^b)^T \cdot B^{-1} \cdot (\mathbf{x}_0 - \mathbf{x}^b)$$

$$+ \frac{1}{2}\sum_{k=1}^{N}(H_k(\mathbf{x}_k) - \mathbf{y}_k)^T \cdot R_k^{-1} \cdot (H_k(\mathbf{x}_k) - \mathbf{y}_k) \qquad (1)$$

$$\mathbf{x}^a = \text{Arg min } \mathcal{J}$$

The first term of the sum quantifies the departure from the background state $\mathbf{x}^b$ at the initial time t_0. The second term measures the distance to the observations $\mathbf{y}_k$, which are taken at N times t_k inside the assimilation window. When assimilating observations only at the initial time t_0, the cost function is known as 3D-Var. Otherwise, when the observations are distributed in time, the cost function is known as 4D-Var. In this study we will focus on 4D-Var. The block-diagonal background error covariance matrix B is built to take into account the spatial correlations of the variables, as well as the periodic boundary conditions. H_k is the observation operator defined at assimilation time t_k. It maps the discrete model state $\mathbf{x}_k \approx \mathbf{x}(t_k)$ to the observation space. R_k is the observations error covariance matrix. The weighting matrices B and R need be synthesized in order to have a fully-defined problem and their quality will influence the accuracy of the analysis.

The efficient numerical minimization of (1) requires the gradient of the cost function or second-order derivative information when available. 4D-Var usually relies on adjoint sensitivity analysis to provide information about the first and second-order derivatives of the objective function. For instance, the gradient of the cost function can be given by the first-order adjoint model, while the second-order information can be computed under the form of a Hessian-vector product from the second-order adjoint. Variational data assimilation is an example of PDE-constrained optimization as it is a problem of nonlinear optimization where the the minimization of the cost function is constrained by a numerical model associated with a set of PDEs. The minimizer of the 4D-Var cost function can only be computed iteratively using gradient-based methods, because an analytical solution is almost impossible to derive. When using 4D-Var to correct initial and boundary conditions in a real-time operational setting, we have a limited number of model runs that can be used before the time window expires. This usually implies the minimization process is halted after a number of iterations, so the global minimum might not be attained. Although the most significant decrease in the cost function usually happens during the first iterations, it is important to take into account that the computed solution might not satisfy the optimality conditions. For more details, see the studies in [3,4].

3 Observation Impact in 4D-Var Data Assimilation

Methods for computing observation impact have been initially developed for the framework of 3D-Var data assimilation in [5,6]. The extension of this theory to address 4D-Var data assimilation was only recently accomplished and the details can be found in [7,8].

If the 4D-Var problem is defined and solved accurately, then the analysis $\mathbf{x}^a$ should provide a better forecast than the background $\mathbf{x}^b$. Since one does not usually have access to the real state of the system (otherwise we would not have to perform data assimilation), the accuracy of the analysis is verified through a forecast score against a solution of higher accuracy. Let M be a nonlinear forecast model that discretizes a system of time-dependent partial differential equations, while we will denote its tangent linear and first-order adjoint models with $\mathcal{M}$ and $\mathcal{M}^T$. A forecast score $e(\mathbf{x}^a)$ is defined on M as a short-range forecast error measure:

$$e(\mathbf{x}^a) = (\mathbf{x}_f^a - \mathbf{x}_f^v)^T C(\mathbf{x}_f^a - \mathbf{x}_f^v) \tag{2}$$

where $\mathbf{x}_f^a = M_{t_0 \to t_f}(\mathbf{x}^a)$ is the model forecast at verification time t_f, $\mathbf{x}_f^v$ is the verifying analysis at t_f and C is a matrix that defines the metric in the state space.

Following the derivations in [7], the equation of forecast sensitivity to observations is:

$$\nabla_{\mathbf{y}} e(\mathbf{x}^a) = R^{-1} H A \nabla_{\mathbf{x}} e(\mathbf{x}^a) \tag{3}$$

where

$$A = [\nabla_{\mathbf{xx}}^2 \mathcal{J}(\mathbf{x}^a)]^{-1} \tag{4}$$

denotes the inverse of the Hessian matrix of the cost at $\mathbf{x}^a$ and it is assumed to be a positive definite matrix.

The same paper [7] provides an approximation to the impact of observation, which is defined as the change the observation brings to the value of the 4D-Var cost function. If we consider the innovation vector as $\delta y = y - h(\mathbf{x}^b)$, then a first-order formula would be

$$\delta e = e(\mathbf{x}^a) - e(\mathbf{x}^b) = (\delta \mathbf{y})^T R^{-1} H A \nabla_{\mathbf{x}} e(\mathbf{x}^a) \tag{5}$$

The elements of the innovation vector are individual observations that get multiplied with their corresponding sensitivity. For those observations that contributed to the decrease of the cost function, the impact (individual value of δe) will be a negative value. The matrix vector product in the matrix A can be computed as a linear system of the form

$$A^{-1} \cdot \mu_0 = \nabla_{\mathbf{x}} e(\mathbf{x}^a) \tag{6}$$

The value of μ_0 is associated with the sensitivity at the initial time. By initializing the tangent linear model with this vector of sensitivities and integrating it from this time to the final time of the assimilation window, we can obtain the sensitivities to observations corresponding to the rest of the assimilation windows.

The problem of computing observation impact has to be solved in real-time since these results are used in applications such as sensor network deployment. Numerical models that simulate the atmosphere are very expensive from a computational standpoint and their adjoint models even more so. Most of the computation bulk is in computing the solution of the linear system $A^{-1} \cdot \mu_0 = \nabla_{\mathbf{x}} e(\mathbf{x}^a)$. This linear system has to be solved with a high degree of accuracy because its solution μ_0 is a vector representing a "supersensitivity". Any errors that it contains can be propagated and amplified by the numerical models used for forecast or sensitivity analysis. At the same time, there is also a constraint on the number of model runs one can execute before the results are due for operational use. The Hessian matrix of a system associated with a numerical model underlined by PDEs is usually sufficiently complicated that an iterative solver generally will not converge to a satisfactory solution in a few iterations unless it is tweaked accordingly. Accuracy of the solution and fast convergence of the solver are two of the main challenges when computing observation impact in data assimilation, along with means of computing the second-order information.

4   The Shallow Water Equations Test Case

We consider the two dimensional Saint-Venant PDE system that approximates fluid flow inside a shallow basin (also known as "Shallow Water Equations" (SWE)):

$$\frac{\partial}{\partial t} h + \frac{\partial}{\partial x}(uh) + \frac{\partial}{\partial y}(vh) = 0$$

$$\frac{\partial}{\partial t}(uh) + \frac{\partial}{\partial x}\left(u^2 h + \frac{1}{2}gh^2\right) + \frac{\partial}{\partial y}(uvh) = 0$$

$$\frac{\partial}{\partial t}(vh) + \frac{\partial}{\partial x}(uvh) + \frac{\partial}{\partial y}\left(v^2 h + \frac{1}{2}gh^2\right) = 0 \ .$$

The spatial domain is square shaped ($\Omega = [-3, \ 3]^2$), and the integration window is $t^0 = 0 \le t \le t^F = 0.1$. Here $h(t, x, y)$ denotes the fluid layer thickness, and $u(t, x, y)$, $v(t, x, y)$ are the components of the velocity field. g is the standard value of the gravitational acceleration. The boundary conditions are periodic in both directions. For ease of presentation, we arrange the n discretized state variables in a column vector

$$x = \begin{bmatrix} \hat{h} \\ \hat{uh} \\ \hat{vh} \end{bmatrix} \in R^n \ . \tag{7}$$

In order to solve these equations, we implemented a numerical model that makes use of a finite volume-type scheme for space discretization and a fourth-order Runge-Kutta scheme for timestepping. This method was introduced by Liska and Wendroff in [9]. In the following, forward solves will be denoted by FWD.

The tangent linear model (TLM), first-order (FOA) and second-order (SOA) adjoints were generated through automatic differentiation of the forward model using TAMC [10,11]. TAMC is a source-to-source translator that generates Fortran code for computing the derivatives of a numerical model. For more details about automatic differentiation see [12]. The FOA and SOA are integrated backwards in time and they require the values computed at each timestep for the FWD and TLM, respectively. There are two ways to generate these numerical values. The first one is to run the FWD and TLM in advance and to store their evolution in time; this approach has the disadvantage of taking up a lot of storage space. The second one is to recompute the FWD and TLM at each timestep where the FOA and SOA need their values. Theoretically this is necessary at each timestep, which means the latter approach saves storage space at the expense of increasing demand for computation.

A question that arises naturally is how large is the overhead of the code generated through automatic differentiation. Table 1 illustrates the CPU times of TLM, FOA and SOA models, normalized with respect to the CPU time of a single forward model run. It is seen that a full SOA integration is about 3.5 times more expensive than a single first-order adjoint run, while this FOA takes 3.7 times longer than the forward run. The time for SOA can be considered to be a large value as it takes longer time to compute the SOA than to approximate it through finite differences (two FOA runs). However, for the purpose of this test they are computationally feasible. Please note that these values apply only for our particular implementation (method of lines) and the overhead can greatly vary with the complexity of the numerical method or the automatic differentiation tool used.

We mentioned that each time the tangent linear model or first-order adjoint are run, the forward model is run beforehand for storing its evolution. Similarly,

Table 1. Ratio between the CPU time of adjoint models and the forward model

FWD	1		
TLM	2.5	FWD + TLM	3.5
FOA	3.7	FWD + FOA	4.7
SOA	12.8	FWD + TLM + FOA + SOA	20.0

in order to compute the second-order adjoint one must first run the forward, first-order adjoint and tangent linear models. As such, the actual timing values are the ones displayed in the right hand side of the table.

5 Data Assimilation Scenario

Prior to computing observation impact, we have to perform the 4D-Var data assimilation experiment with the SWE model presented in the previous section. Following are the parameters of the Data Assimilation System (DAS) we are using:

- The 2-D grid of the SWE model is divided into 40 grid points on each direction (1600 in total). The number of time steps for the model runs is fixed at $N = 1000$.
- The reference solution is generated component-wise: the height component h will be a Gaussian pulse of amplitude $A = 30$ while the velocity fields u and v are assigned a value of 2 at each grid point.
- The background solution $\mathbf{x}^b$ is generated by applying a correlated perturbation on the reference profile for h, u and v.
- The background error covariance B was generated for a standard deviation of 5% with a nondiagonal structure and correlation distance of five grid points. This will help the 4D-Var method to spread the information from observations in each cell based on information passed from its neighbors.
- The model was run with the reference as initial condition in order to generate the synthetic observations $\mathbf{y}_k$. The observation frequency is set to once every 20 time steps. In order to simulate the effect of noise over observations, we apply a normal random perturbation to the perfect synthetic observations.
- The observation error covariance R is a diagonal matrix, based on the assumption that the observational errors are uncorrelated. The standard deviation of these errors was set to 1% of the largest absolute value of the observations for each variable.

The observation operator H is configured to select observations for each variable at each point of the grid, hence we say the observations are dense in space. In an operational setting, the operator H is enforced by the availability of observations provided by meteorological stations and other types of sensors.

In order to minimize the 4D-Var cost function, the L-BFGS-B iterative solver will be employed [13]. The optimization parameters of interest were assigned the following values:

- The stopping criterion is set to $\left\|\nabla\mathcal{J}(\mathbf{x}_{[k]}^0)\right\| < 10^{-6} \cdot \max(1, \left\|\mathbf{x}_{[k]}^0\right\|)$.
- Wolfe conditions parameters are set to $c_1 = 10^{-4}$ and $c_2 = 0.9$.
- A maximum number of 500 iterations is allowed for each optimization.

6   Computing Observation Impact

6.1   Second-Order Information

The matrix of coefficients is the Hessian of the 4D-Var cost function, evaluated at the analysis. For large-scale models like the atmosphere, this matrix cannot be manipulated in its full explicit form due to high computational demands, storage restrictions and I/O bottlenecks. The dimensions of such Hessian matrices, associated with models of many variables, could reach orders of magnitude of 10^6 or more. In practice, one usually tries to evaluate directly the action of the Hessian matrix on a vector, an approach known as "matrix-free".

Evaluating the Hessian or its action on a vector is itself another computational issue. The methodology of adjoint models provides ways for the computation of first-order and second-order derivative information. Second-order adjoint models are considered to be the best approach to computing Hessian-vector products. Unfortunately, they are not that popular because of the tedious process of deriving them. At the same time, the adjoint models usually have higher computational demands than the forward model, unless they are built around a clever strategy to reuse computation from the forward model [4]. When one does not have access to the second-order adjoint but only to the first-order adjoint, Hessian-vector products can be computed by using finite differences. This approach might be more economical in terms of computation than using second-order adjoints, but is also less reliable.

These two arguments indicated to us that there is only a limited class of methods we can use in order to solve, precondition or transform the system. More specifically, we will use those methods that need only matrix-vector products. By elementary analysis, the Hessian of any reasonable system is symmetric. However, this symmetry can be lost due to numerical inaccuracies. At the same time, we cannot guarantee the Hessian is positive definite when the 4D-Var first-order optimality condition is not met and the resulting analysis is far from the true minimizer. While using iterative methods on our linear systems seems to be feasible, it appears to be significantly more difficult to obtain preconditioners, even simple ones based on the diagonal. Also, we can exclude factorizations or direct solve methods since these usually require the full matrix or at least a significant portion of it.

In Figs. 1–4 we present some graphical representations of the explicit form of the Hessian, built through the second-order adjoint methodology from matrix-vector products with the e_i unity vectors. For the scenario presented in Sect. 5 the size of the Hessian is 4800 rows by 4800 columns. The first 1600 rows correspond to the variable h, the next 1600 rows to u and the last 1600 to v. The Hessian thus generated proved to be symmetric. Regarding the minimization of the 4D-Var cost function, we let L-BFGS run for as many iterations as needed to obtain a solution close to the optimum (around 500 iterations). As a result, the Hessian was indeed positive definite. Although not shown here, we can confirm that early termination of L-BFGS after 50 or 100 iterations produced an analysis whose corresponding 4D-Var Hessian was no longer positive definite. As said before, the mere purpose of generating the full Hessian is to analyze it offline. In an operational setting it would not be possible, due to constraints imposed by the computational resources.

The first thing noticed about the Hessian is its somewhat regular structure that looks near block-Toeplitz. Although the diagonals are not constant, their values vary inside a well-defined interval. At the same time, we could notice that these diagonals occur at fixed distance from the main diagonal, spaced at every 40 or 1600 rows/columns. These patterns can be seen in Figs. 3 and 4, which are graphical representations of the upper-left corner of the matrix. The patterns repeat themselves across the whole matrix. Since the 4D-Var Hessian is the inverse of the covariance matrix of the analysis [14], we can attribute the patterns in the Hessian to the covariance of each grid point with the grid points in its discretization scheme stencil. Thus, each point of the grid is correlated to the points to the East and the West (corresponding to adjacent rows/columns in the Hessian) but also to the points found to the North and the South (corresponding to rows/columns situated at a distance of 40) and with the other variables at the same points (corresponding to a distance of 1600 rows/columns). This can be verified in Fig. 3 where the diagonal breaks down after 40 cells and is displaced, due to periodic boundary conditions.

The eigenvalues of the Hessian are displayed in Fig. 5, sorted in ascending order. Based on their values, the condition number for the matrix is of $7.2095e+3$. The magnitude of the condition number does not indicate the matrix is extremely ill-conditioned, but the eigenvalues are not clustered together and many of them are close to zero.

6.2 Solving the Linear System

Next, we will present our results and comments on solving the linear system with Krylov-based iterative methods when no preconditioner is used to accelerate the convergence. For our experiments we used MATLAB and its built-in linear solvers. We chose three algorithms for our comparison of their efficiency at solving the system. These three methods are "generalized minimum residual" (GMRES), "conjugate gradients" (CG) and "quasi-minimum residual" (QMR). There are many other variations of these methods but these three are each representative of one class of such methods. While GMRES and QMR operate on general matrices, CG needs the matrix to be SPD which might not be true all the time for the 4D-Var Hessian.

As we can see from Fig. 6 and Table 2, all three solvers converge to a solution whose norm of the residual is close to zero. GMRES and QMR converge to the same solution and they take the same iterates. This means they are equivalent when applied to this problem. CG converges to a more accurate solution from the perspective of its error norm, but the residual norm is larger. This can

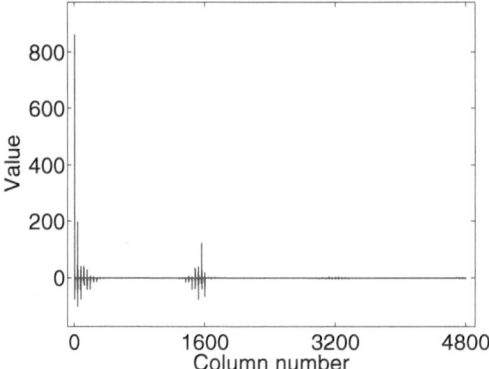

Fig. 1. The values found on the first row of the Hessian matrix

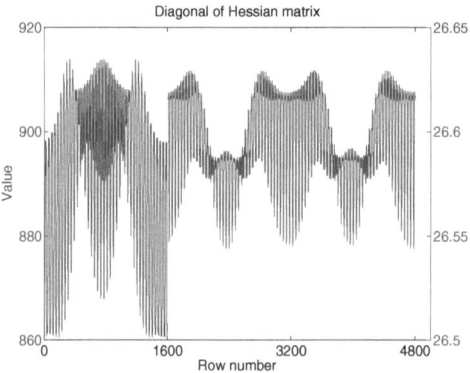

Fig. 2. The values found on the diagonal of the Hessian matrix. (Divided into two scaling groups: first 1600 and last 3200).

cause confusion in practice because we would not have access to the reference solution. Also, the norm of the residual for CG fluctuates, instead of decreasing monotonically. Based on these results, we decided to use GMRES for subsequent tests.

All solvers start with a vector of zeroes as initial solution.

6.3   Preconditioning

The convergence speed of Krylov-based linear solvers depends on the spectrum of the system matrix. These algorithms perform better when the eigenvalues are clustered together, but as seen in Fig. 5 the eigenvalues of our Hessian matrix are scattered across various orders of magnitude. Also, many of these eigenvalues are close to zero which makes them more difficult to be estimated by the solvers. Such a matrix is called "ill-conditioned" and this explains why neither algorithm

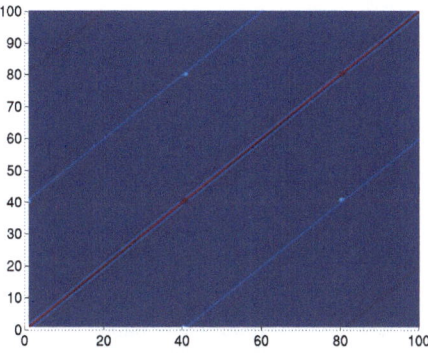

Fig. 3. Density plot of elements in the $(100 \times 100$ minor) of the Hessian matrix

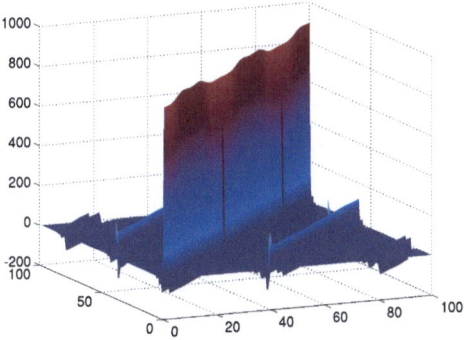

Fig. 4. The values found in the upper-left $(100 \times 100$ minor) of the Hessian matrix

presented in the comparison of the previous section converged in 100 iterations to a solution that is highly accurate.

In order to alleviate slow convergence, a linear system can be preconditioned by multiplying it with a matrix M (called preconditioner) that approximates the inverse of the system matrix or some of its features, but is easier to invert or factorize. One way to use this matrix is to multiply the linear system with it to the left, which leads to an equivalent linear system. This system should be easier to solve if the preconditioner approximates the inverse of the system matrix, since their multiplication will yield a new system matrix that is better conditioned. In the ideal setting when the preconditioner is exactly the inverse of the original system matrix, the new system matrix becomes the identity matrix and the linear solver will converge in one iteration. However, finding the exact inverse of a matrix is usually a more difficult problem then solving the linear system.

The problem of finding a preconditioner for our system is the fact that we do not have access to the full system matrix. This means we have to exclude

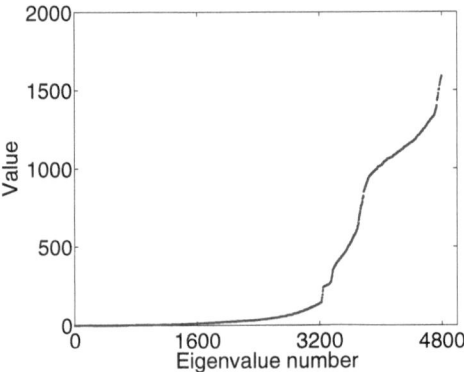

Fig. 5. The computed eigenvalues of the Hessian matrix (in ascending order)

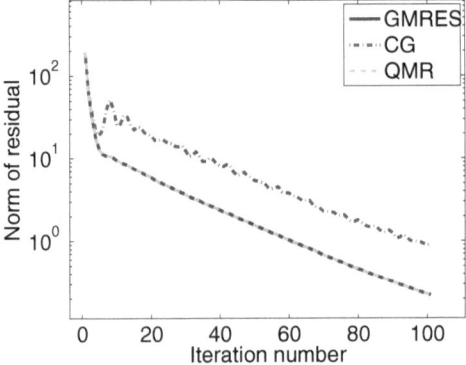

Fig. 6. The evolution of the solution residuals with iteration number for different linear solvers

the possibility of using certain preconditioning techniques such as incomplete factorizations, wavelet-based, or variations of Schur complement. Moreover, we cannot just use matrix-vector products to construct basic preconditioners such as the main diagonal or a diagonal band, unless we construct the full matrix.

Knowing or anticipating the structure of the matrix can be of great help when trying to devise a preconditioner. For instance, we know our Hessian matrix is the inverse of the covariance matrix of the 4D-Var analysis. If the data assimilation process is accurate enough, one can observe a near block-diagonal structure corresponding to each variable. On a 3D grid these blocks correspond to the vertical levels of each variable. This was noted by Zupanski in [15] who hinted at the possibility to approximate the diagonal in a block-wise fashion, where the diagonal values for a certain block are equal to a constant value. We will try to approximate the diagonal by using Hessian-vector products to "probe" the matrix. A straightforward way of accomplishing this for our three-variable model is to run three Hessian-vector products with unity vectors that extract one column

Table 2. Norm of residual and error for solutions computed with different iterative methods (100 iterations)

Solver	GMRES	CG	QMR
Norm of residual	0.222	0.893	0.222
Norm of error	0.357	0.267	0.357

(row) of the Hessian at one time and then use the value corresponding to the diagonal element for all diagonal elements in that block. For example, consider three unity vectors for our 4800 Hessian that have the value 1 at position 1, 1601 and 3201 respectively, and zeros everywhere else. The Hessian-vector product realized with these three vectors will extract the columns number 1, 1601 and 3201 which correspond to the three different variables in our Hessian structure. Building an approximation of the diagonal means using the value found at coordinates 1,1 for the entire first block (so up to coordinates 1600,1600), the value found at coordinates 1601,1601 for the entire second block and so forth. This approximation can be refined by probing for more elements from the same block. If there are many blocks that have to be probed and the computational burden increases significantly, one can employ coloring techniques to probe for more than one element with the same matrix-vector product. We shall refer to this technique as "block-diagonal preconditioner".

Another characteristic of covariance matrices is their diagonally dominant structures and under certain favorable cases, strongly diagonally dominant. In this former case, an approximation of the diagonal can be computed with just one matrix-vector product, where the seed vector is composed only of ones. This is equivalent to computing the sum of each row, which is a good approximation of the diagonal if the diagonal element supersedes the sum of all other elements from the same row. This is the second preconditioner we tried in our experiments.

The Hessian matrix can also be approximated from data collected throughout the minimization process of 4D-Var. Iterative solvers such as L-BFGS work by generating approximations of the Hessian evaluated at the analysis in a certain subspace and minimize the cost function upon it. This approximation is very efficient as it preserves the properties of the matrix (positive definiteness and symmetry) and because it is designed for limited storage. We built a preconditioner by reusing the approximation of the Hessian generated over the last 10 iterations of L-BFGS. Our tests showed this to be as accurate as if using more iterations, a result that confirms the theory.

The norm of the error against the reference solution and that of the residual are shown in Table 3. We also present the results obtained with the exact diagonal, although this will not be available in practice, as stated before. The preconditioner obtained from the sum of elements on each row did not improve the convergence at all. This is due to the fact that in our case, the Hessian was not strongly diagonally dominant so this preconditioner was far from

Table 3. Norm of residual and error for solutions computed with different preconditioners (100 iterations)

Preconditioner	None	Diagonal	Block-diagonal	Row sum	L-BFGS
Norm of residual	0.222	0.009	0.009	62.346	0.222
Norm of error	0.357	0.005	0.005	5.301	0.339

approximating the diagonal. The L-BFGS preconditioner brought a slight improvement in the solution and has the advantage that no extra computation is required. The block-diagonal preconditioner improved significantly the accuracy of the solution and behaved as good as the exact diagonal.

7 Conclusions

This paper presents an adjoint-based framework to compute observation impact in 4D-Var data assimilation. The observation impact calculations need to be performed accurately and rapidly in real-time sensor network deployment problems. The main computational task is solving a linear system whose matrix is the Hessian of the 4D-Var cost function, evaluated at the analysis state. This matrix is typically very large, as each row corresponds to one model state, and it is only accessible through matrix-vector products.

The main contributions of this work are to outline the characteristics of the linear system, to investigate iterative linear solvers, and to propose three efficient preconditioners. Two of the preconditioning methods approximate the diagonal of the Hessian from matrix-vector products, while the third method uses data generated during the 4D-Var minimization to provide a quasi-Newton approximation of the Hessian. Our study shows that these inexpensive preconditioners accelerate convergence. Future work will include the development of improved preconditioners for the linear system, and will study the use of additional iterative methods such as multigrid.

Acknowledgements. This work was supported by National Science Foundation through the awards NSF DMS-0915047, NSF CCF-0635194, NSF CCF-0916493 and NSF OCI-0904397.

References

1. Daley, R.: Atmospheric data analysis. Cambridge University Press (1991)
2. Kalnay, E.: Atmospheric modeling, data assimilation and predictability. Cambridge University Press (2002)
3. Alekseev, A.K., Navon, I.M., Steward, J.L.: Comparison of Advanced Large-scale Minimization Algorithms for the Solution of Inverse Ill-posed Problems. Journal of Optimization Methods & Software 24, 63–87 (2009)
4. Cioaca, A., Alexe, M., Sandu, A.: Second Order Adjoints for Solving PDE-constrained Optimization Problems. Optimization Methods and Software (2011) (to appear)

5. Baker, N.L., Langland, R.H.: Diagnostics for evaluating the impact of satellite observations. In: Park, S.K., Xu, L. (eds.) Data Assimilation for Atmospheric, Oceanic and Hydrologic Applications, pp. 177–196. Springer, Berlin (2009)
6. Baker, N.L., Daley, R.: Observation and Background Adjoint Sensitivity in the Adaptive Observation-targeting Problem. Q.J.R. Meteorol. Soc. 126, 1431–1454 (2000)
7. Daescu, D.N.: On the Sensitivity Equations of Four-dimensional Variational (4D-Var) Data Assimilation. Monthly Weather Review 136(8), 3050–3065 (2008)
8. Daescu, D.N., Todling, R.: Adjoint Sensitivity of the Model Forecast to Data Assimilation System Error Covariance Parameters. Quarterly Journal of the Royal Meteorological Society 136, 2000–2012 (2010)
9. Liska, R., Wendroff, B.: Composite Schemes for Conservation Laws. SIAM Journal of Numerical Analysis 35(6), 2250–2271 (1998)
10. Giering, R., Kaminski, T.: Recipes for Adjoint Code Construction. ACM Transactions on Mathematical Software 24(4), 437–474 (1998)
11. Giering, R.: Tangent Linear and Adjoint Model Compiler, Users Manual 1.4 (1999), http://www.autodiff.com/tamc
12. Griewank, A.: On Automatic Differentiation. In: Iri, M., Tanabe, K. (eds.) Mathematical Programming: Recent Developments and Applications, pp. 83–109. Kluwer Academic Publishers (1989)
13. Zhu, C., Byrd, R.H., Lu, P., Nocedal, J.: L-BFGS-B: A Limited Memory FORTRAN Code for Solving Bound Constrained Optimization Problems. Technical Report, NAM-11, EECS Department, Northwestern University (1994)
14. Gejadze, I., Le Dimet, F.-X., Shutyaev, V.: On Error Covariances in Variational Data Assimilation. Russian Journal of Numerical Analysis and Mathematical Modeling 22(2), 163–175 (2007)
15. Zupanski, M.: A Preconditioning Algorithm for Large-scale Minimization Problems. Tellus (45A), 478–492 (1993)

Discussion

Speaker: Adrian Sandu

Van Snyder: There are two possibilities to assimilate remote sensing data into chemical transport models (CTMs). One is to incorporate the measured quantity, e.g., radiances. The other is to incorporate geophysical quantities that result from separate analyses. The latter is a nonlinear parameter estimation problem, which depends upon having a good starting point. Further, the problem is ill-posed, which requires *a priori* information or Tikhonov for stabilization. If the CTM initializes and stabilizes the remote sensing problem, whose results are then assimilated into the CTM, is there a positive, negative, or neutral effect on the solution quality?

Adrian Sandu: This is a very good question. In principle the direct assimilation of measured quantities is to be preferred, since it does not introduce additional errors/biases in the data. In order to do this, one needs good models of the instrument, e.g., that produce radiances from the concentration fields computed by the CTM.

In practice, there are very many instruments that produce data, and for which good models are not available. For this reason assimilation of geophysical quantities, derived from the raw data through an off line estimation process, is often employed. To the best of my knowledge no comprehensive tests have been carried out to date in order to quantify the impact of assimilating geophysical quantities in lieu of measured quantities.

Interval Based Finite Elements for Uncertainty Quantification in Engineering Mechanics

Rafi L. Muhanna[1] and Robert L. Mullen[2]

[1] School of Civil and Environmental Engineering
Georgia Institute of Technology
Atlanta, GA, USA
`rafi.muhanna@gtsav.gatech.edu`
[2] Department of Civil and Environmental Engineering
University of South Carolina
Columbia, SC, USA
`rlm@sc.edu`

Abstract. This paper illustrates how interval analysis can be used as a basis for generalized models of uncertainty. When epistemic uncertainty is presented as a range and the aleatory is based on available information, or when random variables are assigned an interval probability, the uncertainty will have a Probability Bound (PB) structure. When Interval Monte Carlo (IMC) is used to sample random variables, interval random values are generated. Interval Finite Element Method (FEM) is used to propagate intervals through the system and sharp interval solutions are obtained. Interval solutions are sorted and PBs of the system response are constructed. All relevant statistics are calculated characterizing both aleatory and epistemic uncertainty. The above mentioned sequence is presented in this work and illustrative examples are solved.

Keywords: interval, finite elements, reliability, aleatory, epistemic, probability bounds.

1 Introduction

The presence of uncertainty in all aspects of life is evident. However, quantifying uncertainty is always advancing. There are various ways in which types of uncertainty might be classified. One is to distinguish between "aleatory" (or stochastic) uncertainty and "epistemic" uncertainty. The first refers to underlying, intrinsic variability of physical quantities, and the latter refers to uncertainty which might be reduced with additional data or information, or better modeling and better parameter estimation [1].

Probability theory is the traditional approach to handling uncertainty. This approach requires sufficient statistical data to justify the assumed statistical distributions. Analysts agree that, given sufficient statistical data, probability theory describes the stochastic uncertainty well. However, probabilistic modeling cannot handle situations with incomplete or little information on which to

A. Dienstfrey and R.F. Boisvert (Eds.): WoCoUQ 2011, IFIP AICT 377, pp. 265–279, 2012.

evaluate a probability, or when that information is nonspecific, ambiguous, or conflicting [2], [3],and [4]. Many generalized models of uncertainty have been developed to treat such situations, including fuzzy sets and possibility theory [5], Dempster-Shafer theory of evidence [6], [7], random sets [8], probability bounds [9], [2], and [10], imprecise probabilities [4], convex models [11], and others.

These generalized models of uncertainty have a variety of mathematical descriptions. However, they are all closely connected with interval analysis [12]. For example, the mathematical analysis associated with fuzzy set theory can be performed as interval analysis on different α levels [13] and [14]. Fuzzy arithmetic can be performed as interval arithmetic on α cuts. A Dempster-Shafer structure [6] and [7] with interval focal elements can be viewed as a set of intervals with probability mass assignments, where the computation is carried out using the interval focal sets. Probability bounds analysis [9], [2], and [10] is a combination of standard interval analysis and probability theory. Uncertain variables are decomposed into a list of pairs of the form (interval, probability). In this sense, interval arithmetic serves as the calculation tool for generalized models of uncertainty. A short description of probability bounds is given in the next section.

2   Probability Bounds

Probability Bounds (PB) identifies a specific mathematical framework for analysis when precise discrete probabilities (or PDF) are not completely known [4]. Probability bounds are normally associated with epistemic sources of uncertainty where the available knowledge is insufficient to construct precise probabilities. Calculations on PB can be conducted using Monte Carlo methods combined with an interval finite element method [15], or by using various discretization methods.

The arithmetic for piecewise constant discretization on PB can be found in publications such as [16], [17], [18], [10]. Let the behavior of the system be modeled by a function $y = f(x)$, where $x = (x_1, x_2, \ldots, x_n)$ is the parameter vector, and each x_i is represented by a probability-bounds structure. The CDF of a particular value y^*, $F(y^*)$, is to be determined. The objective of probability-bounds analysis is to obtain an interval to bound the precise probability $F(y^*)$ of interest (in the classical sense). The numerical implementation of probability-bounds analysis can be done using interval analysis and the Cartesian product method. The general procedure can be found in publications such as [19], [20], [21], and [10].

For example, Fig. 1 depicts probability-bounds circumscribing a normal distribution whose mean is sure to lie within the interval $[5.6, 6]$ and whose standard deviation is 1. Such bounds can result from the addition of a normally distributed random variable with a mean of 5.6 and a standard deviation of 1 with a variable bounded between 0 and 0.4.

Probability-bounds structure can also arise by forming probability distributions of intervals. In this context, probability-bounds structure is mathematically analogous to a standard discrete probability distribution except that the

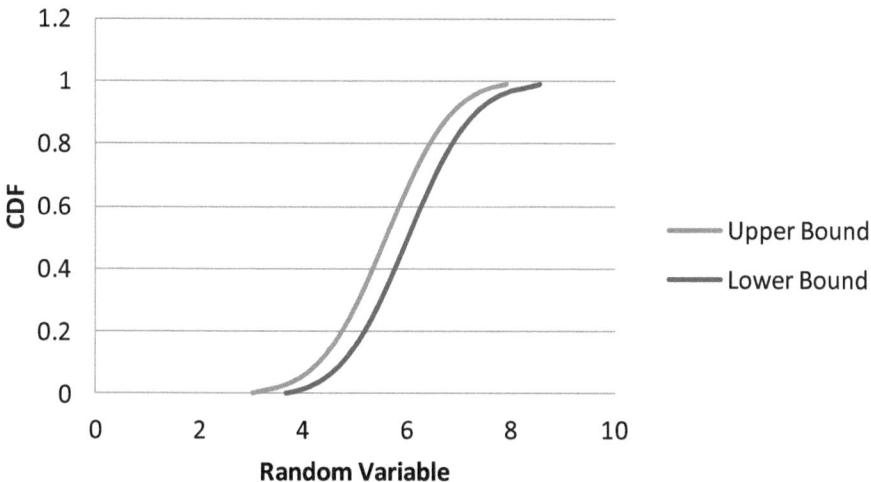

Fig. 1. Probability-bounds associated with normal distribution with mean = [5.6, 6] and standard deviation = 1

probability mass is assigned to an interval rather than to a precise point; thus, the probability-bounds structure can be specified as a list of pairs of the form (interval, probability mass).

The advantage of the probability-bounds approach is that it can capture a wider range of uncertainties than the standard probabilistic approach. On the other hand, the standard probability distribution and interval number are two degenerate cases of probability-bounds structure; thus the probability-bounds approach provides a general framework for handling problems with a mixture of interval-based information and standard probabilistic information.

When and if additional knowledge of a system is obtained, the probability bounds can be refined. Just as interval and scalar calculations can be easily mixed, conventional precise probabilities and probability bounds may be mixed in a single calculation. However, how to accomplish this in a computationally efficient method for finite element analysis is unresolved. In this work interval Monte Carlo (IMC) will be used.

3   Interval Monte Carlo

When a random variable is described in a probability-bound structure, one natural approach for sampling such a random variable is the use of interval Monte Carlo (IMC). A discretization approach can be used as well [15]; however, this approach has not yet been developed for the general case. IMC has been proposed to generate fuzzy numbers in [22] and for the first time for structural reliability in [15].

The first step in the implementation of interval Monte Carlo simulation is the generation of intervals in accordance with the prescribed probability bounds.

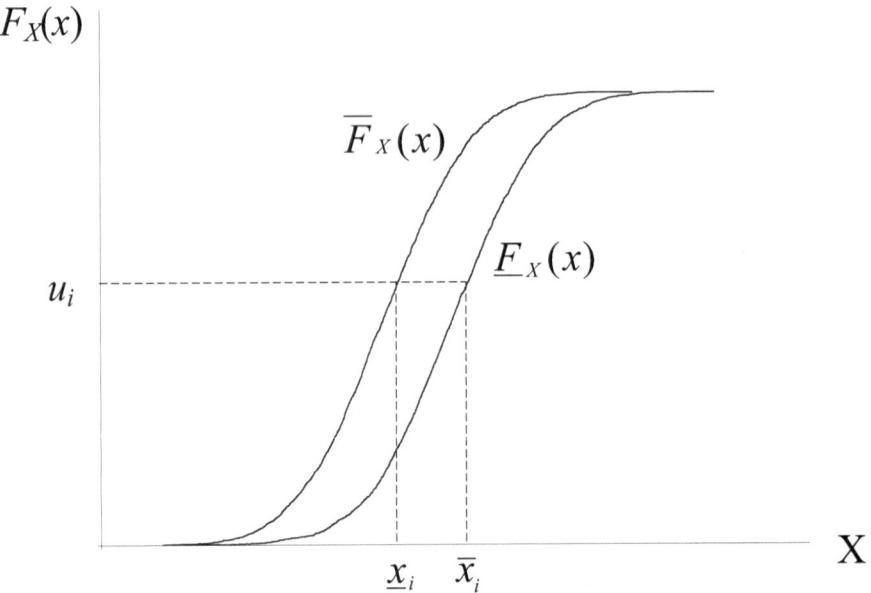

Fig. 2. Generation of random number from distribution with probability-bound structure

The inverse transform method is often used to generate random numbers. Consider a random variable x with CDF $F(x)$. If $(u_1, u_2, \ldots, u_m)$ is a set of values from the standard uniform variate, then the set of values

$$x_i = F_x^{-1}(u_i), \quad i = 1, 2, \ldots, m \tag{1}$$

will have the desired CDF $F(x)$. The inverse transform method can be extended to perform random sampling from a probability bound. Suppose that an imprecise CDF $F(x)$ is bounded by $\underline{F}(x)$ and $\overline{F}(x)$, as shown in Fig. 2. For each u_i in Eq. (1), two random numbers are generated

$$\underline{x}_i = \overline{F}_x^{-1}(u_i) \quad \text{and} \quad \overline{x}_i = \underline{F}_x^{-1}(u_i) \tag{2}$$

Such a pair of $\underline{x}_i$ and $\overline{x}_i$ form an interval $[\underline{x}_i, \overline{x}_i]$ which contains all possible simulated numbers from the ensemble of distributions for a particular u_i. The method is graphically demonstrated in Fig. 2 for the one-dimensional case. The next step is to solve for the generated interval values using interval finite elements described in the next section.

4   Interval Finite Element Methods

One of the main features of interval arithmetic is its capability of providing guaranteed results. However, it has the disadvantage of overestimation if variables have multiple occurrences in the same expression. For example, if x is an interval, evaluation of the function $f(x) = x - x$ using naive rules of interval arithmetic will not return zero but rather an interval that contains zero. Such situations lead to extremely pessimistic results, and have discouraged some researchers of pursuing further developments using interval representations.

The Finite Element Method (FEM) is a numerical method for solving differential and partial differential equations with enormous applications in different fields of the sciences and engineering. Interval Finite Element Methods (IFEM) have been developed for the last 15 years to handle the analysis of systems for uncertain parameters described as intervals. Since the early development of IFEM during the mid-1990s of the last century [23], [24], [25], [26], [27], and [28], researchers have focused, among other issues, on two major problems: the first is how to obtain solutions with reasonable bounds on the system response that make sense from a practical point of view, or in other words, with the least possible overestimation of their bounding intervals; the second is how to obtain reasonable bounds on the derived quantities that are functions of the system response.

The most successful approaches for overestimation reduction are those which relate the dependency of interval quantities to the physics of the problem being considered; details on these developments can be found in the works of the authors and their collaborators [24], [29], [30], and [31].

A brief description of IFEM formulation is presented below, details can be found in the authors' work [31]. The two major issues in this formulation are:

1. Reducing overestimation in the bounds on the system response due to the coupling and transformation in the conventional FEM formulation as well as due to the nature of used interval linear solvers (Muhanna and Mullen, 2001).
2. Obtaining the secondary (derived) variables such as forces, stresses, and strains of the conventional displacement FEM along with the primary variables (displacements) and with the same accuracy of the primary ones. Previous interval methods calculate secondary variable from interval solutions of displacement which result in a significant overestimation.

4.1   Discrete Structural Models

In steady-state analysis, the variational formulation for a discrete structural model within the context of the Finite Element Method (FEM) is given in the following form of the total potential energy functional [32] and [33].

$$\Pi = \frac{1}{2}U^T K U - U^T P \tag{3}$$

with the conditions

$$\frac{\partial \Pi}{\partial U_i} = 0 \quad \text{for all } i \tag{4}$$

where Π, K, U, and P are total potential energy, stiffness matrix, displacement vector, and load vector respectively. For structural problems the formulation will include both direct and indirect approaches. For the direct approach the strain ϵ is selected as a secondary variable of interest, where a constraint can be introduced as $C_2 U = \epsilon$. For the indirect approach constraints are introduced on displacements of the form $C_1 U = V$ in such a way that Lagrange multipliers will be equal to the internal forces. C_1 and C_2 are matrices of orders $m \times n$ and $k \times n$, respectively, where m is the number of displacements constraints, k is the number of strains, and n is the number of displacements degrees of freedom. We note that V is a constant and ϵ is a function of U. We amend the right-hand side of Eq. (3) to obtain

$$\Pi^* = \frac{1}{2} U^T K U - U^T P + \lambda_1^T (C_1 U - V) + \lambda_2^T (C_2 U - \epsilon) \tag{5}$$

where λ_1 and λ_2 are vectors of Lagrange multipliers with dimensions m and k, respectively. Invoking the stationarity of Π^*, that is $\delta \Pi^* = 0$, we obtain

$$\begin{pmatrix} K & C_1^T & C_2^T & 0 \\ C_1 & 0 & 0 & 0 \\ C_2 & 0 & 0 & -I \\ 0 & 0 & -I & 0 \end{pmatrix} \begin{pmatrix} U \\ \lambda_1 \\ \lambda_2 \\ \epsilon \end{pmatrix} = \begin{pmatrix} P \\ V \\ 0 \\ 0 \end{pmatrix} \tag{6}$$

The solution of Eq. (6) will provide the values of the dependent variable U and the derived ones λ_1, λ_2, and ϵ at the same time.

The present interval formulation, which will be introduced in the next section, is an extension of the Element-By-Element (EBE) finite element technique developed in the work of the authors [29].

The main sources of overestimation in the formulation of IFEM are the multiple occurrences of the same interval variable (*dependency problem*), the width of interval quantities, the problem size, and the problem complexity, in addition to the nature of the used interval solver of the interval linear system of equations. While the present formulation is valid for the FEM models in solid and structural mechanics problems, the truss model will be used here to illustrate the applicability and efficiency of the present formulation without any loss of generality.

The current formulation modifies the displacement constraints used in the previous EBE formulation to yield the element forces as Lagrange Multipliers directly and the system strains. *All interval quantities will be denoted by non-italic boldface font.* Following the procedures given in [31] we obtain the interval linear system

$$\begin{pmatrix} \mathbf{K} & C_1^T & C_2^T & 0 \\ C_1 & 0 & 0 & 0 \\ C_2 & 0 & 0 & -I \\ 0 & 0 & -I & 0 \end{pmatrix} \begin{pmatrix} \mathbf{U} \\ \boldsymbol{\lambda}_1 \\ \boldsymbol{\lambda}_2 \\ \boldsymbol{\epsilon} \end{pmatrix} = \begin{pmatrix} \mathbf{P} \\ 0 \\ 0 \\ 0 \end{pmatrix} \tag{7}$$

where $\mathbf{K}$ is an interval matrix of dimension (dof $\times$ dof), where dof is the sum of element degrees of freedom and the free node degrees of freedom. It consists of the individual elements' local stiffness and zeros at the bottom corresponding to the free nodes' degrees of freedom.

The accuracy of the system solution depends mainly on the structure of Eq. (7) and on the nature of the used solver. The solution of the interval system (7) provides the enclosures of the values of dependent variables which are the interval displacements $\mathbf{U}$, interval element forces $\boldsymbol{\lambda}_1$, the multiplier $\boldsymbol{\lambda}_2$, and the element's interval strains $\boldsymbol{\epsilon}$. An iterative solver is discussed in the next section.

4.2 Iterative Enclosures

The best known method for obtaining very sharp enclosures of interval linear system of equations that have the structure introduced in Eq. (7) is the iterative method developed in the work of Neumaier and Pownuk [34]. The current formulation results in the interval linear system of equations given in (7) which can be transformed to have the following general form:

$$(K + BDA)\mathbf{u} = a + F\mathbf{b} \tag{8}$$

where $\mathbf{D}$ is diagonal. Furthermore, defining

$$C := (K + BD_0A)^{-1} \tag{9}$$

where D_0 is chosen to ensure invertablility (often D_0 is selected as the midpoint of $\mathbf{D}$), the solution $\mathbf{u}$ can be written as

$$\mathbf{u} = (Ca) + (CF)\mathbf{b} + (CB)\mathbf{d}. \tag{10}$$

To obtain a solution with tight interval enclosure we define two auxiliary interval quantities,

$$\mathbf{v} = A\mathbf{u} \tag{11}$$
$$\mathbf{d} = (D_0 - \mathbf{D})\mathbf{v}$$

which, given an initial estimate for $\mathbf{u}$, we iterate as follows:

$$\mathbf{v} = \{(ACa) + (ACF)\mathbf{b} + (ACB)\mathbf{d}\} \cap \mathbf{v}, \quad \mathbf{d} = \{(D_0 - \mathbf{D})\mathbf{v}\} \cap \mathbf{d} \tag{12}$$

until the enclosures converge, from which the desired solution $\mathbf{u}$ can be straightforwardly obtained.

This formulation allows obtaining the interval displacement $\mathbf{U}$ and the accompanied interval derived quantities $\boldsymbol{\lambda}_1$, $\boldsymbol{\lambda}_2$, and $\boldsymbol{\epsilon}$ with the same accuracy. A number of examples are introduced in the next section.

5 Examples

Fig. 3 shows a planar truss. The reliability of this structure using single deflection criteria has been presented by Zhang, et al. [15]. Two limit states are considered in this work, the serviceability and the strength. The deflection limit at the mid-span is set to 2.4 cm, and the allowable stress in any member is set 200

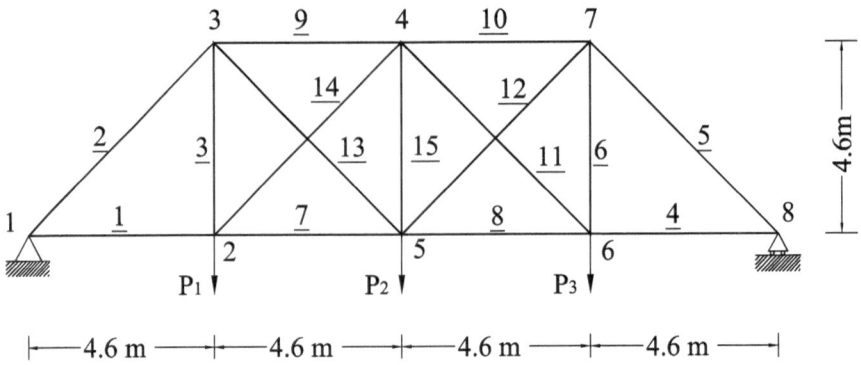

Fig. 3. Truss structure

Table 1. Sample statistics for the basic random variables (truss in Fig. 3)

Variables	Sample mean	Sample standard deviation	No. of samples
A1 – A6 (cm^2)	10.32	0.516	30
A7 – A15 (cm^2)	6.45	0.323	30
Ln P$_1$	3.2122	0.071474	20
Ln P$_2$	3.9982	0.071474	20
Ln P$_3$	3.2122	0.071474	20

MPa. Interval linear elastic analyses are performed. The element stress is calculated using conventional interval methods (stresses are calculated from interval displacements) as well as the improved algorithm outlined in this paper. The cross-sectional areas for the 15 members and the three loads are identified as the basic random variables. All the 18 random variables are assumed to be mutually statistically independent. Assume that based on experience, the cross-sectional areas can be modeled by normal distributions, and the loads modeled by lognormal distributions. Suppose the statistics for the random variables were estimated from limited samples of data. Table 1 gives the available sample statistics for the cross-sectional areas and the logarithm of the loads (Ln P). The Youngs modulus is assumed deterministic (200 GPa).

From the sample size, one can calculate confidence bounds on the mean of the random variables. We will use these bounds to construct the bounding functions defining the probability bounds for the random variables. Two cases are considered (1) the uncertain means for the (logarithm of) loads only and (2) uncertain cross-sectional area as well as loading. The interval Monte Carlo method is used to obtain probability bounds information on the variables associated with the limit states as well as interval estimates for the failure probability. All results are calculated using 10,000 realizations.

Fig. 4 presents the probability bounds for the central deflection of the truss subject to uncertain loading. Using a limit state for deflection of 2.4 cm, the

Table 2. Probability bounds for the basic random variables (truss in Fig. 3)

Variables	Population means 90% confidence interval bounds	Population standard deviation
A1 − A6 (cm^2)	[10.160, 10.480]	0.516
A7 − A15 (cm^2)	[6.3498, 6.5502]	0.323
Ln P$_1$	[3.1846, 3.2398]	0.071474
Ln P$_2$	[3.9706, 4.0259]	0.071474
Ln P$_3$	[3.1846, 3.2398]	0.071474

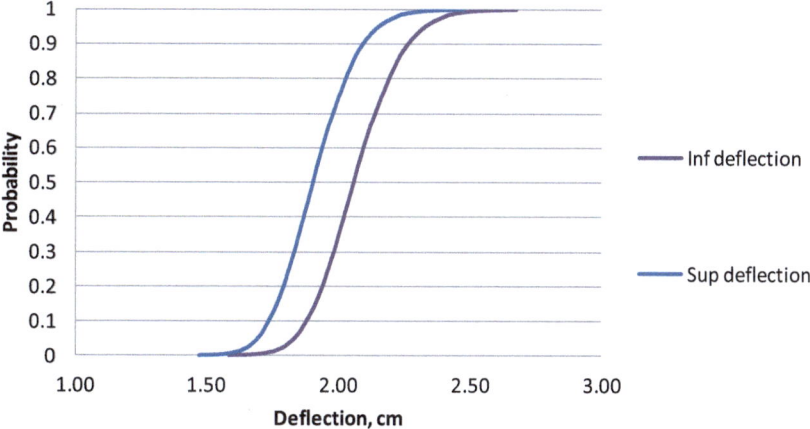

Fig. 4. Probability bounds for central deflection, uncertain load

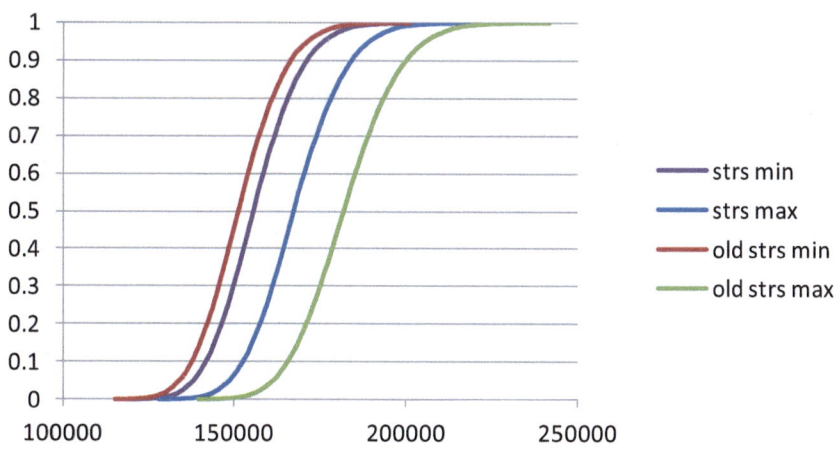

Fig. 5. Probability bounds of maximum absolute stress in the structure, uncertain load

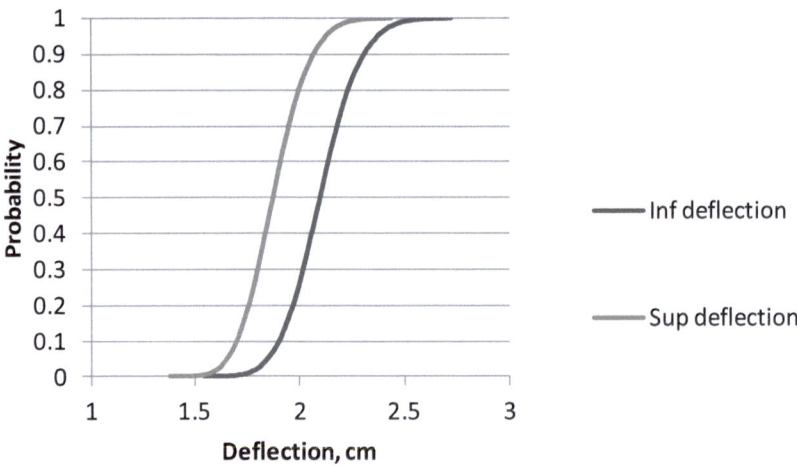

Fig. 6. Probability bounds of central deflection for uncertain loads and element cross-sectional areas

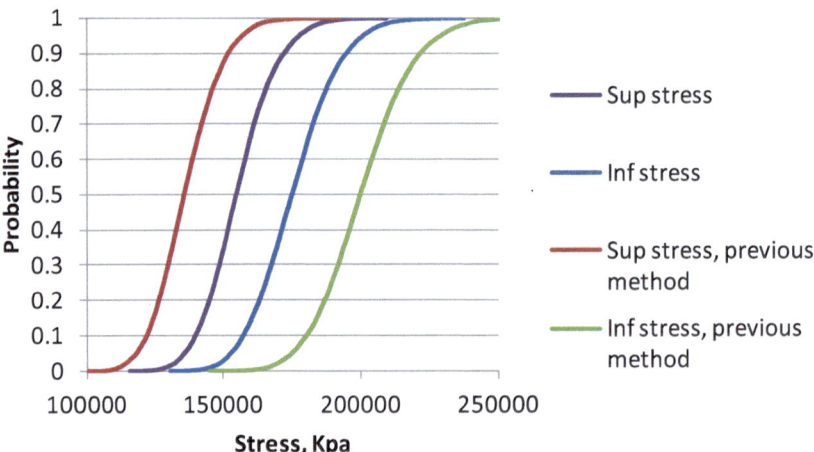

Fig. 7. Probability bounds of maximum absolute stress in the structure, uncertain load and element cross-sectional area

bounds on the probability of survival is given by the intersection of a vertical line at 2.4 cm and the probability bounds. The failure probability bounds are then calculated from the survival bounds and are $[1.34, 1.93]\%$. Fig. 5 presents the probability bounds results for the maximum absolute value of the stress the structure. Two different bounds are presented; the wider bounds are calculated using standard interval methods while the narrower bounds result from the current method. The calculated failure probabilities for the strength-based limit state are: $[0.01, 0.76]\%$ using the current method and $[< 0.01, 9.95]\%$ using the previous method. While both results bound the failure probability, the previous method significantly overestimates the bounding values.

Fig. 6 presents the probability bounds for the central deflection of the truss subject to uncertain load and element cross-sectional areas. Using the same 2.4 cm limit state, the bounds on the probability of failure increase to $[0.11, 5.39]\%$. It should be expected that increasing the uncertainty associated with the analysis will increase the width of the bounds on failure probability. Fig. 7 presents the probability bounds results for the maximum absolute value of the stress in the structure. Again, two different bounds are presented: the new method of this work and previous interval method. Using the same limit state, the failure probabilities bound are: $[0.04, 3.71]\%$ using the new method and $[< 0.01, 48.88]\%$ using the previous method. The overestimation using previous method of calculations of element stress clearly renders the previous method useless in engineering design.

6   Conclusion

An interval Monte Carlo finite element method with improved calculation of the bounds of secondary quantities is presented. Using element stress as an example, the improved sharpness of the bounds is illustrated. This work resolves the issue of engineering design with limit states calculated from secondary quantities (stress) that existed in previous interval Monte Carlo finite element analyses.

References

1. Melchers, R.E.: Structural Reliability Analysis and Prediction, 2nd edn. John Wiley & Sons, West Sussex (1999)
2. Ferson, S., Ginzburg, L.R.: Different Methods Are Needed to Propagate Ignorance and Variability. Reliab. Engng. Syst. Saf. 54, 133–144 (1996)
3. Sentz, K., Ferson, S.: Combination of Evidence in Dempster-Shafer Theory. Technical Report SAND2002-0835, Sandia National Laboratories (2002)
4. Walley, P.: Statistical Reasoning with Imprecise Probabilities. Chapman and Hall, London (1991)
5. Zadeh, L.A.: Fuzzy Sets as a Basis for a Theory of Possibility. Fuzzy Sets and Systems 1, 3–28 (1978)
6. Dempster, A.P.: Upper and Lower Probabilities Induced by a Multi-Valued Mapping. Ann. Mat. Stat. 38, 325–339 (1967)
7. Shafer, G.: A Mathematical Theory of Evidence. Princeton University Press, Princeton (1976)
8. Kendall, D.G.: Foundations of a Theory of Random Sets. In: Harding, E., Kendall, D. (eds.) Stochastic Geometry, New York, pp. 322–376 (1974)
9. Berleant, D.: Automatically Verified Reasoning with Both Intervals and Probability Density Functions. Interval Computations 2, 48–70 (1993)
10. Ferson, S., Kreinovich, V., Ginzburg, L., Myers, D.S., Sentz, K.: Constructing Probability Boxes and Dempster-Shafer Structures. Technical Report SAND2002-4015, Sandia National Laboratories (2003)

11. Ben-Haim, Y., Elishakoff, I.: Convex Models of Uncertainty in Applied Mechanics. Elsevier Science, Amsterdam (1990)
12. Moore, R.E.: Interval Analysis. Prentice Hall, Englewood Cliffs (1966)
13. Lodwick, W.A., Jamison, K.D.: Special Issue: Interface between Fuzzy Set Theory and Interval Analysis. Fuzzy Sets and Systems 135, 1–3 (2002)
14. Muhanna, R.L., Mullen, R.L.: Development of Interval Based Methods for Fuzziness in Continuum Mechanics. In: Proc. ISUMA-NAFIPS 1995, pp. 23-45 (1995)
15. Zhang, H., Mullen, R.L., Muhanna, R.L.: Interval Monte Carlo Methods for Structural Reliability. Structural Safety 32, 183–190 (2010)
16. Yager, R.R.: Arithmetic and Other Operations on Dempster-Shafer Structures. International Journal of Man-Machine Studies 25, 357–366 (1986)
17. Williamson, R., Downs, T.: Probabilistic Arithmetic I: Numerical Methods for Calculating Convolutions and Dependency Bounds. International Journal of Approximate Reasoning 4, 89–158 (1990)
18. Berleant, D., Goodman-Strauss, C.: Bounding the Results of Arithmetic Operations on Random Variables of Unknown Dependency Using Intervals. Reliable Computing 4(2), 147–165 (1998)
19. Tonon, F., Bernardini, A., Mammino, A.: Reliability Analysis of Rock Mass Response by Means of Random Set Theory. Reliab. Engng. Syst. Saf. 70, 263–282 (2000)
20. Tonon, F.: A Search Algorithm for Calculating Validated Reliability Bounds. In: Proceedings of NSF Workshop on Reliable Engineering Computing (2004), http://savannah.gatech.edu/workshop/rec04/
21. Joslyn, C., Kreinovich, V.: Convergence Properties of an Interval Probabilistic Approach to System Reliability Estimation. Report No. LA-UR-02-6261, Los Alamos National Laboratory, Los Alamos, NM (2002)
22. Ferson, S., Ginzburg, L.: Hybrid Arithmetic. In: IEEE Proceedings of ISUMA – NAFIPS 1995 The Third International Symposium on Uncertainty Modeling and Analysis and Annual Conference of the North American Fuzzy Information Processing Society, pp. 619–623 (1995)
23. Koyluoglu, U., Cakmak, S., Ahmet, N., Soren, R.K.: Interval Algebra to Deal with Pattern Loading and Structural Uncertainty. Journal of Engineering Mechanics 121(11), 1149–1157 (1995)
24. Muhanna, R.L., Mullen, R.L.: Development of Interval Based Methods for Fuzziness in Continuum Mechanics. In: IEEE Proceedings of ISUMA – NAFIPS 1995 The Third International Symposium on Uncertainty Modeling and Analysis and Annual Conference of the North American Fuzzy Information Processing Society, pp. 145–150 (1995)
25. Rao, S.S., Sawyer, P.: Fuzzy Finite Element Approach for Analysis of Imprecisely Defined Systems. American Institute of Aeronautics and Astronautics Journal 33(12), 2364–2370 (1995)
26. Nakagiri, S., Yoshikawa, N.: Finite Element Interval Estimation by Convex Model. In: Proceedings of 7th ASCE EMD/STD Joint Specialty Conference on Probabilistic Mechanics and Structural Reliability, pp. 278–281. WPI, Worcester (1996)
27. Rao, S.S., Berke, L.: Analysis of Uncertain Structural Systems using Interval Analysis. American Institute of Aeronautics and Astronautics Journal 35(4), 727–735 (1997)
28. Rao, S.S., Chen, L.: Numerical Solution of Fuzzy Linear Equations in Engineering Analysis. International Journal of Numerical Methods in Engineering 43, 391–408 (1998)

29. Muhanna, R.L., Mullen, R.L.: Uncertainty in Mechanics Problems–Interval-based Approach. Journal of Engineering Mechanics 127(6), 557–566 (2001)
30. Zhang, H.: Nondeterministic Linear Static Finite Element Analysis: An Interval Approach. PhD Dissertation, School of Civil and Environmental Engineering, Georgia Institute of Technology, Atlanta, GA (2005)
31. Rao, M.V.R., Mullen, R.L., Muhanna, R.L.: A New Interval Finite Element Formulation with the Same Accuracy in Primary and Derived Variables. Int. J. Reliability and Safety 5(3/4) (2011)
32. Bathe, K.: Finite Element Procedures. Prentice Hall, Englewood Cliffs (1996)
33. Gallagher, R.H.: Finite Element Analysis Fundamentals. Prentice Hall, Englewood Cliffs (1975)
34. Neumaier, A., Pownuk, A.: Linear Systems with Large Uncertainties, with Applications to Truss Structures. Reliable Computing 13(2), 149–172 (2007)

Discussion

Speaker: Rafi Muhanna

Richard Hanson: Does the use of interval arithmetic in finte element methods scale to the size of the structure show in your slide (large!)?

Rafi Muhanna: Yes. We have tested that and the results are reported in following paper: Muhanna, R., Mullen, R., and Zhang, H., "Interval Finite Element as a Basis for Generalized Models of Uncertainty in Engineering Mechanics," Reliable Computing Journal, Springer Netherlands, Vol. 13, No. 2, pp. 173-194, April 2007, and it has been found that the computer time scaled cubically to the number of interval variables which is similar to the conventional finite elements.

William Kahan: This talk illustrates that the use of interval arithmetic to get bounds not grossly excessive requires close analysis and exploitation of a problem's properties. Here the ideas that work on static analysis by finite elements of elastic structures also works on elliptic partial differential equations. But the methods presented here do not always succeed on dynamical systems, nor on solutions of nonlinear systems of equations with interval coefficients. The right person to ask about interval arithmetic's successes is Professor Ulrich Kulisch in attendence here.

Ulrich Kulisch: Yes, for dynamical systems double precision interval and floating-point arithmetic are not very successful. Here long interval arithmetic is the appropriate tool. For the logistic equation $x_{n+1} := 3.75x_n(1 - x_n), n \geq 0$, and the initial value $x_0 = 0.5$ double precision floating-point and interval arithmetic totally fail (no correct digit) after 30 iterations. Long interval arithmetic still computes correct digits of a guaranteed enclosure after 2790 iterations.

Rafi Muhanna: Note also that nonlinear structural problems have been successfuly solved and reported in: ICASP'11: Applications of Statistics and Probability in Civil Engineering, Faber, Köhler and Nishijima (eds.), Taylor & Francis Group, London (2011), with the title: "Interval finite elements for nonlinear material problems." Linear dynamic problems are addressed in: Modares, M., Mullen, R., L. and Muhanna, R. L. "Natural Frequency of a Structure with Bounded Uncertainty," Journal of Engineering Mechanics, ASCE, Vol. 132, No. 12, pp. 1363 1371, 2006.

Ronald Boisvert: Have you written your own interval arithmetic software infrastructure, or are you using tools developed elsewhere?

Rafi Muhanna: The results in this presentation are calculated using MATLAB with the INTLAB toolbox.

Jeffrey Fong: Log normal or normal distributions are not suitable for the modeling of ultimate tensile strength data because they do not predict a minimum strength. A 3-parameter Weibull distribution with a positive location parameter is a better model for predicting a minimum strength. Do you have experience using Weibull in interval finite elements?

Rafi Muhanna: In this work we did not model the material strength. The authors have studied probability bounds using a 3-parameter Weibull distribution. The results are not yet published.

Reducing the Uncertainty When Approximating the Solution of ODEs

Wayne H. Enright*

Department of Computer Science
University of Toronto
Canada

Abstract. One can reduce the uncertainty in the quality of an approximate solution of an ordinary differential equation (ODE) by implementing methods which have a more rigorous error control strategy and which deliver an approximate solution that is much more likely to satisfy the expectations of the user. We have developed such a class of ODE methods as well as a collection of software tools that will deliver a piecewise polynomial as the approximate solution and facilitate the investigation of various aspects of the problem that are often of as much interest as the approximate solution itself. We will introduce measures that can be used to quantify the reliability of an approximate solution and discuss how one can implement methods that, at some extra cost, can produce very reliable approximate solutions and therefore significantly reduce the uncertainty in the computed results.

Keywords: Numerical methods, initial value problems, ODEs, reliable methods, defect control.

1   Introduction

In the numerical solution of ODEs, it is now possible to develop efficient techniques that compute approximate solutions that are more convenient to interpret and understand when used by practitioners who are interested in accurate and reliable simulations of their mathematical models. When implementing numerical methods for ODEs, there is inevitably a trade-off between efficiency and reliability that must be considered and most methods that are widely used are designed to provide reliable results most of the time. The methods we develop in this paper are designed so that the resulting piecewise polynomial will satisfy a perturbed ODE with an associated defect (or residual) that is *reliably* controlled. We also show how these methods can be the basis for implementing effective tools for visualizing an approximate solution, and for performing key tasks such as sensitivity analysis, global error estimation and parameter fitting. Software implementing this approach will be described for systems of IVPs, BVPs, DDEs, and VIEs.

---

* This research was supported by the Natural Science and Engineering Research Council of Canada.

A. Dienstfrey and R.F. Boisvert (Eds.): WoCoUQ 2011, IFIP AICT 377, pp. 280–293, 2012.
© IFIP International Federation for Information Processing 2012

Numerical results will be presented which quantify the improvement in reliability that can be expected with the methods we have developed. We will also show an example of the use of a related software tool for estimation of the underlying mathematical conditioning of a problem and the global error of the approximate solution.

Consider an IVP defined by the system

$$y' = f(x, y), \quad y(a) = y_0, \quad \text{on } [a, b]. \tag{1}$$

When approximating the solution of this problem, a numerical method will introduce a partitioning $a = x_0 < x_1 < \cdots < x_N = b$ and determine corresponding discrete approximations $y_0, y_1 \cdots y_N$ where $y_i \approx y(x_i)$. The number of and the distribution of the meshpoints, x_i, are determined adaptively as the method attempts to satisfy an accuracy that is consistent with an accuracy parameter, TOL, that is specified as part of the numerical problem associated with (1).

For many applications it is now recognized that an accurate discrete approximation is not enough and most numerical methods now provide an accurate approximation to the solution of (1) that can be evaluated at any value of $x \in [a, b]$. For a discussion of how this is done and how such methods are used see [10], [3] and [7]. In particular Figures 1 and 2 show the advantage such a method has when it is used to display (or visualize) the solution of an IVP. [Note that the particular problem visualized here will be defined and investigated in more detail in section 3.2.1.] These methods are often called continuous methods (in contrast to the more traditional discrete methods discussed above). [This name can be confusing as the approximate solution provided by a continuous method may not produce an approximate solution $S(x)$ that is in $C^0[a, b]$.] In this investigation we will consider a class of numerical methods which produce a computeable approximation $S(x) \approx y(x)$ for any $x \in [a, b]$ and where the reliability and accuracy of such methods will be quantified in terms of how accurately and reliably $S(x)$ agrees with $y(x)$.

In the next section we will introduce and justify a class of continuous explicit Runge-Kutta methods (SDC-CRKs) that have a rigorously justified error control strategy and are designed to be very reliable when applied to non-stiff IVPs. We will introduce suitable measures that can be used to quantify the reliability of the performance of a CRK method when applied to a particular problem. We will then use these measures to assess the performance of three methods we have implemented (of orders five, six and eight) on a standard collection of 25 non-stiff test problems.

In the third section we will discuss how the approach we have introduced for IVPs has been used to develop reliable CRK-based methods for boundary value problems (BVPs), delay differential equations (DDEs) and Volterra integro-differential equations (VIDES). We also discuss how these CRK method can be used to develop effective software tools to investigate important properties of the problem and/or its approximate solution when the problem belongs to one of these classes. As an example we will show how this approach can be used to develop an effective technique to estimate the mathematical conditioning of

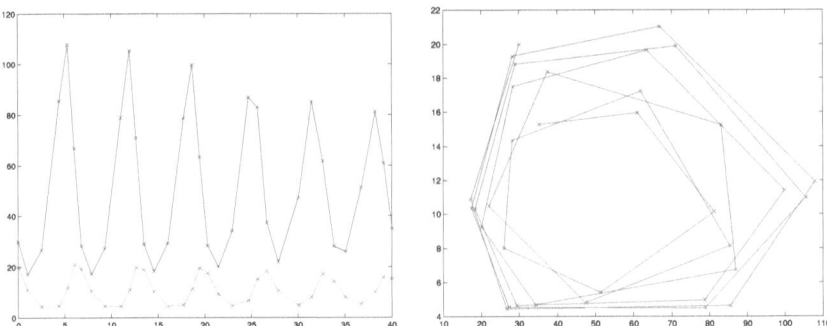

Fig. 1. Visualizing the approximate solution using an accurate discrete approximation. A standard solution plot of each component is displayed on the left while a phase plot of $y_1(t)$ vs $y_2(t)$ is displayed on the right

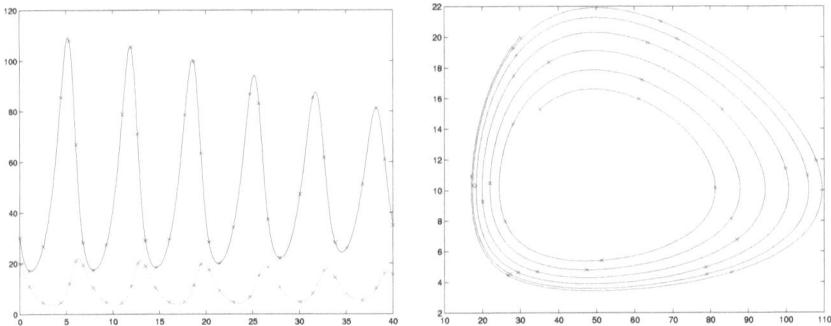

Fig. 2. Visualizing the approximate solution using an accurate continuous approximation. A standard solution plot of each component is displayed on the left while a phase plot of $y_1(t)$ vs $y_2(t)$ is displayed on the right

an IVP as well as an estimate of the global error of an approximate solution. This technique will be illustrated by applying it to two problems.

In the final section we will make some general observations and discuss some ongoing and future work that extends the techniques discussed in this paper to other classes of problems.

2   Continuous Runge-Kutta Methods

A classical, explicit, p^{th}-order, s-stage, discrete Runge-Kutta formula is defined by the vectors $(c_1, c_2, \ldots c_s), (w_1, w_2, \ldots w_s)$ and the lower triangular matrix $(a_{i,j}), i = 1, 2 \ldots s, j = 1, 2 \ldots i - 1$. When approximating the solution of (1), after $y_0, y_1, \ldots y_{i-1}$ have been generated, the formula determines,

$$y_i = y_{i-1} + h_i \sum_{j=1}^{s} w_j k_j,$$

where $h_i = x_i - x_{i-1}$ and the jth stage is defined by,

$$k_j = f(x_{i-1} + h_i c_j, y_i + h_i \sum_{r=1}^{j-1} a_{jr} k_r).$$

Let $z_i(x)$ be the solution of the local IVP associated with the ith step,

$$z_i' = f(x, z_i(x)), \quad z_i(x_{i-1}) = y_{i-1}, \quad \text{for } x \in [x_{i-1}, x_i].$$

A Continuous extension (CRK) of this discrete RK formula is determined by adding $(\bar{s} - s)$ additional stages on step i to obtain an order p approximation for $x \in (x_{i-1}, x_i)$

$$u_i(x) = y_{i-1} + h_i \sum_{j=1}^{\bar{s}} b_j \left(\frac{x - x_{i-1}}{h_i} \right) k_j,$$

where $b_j(\tau)$ is a polynomial of degree at least p and $\tau = \frac{x - x_{i-1}}{h_i}$.

The set of polynomials, $[u_i(x)]_{i=1}^{N}$, define a piecewise polynomial $U(x)$ for $x \in [a, b]$. We consider $U(x)$ to be the numerical solution generated by the CRK method. The particular class of $O(h^p)$ extensions considered here, were introduced in [4]. They satisfy,

$$u_i(x) = y_{i-1} + h_i \sum_{j=1}^{\bar{s}} b_j(\tau) k_j = z_i(x) + O(h_i^{p+1}).$$

$U(x) \in C^0[a, b]$ and will interpolate the underlying discrete RK values, y_i, if $b_j(1) = w_j$ for $j = 1, 2 \cdots s$ and $b_{s+1}(1) = b_{s+2}(1) = \cdots b_{\bar{s}}(1) = 0$. If $k_1 = f(x_{i-1}, y_{i-1})$ and $k_{s+1} = f(x_i, y_i)$, a similar set of constraints on the $\frac{d}{d\tau}(b_j(\tau))$ will ensure $U'(x)$ interpolates $f(x_i, y_i), f(x_{i-1}, y_{i-1})$ and therefore $U(x) \in C^1[a, b]$. All the CRK extensions we consider in this investigation are in $C^1[a, b]$.

2.1 Defect Error Control for CRK Methods

When applied to (1) a CRK method will determine an approximate solution, $U(x)$. This approximate solution has a defect (or residual) defined by,

$$\delta(x) = f(x, U(x)) - U'(x). \tag{2}$$

It can be shown (see [1] for details) that, for such a CRK and $x \in (x_{i-1}, x_i)$),

$$\delta(x) = G(\tau) h_i^p + O(h_i^{p+1}),$$

$$G(\tau) = \tilde{q}_1(\tau) F_1 + \tilde{q}_2(\tau) F_2 + \cdots + \tilde{q}_k(\tau) F_k, \tag{3}$$

Table 1. Cost per step of the explicit SDC CRK formulas we have implemented

Formula p	s	$\bar{s}$
SDC5 5	6	12
SDC6 6	7	15
SDC8 8	13	27

where $k \geq 1$ depends on the particular CRK formula and the $\tilde{q}_j's$ are polynomials in τ that depend only on the coefficients defining the CRK formula, while the $F_j's$ are constants (elementary differentials) that depend only on the problem.

CRK Methods can be implemented to adjust h_i in an attempt to ensure that the maximum magnitude of $\delta(x)$ is bounded by TOL on each step (see [11] and [2] for details). The quality of an approximate solution can then be described in terms of the maximum value of $\|\delta(x)\|/TOL$. From (3) it is clear that, as $h_i \to 0$, the defect will look like a linear combination of the $\tilde{q}_j(\tau)$ over $[x_{i-1}, x_i]$. Then the maximum defect will be easier to estimate if $k = 1$, in which case the maximum should occur (as $h_i \to 0$) at $\tau = \tau^*$ where τ^*, is the location in $[0, 1]$ of the local maximum of $\tilde{q}_1(\tau)$. In this case we call the defect control strategy **Strict Defect Control (SDC)** and CRK methods that implement this strategy are called SDC CRK methods. Figure 3 shows how the defect of an SDC method has a consistent shape when applied to a typical non-stiff IVP. We will consider only SDC extensions, $u_i(x)$,

$$SDC: \quad u_i(x) = y_{i-1} + h_i \sum_{j-1}^{\bar{s}} b_j(\tau)k_j = z_i(x) + O(h_i^{p+1}).$$

In the next section we will discuss how, for a given discrete RK formula, we can identify a suitable continuous extension. SDC methods SDC5, SDC6 and SDC8 have been implemented at a cost per step that is given in table 1. We will report on how well these methods are able to provide reliable and consistent control of the size of the defect on non-stiff problems over a range of prescribed values of TOL.

2.2   Optimal SDC Extensions of a Discrete RK Formula

For a particular discrete explicit RK formula, we generally have a family of possible continuous extensions and we are interested in a continuous extension with the lowest cost per step (the smallest value of $\bar{s}$).

In selecting an optimal continuous extension, one should also attempt to avoid potential difficulties which can arise. Each SDC extension satisfies,

$$\delta(x) = \tilde{q}_1(\tau)F_1 h_i^p + (\hat{q}_1(\tau)\hat{F}_1 + \hat{q}_2(\tau)\hat{F}_2 + \cdots\cdots\hat{q}_{\hat{k}}(\tau)\hat{F}_{\hat{k}})h_i^{p+1} + O(h_i^{p+2})$$

and a particular extension might be inappropriate for two reasons,

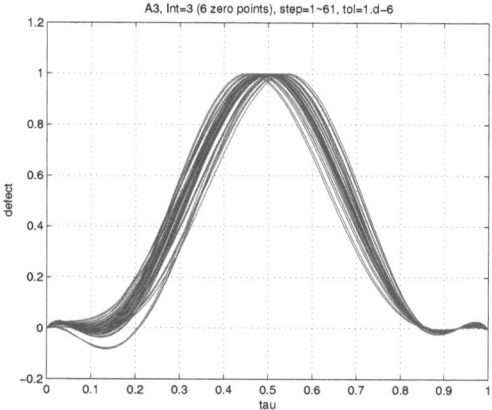

Fig. 3. Plot of scaled defect vs τ (ie. $\delta(\tau)/\delta(\tau^*)$ vs τ) for each step required to solve a typical problem with SDC CRK6 and $TOL = 10^{-6}$. Note that all components of the defect have a similar "shape" on each problem.

- $\tilde{q}_1(\tau)$ may have a large maximum (It is straightforward to show that, for the SDC extensions we are considering, $\tilde{q}_1(0) = \tilde{q}_1(1) = 0$ and its "average' value must be one, for $\tau \in (0,1)$).
- The $\hat{q}_j(\tau)$ may be large in magnitude relative to $\tilde{q}_1(\tau)$ (and therefore h_i would have to be small before the estimate is justified). (That is, before $|h_i\hat{q}_j(\tau)| << |\tilde{q}_1(\tau)|$.)

For each p we have identified a particular SDC-CRK that minimizes these difficulties and uses the fewest number of additional stages, $\bar{s}$. Note that if $|F_1|$ is zero or very small on isolated steps then the associated error control may still be unreliable. Figure 4 shows plots of the polynomials $\tilde{q}_1(\tau)$ and $\hat{q}_j(\tau), \ldots \hat{q}_{\hat{k}}(\tau)$ for the particular order 6 SDC extension we have chosen to implement.

2.3   Quantifying Reliability of a SDC Method

Consider two measures of reliability of a CRK method:

- How well does the **Method** control the maximum magnitude of the defect? We can measure the ratio of the max defect to TOL on each step (DMAX) and the fraction of steps where this ratio is greater than 1 (Frac-D).
- How well does the **Estimate** of the max defect reflect its true value? We can measure both the ratio of the true maximum defect (on a successful step) to its estimated value (R-Max) and the fraction of attempted steps where the estimated maximum is within <u>one percent</u> of the true maximum (Frac-G).

We will use these measures of reliability to demonstrate that SDC error control significantly reduces the uncertainty of approximate solutions to ODE problems.

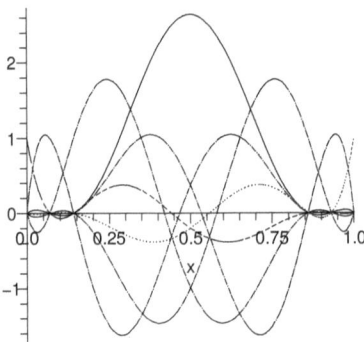

Fig. 4. Plots of $\tilde{q}_1$ and $\hat{q}_2 \cdots \hat{q}_7$ for SDC CRK6. $\tilde{q}_1$ is represented by the solid line and has the highest magnitude

We have implemented SDC RK methods of orders five, six and eight (SDC5, SDC6 and SDC8) and have run each of these methods on the 25 IVP test problems of DETEST [6] (all non-stiff), at 9 tolerances from 10^{-1} to 10^{-9}. The performance of the methods on the 25 test problems on a subset of the tolerances is summarized in Table 2, where we report the above reliability measures, the total number of steps (NSTP) and the total number of function evaluations (NFCN) for all the problems.

Table 2. Numerical Results for SDC CRKs on the 25 problems of DETEST

TOL	CRK	NSTP	NFCN	DMAX	Frac-D	R-Max	Frac-G
	SDC5	625	11709	0.97	.000	1.05	.67
10^{-2}	SDC6	549	12300	1.00	.000	1.43	.71
	SDC8	333	12793	1.01	.003	1.65	.35
	SDC5	1065	19033	1.01	.001	1.12	.78
10^{-4}	SDC6	931	19819	1.00	.001	1.08	.87
	SDC8	465	17319	1.05	.004	1.47	.45
	SDC5	2099	35703	1.01	.002	1.08	.86
10^{-6}	SDC6	1748	35073	1.01	.001	1.08	.96
	SDC8	712	26253	1.02	.001	1.34	.59
	SDC5	4566	66937	1.01	.001	1.07	.95
10^{-8}	SDC6	3547	65148	1.01	.001	1.07	.98
	SDC8	1081	38251	1.12	.007	2.60	.62

3   SDC RK Methods for Other Classes of ODEs

In addition to reliable methods for IVPs, we have developed (or are actively developing) effective and very reliable SDC methods for other important classes of differential equations. These include,

– BVPs ([5]):
$$y' = f(x, y), \quad x \in [a, b],$$

with

$$g(y(a), y(b)) = 0, \quad g : \Re^n \times \Re^n \to \Re^n.$$

– DDEs (both retarded and neutral problems) ([12]):
$$y' = f(x, y(x), y(x - \sigma_1) \cdots y(x - \sigma_k), y'(x - \sigma_{k+1}),$$
$$\cdots y'(x - \sigma_{k+\ell})), \quad \text{for } x \in [a, b],$$

where $y(x) \in \Re^n$ and,
$$y(x) = \phi(x), \quad y'(x) = \phi'(x), \quad \text{for } x \leq a,$$

$$\sigma_i \equiv \sigma_i(x, y(x)) \geq 0 \quad \text{for } i = 1, 2 \cdots k + \ell.$$

– VIDEs (with a time dependent delay) ([9]):

$$y'(x) = f(x, y(x)) + \int_{x-\sigma(x)}^{x} K(x, s, y(s), y'(s)) ds, \tag{4}$$

for $x \in [a, b]$, $f : \Re \times \Re^n \to \Re^n$ and $K : \Re \times \Re \times \Re^n \times \Re^n \to \Re^n$ and $y(x) = \phi(x)$ for $x \leq a$.

For each Class of ODEs we are not only interested in providing effective SDC methods to approximate the solution of the ODE, but we are also developing effective software tools to investigating important properties of the problem and its approximate solution. For example:

– Detecting, Locating and Coping with <u>Discontinuous</u> Problems
– Estimating the Global Error and the Mathematical Conditioning of the Problem
– Computing a sensitivity analysis of the solution (eg., $\frac{\partial y_i(x)}{\partial p_j}$).
– Solving Problems which depend on parameters and parameter determination.

3.1 Global Error Estimates and Condition Number of an IVP

Assume $y(x)$ satisfies (1) and the computed approximate solution $U(x)$ satisfies (from (2)) the perturbed IVP,

$$U' = f(x, U) - \delta(x), \quad U(x_0) = y_0 \quad \text{on } [a, b], \quad \text{with } \|\delta(x)\| \leq TOL.$$

Let $\epsilon(x) = y(x) - U(x)$. From the variation of constants formula, (see for example [8]), one can show,
$$\|\epsilon(x)\| \leq K(x) \, TOL,$$

where $K(x)$ reflects the sensitivity of $y(x)$ to perturbations. Then

$$\bar{K} \equiv \max_{x \in [a, b]} K(x),$$

can be viewed as the condition number of this IVP. From the definition of $\bar{K}$ we can determine a lower bound, $\hat{K}$,

$$\hat{K} \equiv \max_{x \in [a,b]} \|\epsilon(x)\|/TOL.$$

If we compute an accurate approximation $E(x)$, to $\epsilon(x)$, (and the inequality $\|\delta(x)\| \le TOL$ is almost sharp), then an effective estimate of the conditioning of the IVP is,

$$\tilde{K} \equiv \max_{x \in [a,b]} \|E(x)\|/TOL. \tag{5}$$

We know that, $\epsilon(x) = y(x) - U(x)$, is the exact solution of the IVP,

$$\begin{aligned} \epsilon' &= f(x,y) - f(x,U) - \delta(x), \\ &= f(x,U(x) + \epsilon(x)) - f(x,U) - \delta(x), \\ &= f(x,U(x) + \epsilon(x)) - U'(x), \\ &\equiv g(x,\epsilon). \end{aligned}$$

Therefore if we solve this 'companion' IVP using the same SDC method used to determine $U(x)$, we can determine an inexpensive estimate $E(x)$ of the global error and use this to obtain (from (5)) an estimate of the conditioning of the IVP. Note that this computed $E(x)$ will satisfy the IVP,

$$E' = g(x,E) + \delta_2(x), \quad \text{where } \|\delta_2(x)\| \le TOL_2.$$

We can also use this estimate of the global error to improve the accuracy of the numerical solution since $U_1(x) = U(x) + E(x)$ satisfies the perturbed IVP:

$$\begin{aligned} U_1'(x) &= U'(x) + E'(x), \\ &= f(x,U) + \delta(x) + g(x,E) + \delta_2(x), \\ &= f(x,U) + \delta(x) + f(x,U(x) + E(x)) - U'(x) + \delta_2(x), \\ &= f(x,U(x) + E(x)) + \delta_2(x), \\ &= f(x,U_1(x)) + \delta_2(x), \end{aligned}$$

where $\|\delta_2(x)\| \le TOL_2$ and TOL_2 can be determined by sampling $\|\delta_2(\tau^*)\|$ on each step.

3.2   Two Sample Problems

Predator – Prey Problem:.
This is a well known system that models (over time) the populations of two computing species in an isolated environment. It is a well conditioned problem.

$$\begin{aligned} y_1' &= y_1 - 0.1y_1y_2 + 0.02x, \\ y_2' &= -y_2 + 0.02y_1y_2 + 0.008x, \end{aligned}$$

with $y_1(0) = 30, \quad y_2(0) = 20, \quad \text{and } x \in [0,4].$

Lorenz Problem:

This is a standard example often cited in the literature on dynamical systems as a system which can exhibit chaotic behaviour. The condition number is exponential in the length of the integration interval.

$$y_1' = 10(y_2 - y_1),$$
$$y_2' = y_1(28 - y_3) - y_2,$$
$$y_3' = y_1 y_2 - \frac{8}{3} y_3,$$

with $y_1(0) = 15$, $y_2(0) = 15$, $y_3(0) = 36$, and $x \in [0, 15]$.

For each method we monitor performance on these problems over a range of tolerances and report, in Table 3 and Table 4, the following:

- NS – The number of steps to determine $U(x)$.
- NSE – The number of steps to determine $E(x)$.
- DEFUM – The maximum magnitude of the defect $\delta(x)$, (associated with $U(x)$), in units of TOL. This is determined by evaluating the defect at several sample points per step.
- G-ERRM – The maximum global error associated with $U(x)$ in units of TOL. This is determined by computing the true global error at 100 sample points per step.
- K-ESTM – The estimate of the conditioning corresponding to the maximum observed value of $\|E(x)\|/TOL$ measured over 100 sample values per step.
- DEFEM – The maximum magnitude of the defect $\delta_2(x)$, (associated with $E(x)$), in units of TOL.
- GE(U+E) – The The maximum global error associated with the improved solution $U(x) + E(x)$ in units of TOL.

4   Observations and Future Work

The results presented in Table 2 demonstrate the strong reliability of the SDC IVP methods we have implemented. In particular these tables show, that over a wide range of non-stiff problems and accuracy requests, the computed approximate solution will almost always satisfy a perturbed ODE with the norm of the perturbation bounded by the requested accuracy parameter, TOL. Furthermore, our analysis in the previous section shows that, as a result of this strong reliability property, the maximum global error will be proportional to the tolerance and the proportionality constant will be insensitive to the order of the SDC method. The results reported in Table 3 and Table 4 confirm that this is true for our two test problems. This allows us to implement and justify a rigorous and inexpensive measure of the underlying mathematical conditioning. For example, in the case of the Lorenz problem, which is known to be badly conditioned, Table 4 shows, that in order to compute an approximate solution with an accuracy of two significant figures, one must specify a value for TOL that is less than 10^{-7}.

Table 3. Reliability of Error Control and Validity of the Estimate of Conditioning for SDC on the pred-prey problem

Method	TOL :	10^{-2}	10^{-4}	10^{-6}	10^{-8}
SDC5:	NS	70	148	315	705
	NSE	147	307	644	1412
	DEFUM	1.8	1.1	1.2	1.2
	G-ERRM	3.7	7.3	11.4	14.4
	K-ESTM	3.7	7.3	11.4	14.6
	DEFEM	.009	.009	.011	.034
	GE(U+E)	.002	.009	.004	.041
SDC6:	NS	65	134	277	585
	NSE	132	265	551	1168
	DEFUM	1.3	1.0	1.0	1.2
	G-ERRM	2.2	4.6	2.5	3.5
	K-ESTM	2.2	4.6	2.5	3.6
	DEFEM	.009	.005	.007	.013
	GE(U+E)	.0006	.001	.001	.008
SDC8:	NS	34	53	83	127
	NSE	65	104	177	262
	DEFUM	1.3	1.1	0.9	2.1
	G-ERRM	9.5	6.1	6.1	14.4
	K-ESTM	9.5	6.1	6.1	13.9
	DEFEM	.012	.010	.018	1.9
	GE(U+E)	.0009	.002	.003	2.0

It must be acknowledged that the analysis and methods developed in this paper apply to the usual case where truncation error of the RK formulas dominates the affects of rounding errors when approximating the solution of an ODE. If one is interested in satisfying severe accuracy requirements and using a high order SDC method then round-off error can become significant and reduce the reliability of the computed results. In such cases, an SDC method can (at a small amount of extra work) detect that the defect estimates are adversely affected by round-off error (see [4] for details) and signal that this is the case. The remedy, in this case, would be to use higher precision (if it is available) or to use a lower order SDC method which is not as sensitive to round-off errors.

The SDC methods investigated in this paper are suitable for non-stiff problems. We are currently implementing and testing continuous extensions of implicit RK methods that could be suitable for stiff problems. The derivation of these extensions is straightforward, but the development of an effective adaptive stepsize control strategy for stiff problems remains a challenge. We are considering some alternative techniques related to defect control for use on these problems. We are also considering how to best develop accurate continuous extensions and reliable defect control for multistep methods.

Table 4. Reliability of Error Control and Validity of the Estimate of Conditioning for SDC on the Lorenz problem

Method	TOL :	10^{-2}	10^{-4}	10^{-6}	10^{-8}
SDC5:	NS	356	751	1738	4304
	NSE	834	1591	3470	4306
	DEFUM	1.3	1.4	1.4	1.4
	G-ERRM	$4.4 \cdot 10^3$	$1.9 \cdot 10^5$	$1.9 \cdot 10^5$	$1.8 \cdot 10^6$
	K-ESTM	$4.6 \cdot 10^3$	$1.9 \cdot 10^5$	$1.9 \cdot 10^5$	$1.9 \cdot 10^6$
	DEFEM	.016	.018	.020	.14
	GE(U+E)	$.47 \cdot 10^3$	$.50 \cdot 10^2$	$.17 \cdot 10^3$	$.68 \cdot 10^4$
SDC6:	NS	316	642	1339	2865
	NSE	731	1326	2678	2865
	DEFUM	1.4	1.3	1.3	1.2
	G-ERRM	$4.2 \cdot 10^3$	$2.8 \cdot 10^5$	$1.5 \cdot 10^5$	$1.5 \cdot 10^5$
	K-ESTM	$4.2 \cdot 10^3$	$2.8 \cdot 10^5$	$1.5 \cdot 10^5$	$1.3 \cdot 10^5$
	DEFEM	.011	.011	.004	.20
	GE(U+E)	$.29 \cdot 10^3$	$.31 \cdot 10^3$	$.32 \cdot 10^3$	$.20 \cdot 10^5$
SDC8:	NS	145	228	371	634
	NSE	292	454	803	1349
	DEFUM	1.3	1.4	1.6	1.4
	G-ERRM	$5.5 \cdot 10^3$	$.16 \cdot 10^5$	$.14 \cdot 10^5$	$.20 \cdot 10^5$
	K-ESTM	$5.5 \cdot 10^3$	$.16 \cdot 10^5$	$.15 \cdot 10^5$	$.70 \cdot 10^5$
	DEFEM	.013	.003	.076	9.0
	GE(U+E)	$.70 \cdot 10^2$	$.82 \cdot 10^1$	$.42 \cdot 10^3$	$.48 \cdot 10^5$

References

1. Enright, W.: A new error-control for initial value solvers. App. Math. Comp. 31, 288–301 (1989)
2. Enright, W.: The relative efficiency of alternative defect control schemes for high-order continuous Runge-Kutta formulas. SIAM Journal on Numerical Analysis 30(5), 1419–1445 (1993)
3. Enright, W., Jackson, W., Nørsett, S., Thomsen, P.: Interpolants for Runge-Kutta formulas. ACM Transactions on Mathematical Software 12(3), 193–218 (1986)
4. Enright, W., Yan, L.: The Reliability/Cost trade-off for a class of ODE solvers. Numerical Algorithms 53(2), 239–260 (2009)
5. Enright, W., Muir, P.: New Interpolants for Asymptotically Correct Defect Control of BVODEs. Numerical Algorithms 53(2), 219–238 (2009)
6. Enright, W., Pryce, J.: Two FORTRAN packages for assessing initial value methods. ACM Transactions on Mathematical Software 13(1), 1–27 (1987)
7. Gladwell, I., Shampine, L., Baca, L., Brankin, R.: Practical aspects of interpolation in Runge-Kutta codes. SIAM Journal of Scientific and Statistical Computing (8), 322–341 (1987)
8. Hairer, E., Nørsett, S., Wanner, G.: Solving Ordinary Differential Equations I: Nonstiff Problems. Springer, Berlin (1987)
9. Shakourifar, M., Enright, W.: Reliable Approximate Solution of Systems of Volterra Integro-Differential Equations with Time Dependent Delays. SIAM Journal of Sc. Comp. 33, 1134–1158 (2011)

10. Shampine, L.: Interpolation for Runge-Kutta methods. SIAM Journal of Numererical Analysis (22), 1014–1027 (1985)
11. Shampine, L.: Solving ODEs and DDEs with residual control. Applied Numererical Mathematics (52), 113–127 (2005)
12. Zivaripiran, H., Enright, W.: An Efficient Unified Approach for the Numerical Solution of Delay Differential Equations. Numerical Algorithms 53(2), 397–417 (2009)

Discussion

Speaker: Wayne Enright

Bill Oberkampf: Is the advantage of continuous Runge-Kutta methods over traditional Runge-Kutta methods that you can relax the assumption on the solution from C^1 to C^0?

Wayne Enright: No. If the solution is not not differentiable at $\bar{x} \in [a, b]$ then, for any numerical method to be effective, it must locate all such points and force these points to be meshpoints. This can be done automatically by CRK method which detect such points by observing sudden increases in the magnitude of the defect. The main advantage of CRK methods is that they provide accurate approximations to the solution for any value of $x \in [a, b]$, (not just at the meshpoints, x_i).

The term continuous Runge Kutta method can be misleading. It would perhaps be better to refer to this class of Runge Kutta methods as continuous-output Runge Kutta methods (CORK), or dense-output Runge Kutta methods (DORK).

Van Snyder: Can the ideas underlying continuous Runge-Kutta methods be applied to Adams method?

Wayne Enright: This is an extension that we have thought about for some time. The main difficulty in extending the approach is that, for the most natural piecewise polynomial approximations, the associated defect would depend on past stepsizes as well as on the current stepsizes. This would make it particularly challenging to define local interpolants that permit an asymptotically correct estimate of the maximum defect.

Uncertainties in Predictions of Material Performance Using Experimental Data That Is Only Distantly Related to the System of Interest

Wayne E. King[1], Athanasios Arsenlis[1], Charles Tong[2],
and William L. Oberkampf[3]

[1] Physical and Life Sciences Directorate
[2] Computations Directorate
Lawrence Livermore National Laboratory
Livermore, CA, USA
{weking,arsenlis,tong10}@llnl.gov
[3] Consulting Engineer
Georgetown, TX, USA
wloconsulting@gmail.com

Abstract. There is a need for predictive material "aging" models in the nuclear energy industry, where applications include life extension of existing reactors, the development of high burnup fuels, and dry cask storage of used nuclear fuel. These problems require extrapolating from the validation domain, where there is available experimental data, to the application domain, where there is little or no experimental data. The need for predictive material aging models will drive the need for associated assessments of the uncertainties in the predictions. Methods to quantify uncertainties in model predictions, using experimental data that is only distantly related to the application domain, are discussed in this paper.

Keywords: uncertainty quantification, model form uncertainty, model uncertainty, nuclear energy, neutron damage, ion damage, irradiation effects scaling, extrapolation.

1 Introduction

There is a growing need to make predictions of material performance in extreme environments and over very long periods of time where there is little or no experimental data. This is particularly the case in the prediction of the effects of "aging" on material performance where desired material lifetimes can exceed by a large margin what is practical for validation under normal application conditions. In large, complex engineering systems, the costs are often too high and/or the times too long to carry out desired validation experiments under actual operating conditions. In the case of aging, the required extrapolations can be orders of magnitude beyond the validation domain.

A. Dienstfrey and R.F. Boisvert (Eds.): WoCoUQ 2011, IFIP AICT 377, pp. 294–311, 2012.

Further, model development can require reliance on accelerated experiments. Experiments can be accelerated by changing temperature to take advantage of Arrhenius behavior or the rate of application of the experimental forcing function can be increased. Results from accelerated experiments, which are also outside of the application domain, require extrapolation, perhaps over orders of magnitude in rate, to the actual operating conditions, by way of a model.

The need for predictive models is particularly acute in the nuclear energy industry, where applications include (i) the desire for life extension of existing reactors to 80 years, (ii) the development of high burnup fuels, and (iii) the imperative for dry cask storage possibly to hundreds of years.

The nuclear materials community has adopted an experimental strategy involving experiments that are not conducted in commercial power plants. The nuclear industry and regulators require a "sound and defensible case for the relevance of these techniques to actual service conditions" [1]. This implies the ability (i) to make predictions across diverse irradiation energy spectra, irradiation rates, and irradiating particles including thermal neutrons, fast neutrons, and energetic ions, (ii) to scale the prediction to the relevant reactor conditions, and (iii) then to extrapolate to the application domain of interest.

Predictions are most useful in the presence of quantified uncertainties. In an engineered system, high uncertainty can lead to excessive conservatism and thus necessarily increased margins, which can adversely affect cost, schedule, and system performance. Thus, in addition to the predictive model, estimates of uncertainties in predictions are also required. Uncertainty quantification (UQ) provides a framework within which uncertainties for predictions can be estimated. The case that we are interested in here is uncertainties due to prediction in domains where there is little or no experimental data. This is generally referred to as model extrapolation.

Extrapolations using physics-based models differ from extrapolations using regression curve fits of the system responses quantities of interest. When making an assessment of uncertainties, there are several sources that must be considered, including model inputs and numerical solution approximation. When making an extrapolation using a model, the assumptions associated with the mathematical model itself result in a source of uncertainty usually referred to as model form uncertainty[1], and it must also be considered [2,3]. The inclusion of model form uncertainty represents a specialized field within UQ [3]. Approaches to dealing with uncertainties in model inputs are well established and can be implemented. The approach to uncertainties associated with model form is less well established, particularly for cases where models are assessed in a validation domain that does not fully overlap or overlap at all with the intended application domain.

In this paper, we discuss a predictive modeling problem in the nuclear energy area that requires large extrapolation. We also review possible approaches to the extrapolation problem. Throughout, we focus on the special case of predictions of models validated with experimental data that are only distantly related to the system of interest.

---

[1] Model form uncertainty is referred to as model uncertainty, model bias, or structural uncertainty.

2   Background

2.1   Accelerated Experimentation for Nuclear Energy Materials Applications [4]

The nuclear industry needs models that predict the time dependence of microstructural and fission product evolution in structural materials and fuels. The most challenging extreme environment to study is that of high irradiation dose. Models developed to address this extreme are difficult to validate because of the inability to reach these doses using existing neutron-irradiation facilities in reasonable amounts of time and at modest costs. Furthermore, reactor facilities are problematic experimental venues for combining the various aspects of the extreme environments into a quantitative in situ study of material behavior.

Understanding radiation damage using ion irradiation is not a new idea. It has a long history of significant contributions spanning several decades. In fact, much of our understanding of material behavior under irradiation comes from well-controlled ion-irradiation experiments.

However, a key challenge is the scaling, or extension, of ion irradiation experiments and data to actual in-service conditions. Fig. 1 illustrates the particular case of scaling and extrapolation for the damage rate parameter. Scaling refers to use of models to bridge two unconnected validation domains. Extrapolation refers to the use of models to project into an application domain where there is no experimental data. The plot shows the range of damage expected for advanced reactors, GEN IV reactors, GEN III reactors, and light water reactors (LWRs). The plot further shows the damage levels that could be obtained in a 5-year irradiation experiment in a number of test reactors and the damage levels that could be obtained by ion irradiation in 5 days [5]. Consequently, a scientifically defensible argument for the applicability of models developed using accelerated experiments to neutron irradiation environments is critically needed. This should include rate scaling, effects of recoil energy spectra, and the ability to extrapolate to dose regimes not explored by neutrons. This irradiation-effects scaling is identified as a priority research direction in the Science for Energy Technology Workshop Report [1]. By definition, models developed for materials under neutron irradiation conditions at the extreme of high irradiation dose cannot be validated because little or no neutron irradiation data exists in that domain. Consequently, UQ will be particularly important in dose regimes that have not been explored using neutrons.

2.2   Other Relevant Science and Technology Application Areas

"Aging" of materials for nuclear energy applications is not the only area requiring scaling across and extrapolation outside of the validation domains. Fusion energy applications have a similar problem. In fusion machines, materials that can withstand very high levels of radiation damage are required. Without a fusion-relevant neutron source, the fusion materials community has adopted a research strategy similar to the fission community.

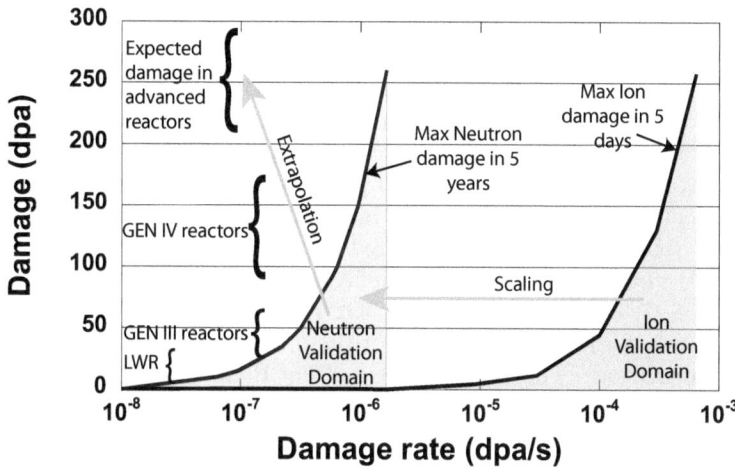

Fig. 1. Plot of radiation damage, measured in displacements per atom (dpa), as a function of the rate at which the damage is produced [5]

Uncertainty in the calculation of the depletion of nuclear fuels is composed of a number of individual components. Some of these include uncertainty in the cross section at a given neutron energy, uncertainty in core composition and other externally driven parameters such as power level and temperatures, and model approximations made to accommodate computer modeling capabilities.

These uncertainties will affect the prediction of the evolution of the fuel isotopic components in time. The isotopic component distribution at any given time represents the integration of the depletion conditions and uncertainties over all previous time. This effect can compound the effects of uncertainty or approximation. This compounding effect may also be limited by compensating effects. A common figure of merit for depletion is the energy extraction per mass of fuel, which is quantified as giga-watt-days per metric tonne of initial heavy metal (GWD/t). Present commercial fuel is depleted to about 60 GWD/t and the effects of uncertainties are well benchmarked for this range of operation. Advanced fuels may go beyond today's 60 GWd/t burnup. To operate outside the range of experience will require identifying the compounding effects for a given fuel type at a specific higher burnup so that they can be applied to the known uncertainties derived from inside the benchmarked region.

The effects of aging of materials are also of interest in both intermediate and long-term storage of nuclear waste. Analyses of the possible behavior of radioactive waste in a repository at Yucca Mountain (YM), Nevada, were conducted between 1982 and 2008. Early analyses (termed performance assessments or PAs) were for selecting the site and determining the feasibility of the disposal concept. Numerous parameter values for the numerical models were required and were mostly assigned by individual analysts and scientists for the early analyses. However, in 1987, Congress asked the YM Project (YMP) to evaluate the viability of a repository at YM [6]. For this viability assessment (PA-VA), the

YMP formed five panels to examine: (1) groundwater flow in the unsaturated zone, (2) groundwater flow and radionuclide transport in the saturated zone, (3) the near-field effects of heat on the region around the engineered barrier system, (4) waste form degradation, and (5) waste package (WP) degradation. These five panels assigned parameter values by aggregating disparate data available in the literature, prior to completion of project experiments, and estimated the uncertainty present as literature-based information was often for conditions and spatial and temporal scales that differed from those required for the PA-VA. The analysis underlying the license application for the Yucca Mountain repository in 2008 [7] considered a total of 392 uncertain analysis inputs (see Ref. [8], Table K3-3, for a complete listing of these inputs and additional sources of detailed information). The Waste Isolation Pilot Plant (WIPP), a repository for transuranic radioactive waste in southern New Mexico, also had similar needs [9,10,11].

Another closely related application is the Qualification Alternatives to the Sandia Pulsed Reactor (QASPR) project at Sandia National Laboratories [12,13]. In this case, pulsed ion beams are being used to understand radiation effects in semiconducting materials. The challenge is to extrapolate those results to the relevant pulsed neutron environment in the absence of a relevant neutron source.

There is a related field known as accelerated testing. Accelerated tests are used to obtain timely information on product-life or performance degradation over time [14]. Ref. [15] provides a comprehensive discussion of useful models and statistical methods for accelerated testing. Ideally, predictions from accelerated tests should be based on models of physics of failure. In practice, however, users of accelerated tests often use a combination of past experience and empirical fitting of data to statistical models. Although these procedures seem to have been adequate in the past, it is generally recognized that the path forward for large extrapolations must be based on physics-based modeling.

2.3   The Problem Recast in More General Terms

One of the goals of UQ is to provide a means to evaluate a model's predictive capability. Fig. 2(a) schematically illustrates the synergistic use of modeling, experiments, and UQ to make a prediction [3]. In this case, the system response quantity of interest is shown as a function of two system or environmental parameters. The application domain is highlighted in light brown, is the range of parameters #1 and #2 that are of interest in the application. The validation domain, highlighted in burgundy, is the range of parameters #1 and #2 where validation experimentation are carried out. The response surface is also shown with the application domain highlighted. In this example, the validation domain fully contains the application domain and predictions could be obtained through various types of interpolation over the validation domain. Likewise, uncertainties in predictions could be quantified by interpolation of uncertainties over the validation domain, or by direct calculation using the input uncertainties in the model.

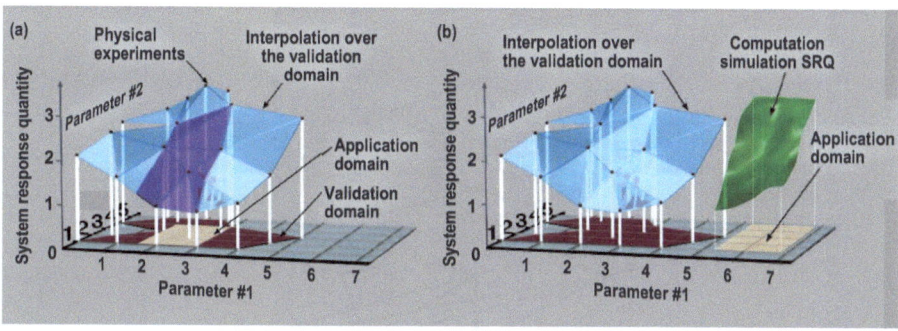

Fig. 2. Possible relations of the validation domain to the application domain. (a) Complete overlap and (b) no overlap [3]

One of the main reasons why we rely on modeling and simulation is to make predictions in domains where there is little or no experimental or observational data. Fig. 2(b) shows the situation where the application domain is no longer within the validation domain and there is no overlap of the application domain with the validation domain. While there is a clear path to quantify uncertainties in a prediction when the application domain is inside the validation domain, the approach to quantifying uncertainties for the case shown in Fig. 2(b) is less clear.

Although Fig. 2 is illustrative, it fails to capture the complexity inherent in today's multiphysics simulations. Another view of the extrapolation issue is given in Fig. 3, which illustrates the validation hierarchy for a complex engineering system model [16]. At each level of the hierarchy, the problem is broken down into smaller and smaller pieces until it is reduced to the unit problem (or unit mechanism) level. This is a more physics-based, or system-based, perspective than in Fig. 2. Each box at the unit problem level contains individual physics models that can be validated using targeted experiments that may not be in the same conditions as would apply to the next higher level in the hierarchy. At each higher level, the individual effects are brought together forming coupled physics and coupled subsystems and systems interactions.

In the case of small extrapolations, we would expect this physics-based approach to enable extrapolation of uncertainties, including model form uncertainty, to the application domain. However, for large extrapolations, it is possible that new physics could appear at higher levels, or that unexpected coupling could emerge at the higher levels. These possibilities are present in the nuclear materials aging problem, where (i) scaling is required across diverse irradiation energy spectra, irradiation rates, and irradiating particles and (ii) large extrapolation is required to the application domain. Extrapolation of uncertainty includes both (a) extrapolation to the application domains where there is little or no experimental data, Fig. 2, and (b) effects present in higher levels of system complexity in the validation hierarchy, Fig. 3, that are not anticipated in the modeling [17].

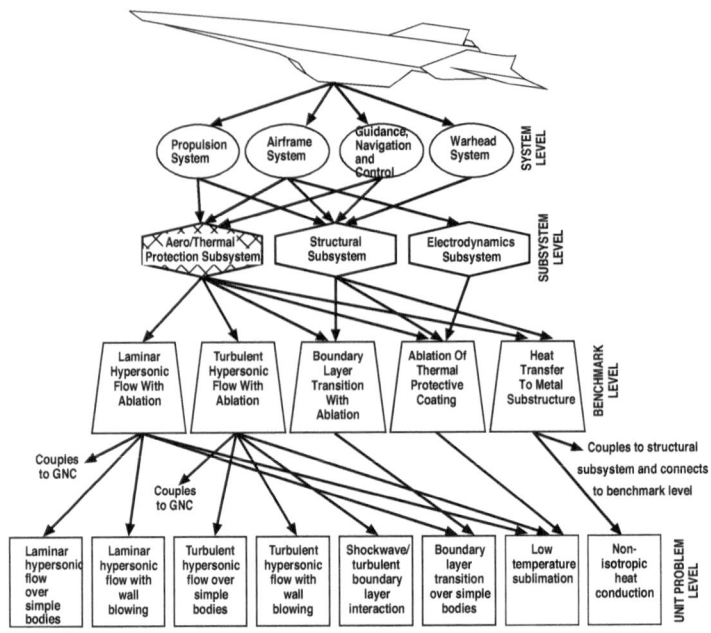

Fig. 3. A hierarchical validation structure for the hypersonic cruise missile (from [3])

2.4 Prediction-Coupling of Accelerated Experiments with Physics Models [4]

It takes a long time to develop a new material or investigate the properties of materials that undergo low-dose-rate irradiation, such as the pressure vessel or core internals including the fuel and cladding. Experiments using neutron irradiation can take up to 7 years, including the irradiation time, the radioactive cool-down time, and the post-irradiation examination. Incorporating the effects of high temperature, stresses and a corrosive environment along with irradiation make the problem multidimensional and extremely complicated. Translating that into a program to satisfy a regulatory requirement for a new material or new fuel design can lead to a multi-decadal process. Such a timescale is unacceptable to efficient progress, and yet, it is the present-day norm.

One pathway to accelerate this process is to carry out accelerated experiments either inside the reactor core (in the case of the low-dose-rate regimes for the pressure vessel) or using external radiation sources such as ion beams (in the case of core internals that would see high neutron doses over their lifetimes). Using ion beams, one can investigate a large parameter space (in terms of external forcing functions) for irradiation effects on microstructure and macroscale properties. The phase space includes temperature, ion type, ion dose rate, ion energy, and total dose. It also includes the ability to apply in situ mechanical loading, chemical environments, and coolant fluids.

In some special cases, it is possible to create microstructures and material properties that are very similar to those that would be found in a particular nuclear reactor irradiation experiment. However, microstructures, mechanical properties, or other physical properties that deviate from neutron irradiations significantly, will also be observed. The challenge is to employ all of those observations to develop a science-based understanding of material degradation and performance, specifically, to establish a scientific basis for the key mechanisms of material performance.

The best approach to quantifying such scientific understanding is to build a model that captures all of the relevant physics, which is where modeling and simulation come into play. One seeks to understand the ion-beam forcing function and the material response to that forcing function. With that understanding, if a different boundary condition was applied (in terms of say temperature, ion type, or dose rate), it is reasonable to expect to be able to predict the material response. That is, one could have sufficient confidence that, with this robust model, interpolation and reproduction of an experimental result is possible.

The question then becomes: How does one extrapolate, given a model, to high-dose neutron-irradiation environments? Compared with ion irradiations, neutron dpa rates are much, much lower ($\sim 10^2 - 10^3$ lower). In addition, physical mechanisms, such as transmutation and chemistry changes, occur simultaneously with the neutron bombardment and displacements in a material. With a robust model, the boundary conditions can be altered, while not perturbing any of the model internals, and an extrapolation can be made to project material performance to the neutron-irradiation environment. To make that extrapolation, researchers must also quantify the quality or accuracy of the extrapolation. This forward extrapolation and the qualification of the quality of the predicted extrapolation is where uncertainty quantification becomes important.

3   Approaches to Extrapolation of Uncertainty

3.1   Calibrate Model Parameters over the Validation Domain and Ignore Model Form Uncertainty

This approach uses the physics model as it is (assuming no model form uncertainties) together with available experimental data for validation and calibration (given a number of model input parameters). The posterior distributions of the calibration parameters are then used in forward uncertainty propagation (through the computational model) to predict the extrapolated configuration and to estimate the corresponding uncertainties. To speed up calibration, response surface or surrogate modeling using Gaussian process, polynomial regression or polynomial chaos is often used. This approach has the inherent assumption that the computational model has captured all essential physics (except that there are uncertainties about some physics parameters that can be estimated using data) and the surrogate models are adequate for the extrapolated regime.

A simple example of this method is shown in Fig. 4, where a prediction and an associated uncertainty are required for the time required for an object to drop

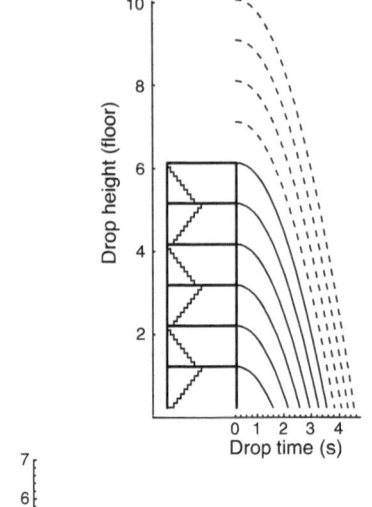

Experimental data:
The time it takes an object to drop from each of 6 floors of a tower is recorded. There is an uncertainty in the measured drop times of about ±0.2 seconds. Predictions for times are desired for drops from floors 7 through 10, which do not exist yet.

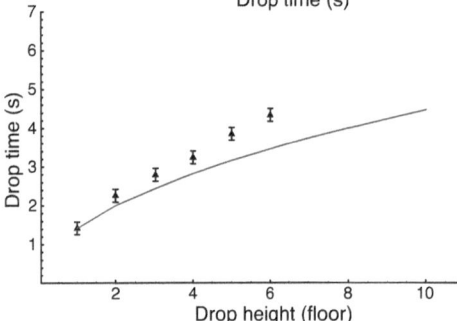

Simulated drop times:
A (trivial) computational model is developed to predict the drop times as a function of height. The simulated drop times (red line) are systematically too low when compared with the experimental data (symbols). the error bars around the observed drop times show the observation uncertainty.

Fig. 4. A simple UQ example using experimental data and a computational model to predict drop times from new heights [18]

from yet unconstructed floors of a building [18]. Measurements are made of the time to drop the object from the first six floors and a model is constructed. Using the available experimental data from the drops, a Bayesian inference methodology is used to update/calibrate the uncertain input parameters in the model. Once the model is calibrated it is then used for prediction outside the calibration range. This is the most commonly used form of extrapolation: the model form uncertainty is assumed to be zero.

Continuing with the example in Fig. 4, when comparing the model against experimental data, systematic errors are observed which cannot be resolved satisfactorily by calibration. Model bias (or discrepancy) and its dependence on the drop height are evident. As shown in Fig. 5, an extrapolation would have led to a predicted time and uncertainty that, in fact, would not have predicted what would have been observed had an experiment been carried out. This illustrates that in this case, as in many others, the model form uncertainty can dominate the extrapolated uncertainties. The experiments revealed the presence of physics that was not accounted for in the original model.

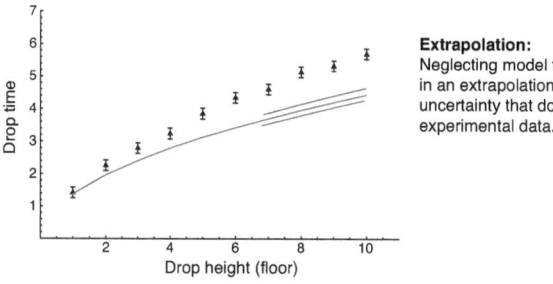

Fig. 5. The impact of ignoring model form uncertainty on the extrapolation and uncertainties [18]

3.2 Calibrate Parameters and Identify a Discrepancy Function to Characterize Model Inadequacy

This approach explicitly assumes a functional form for the discrepancy between the simulation and the actual physical process. The full method estimates from data simultaneously both the posteriors for the calibration parameters as well as the parameters in the discrepancy function. Some simplifications were suggested to reduce the complexity of the estimation process (see [19] for details). This approach assumes both the functional forms of the simulation model and the discrepancy stay the same in the extrapolated regime.

Kennedy and O'Hagan proposed a Bayesian approach that represents the model form uncertainty using a discrepancy function (in terms of some input parameters) to characterize model inadequacy [19]. This discrepancy function is created using experimental data as well as a selected regression or statistical emulator such as Gaussian process. Predictions are performed by incorporating the discrepancy function evaluated at the extrapolated points, in addition to the uncertainty due in posterior distributions of the calibrated parameters. This approach has the inherent assumption that the discrepancy function essentially captures the misrepresented physics.

In their approach, the system is modeled by

$$z = \zeta(\mathbf{x}) + e = \rho\eta(\mathbf{x}, \mathbf{\theta}) + \delta(\mathbf{x}) + e \tag{1}$$

where $\mathbf{x}$ is a vector of input parameters; $\mathbf{\theta}$ is a vector of calibration parameters; $\eta(\mathbf{x}, \mathbf{\theta})$ denotes the function for the simulation model; z is the observation; e is the observation error (independent normal distribution); $\zeta(\mathbf{x})$ is the true value of the process being modeled; ρ is an unknown regression parameter to be determined; and $\delta(\mathbf{x})$ is a function to describe model inadequacy, which is independent of $\eta(\mathbf{x}, \mathbf{\theta})$.

The full Bayesian calibration of this system is very complicated. Instead, Kennedy and O'Hagan proposed a multi-step approach:

– Build a Gaussian process model for $\eta(\mathbf{x}, \mathbf{\theta})$ based on sampling different values of $\mathbf{x}$ and $\mathbf{\theta}$ (that is, to estimate the "hyper-parameters" used to describe a Gaussian process model).

- Use data ($\{z\}$) to estimate the regression parameter ρ; the standard deviation of the observation error e; the hyper-parameters in the Gaussian process model of the model inadequacy function $\delta(\mathbf{x})$.
- Use the data ($\{z\}$) and the model ($z = \rho\eta(\mathbf{x}, \boldsymbol{\theta}) + \delta(\mathbf{x}) + e$) for calibration to get the posterior distribution of $\boldsymbol{\theta}$.
- Use the model ($z = \rho\eta(\mathbf{x}, \boldsymbol{\theta}) + \delta(\mathbf{x}) + e$) and the posteriors of $\boldsymbol{\theta}$ for prediction and uncertainty analysis.

Model inadequacy is defined as "the difference between the true mean value of the real world process and the code output at the true value of the inputs" [19]. There is debate across the field regarding the use of this definition of model inadequacy to describe the difference between a simulation and an experiment (see Sects. 2.2.3, 2.4, and Sects. 12.1-12.3 in [3].

This method is suitable if there is (a) sufficient experimental data from different input parameter configurations to characterize the discrepancy function, (b) reason to believe that the assumption concerning the discrepancy functional form will be valid in the extrapolation regime, and (c) that the discrepancy function is more significant than the observation error.

An alternative approach is where the candidate response surface methods for $\eta(\mathbf{x}, \boldsymbol{\theta})$ in Eq. (1) and the discrepancy functions are based on generalized polynomial chaos (or stochastic collocation) methods [20]. The advantages of polynomial chaos UQ methods are their efficiency and their utility for representing and propagating large uncertainties through complex models [21]. Both intrusive and non-intrusive applications of the polynomial chaos method are reviewed in [21].

This method is suitable if there is sufficient experimental evidence that the discrepancy function has polynomial form in the input parameters, and that this form will also be valid in the extrapolation regime. Moreover, another requirement is that the discrepancy function be more significant than the observation error.

3.3   Validation Metric Approach [2,3]

Oberkampf and Roy argue that model form uncertainty should be estimated as part of the process of model validation. They estimate the model form uncertainty in the validation domain using a validation metric which they define as "a mathematical operator that requires two inputs: the experimental measurements of the system response quantities of interest and the prediction of the system response quantities at the conditions used in the experimental measurements" [2]. Oberkampf and Roy in Sects. 13.2, 13.4, and 13.5 of [3] and in [22], describe two validation metrics: the confidence interval approach and the method of comparing cumulative distribution functions (CDFs) from the model and the experiment. In the confidence interval approach, they define the validation metric for model form uncertainty as the difference between the mean of model prediction and the estimated mean of the experimental data. In the CDF method, the validation metric is defined as the area between the experimental and simulation CDFs.

Once a validation metric is estimated over the validation domain, the critical issue is how this error structure should be extrapolated to the application conditions of interest. One simple method for extrapolation is to construct a regression fit of the error structure over the validation domain using a low degree polynomial function [2,3,22]. The regression function is then evaluated at the application conditions, along with the statistical prediction interval at those conditions. The estimate of the model form uncertainty is increased by the prediction interval not only because of the imprecision of the regression function to fully represent the model form uncertainty, but also because of the random measurement uncertainty that is present in each experimental measurement. A level of statistical confidence is chosen for the prediction interval, say 90 or 95%, and the upper bound on the prediction interval is used as the estimate of the model form uncertainty at the application conditions of interest.

This estimated model form uncertainty is considered as an epistemic uncertainty, i.e., an uncertainty whose source is lack of knowledge as opposed to randomness, for the prediction of the system response quantities of interest at the application conditions. It has been found [2,3,22] that even if the model form uncertainty is relatively small over the validation domain, but the magnitude of the extrapolation is large in the multi-dimensional input space over which data are available, the estimate of model form uncertainty is typically quite large at the application conditions of interest. The model form uncertainty is clearly represented to the user of the simulation results, e.g., a designer or decision maker, as a probability-box, or p-box. The p-box is an interval-valued CDF, where the range of possible probabilities of the system response quantity reflects the epistemic uncertainty due to the model form.

3.4   Method of Alternate Plausible Models [3]

An approach for assessing uncertainty, both model form uncertainty and parametric uncertainty, is to compare predictions from alternative plausible models. This method is also referred to as the method of competing models. While simple in concept, it is not commonly used because of the time and expense of developing multiple models for a system. Examples of applications of this method include hurricane forecasting (Fig. 6) (e.g., see [23,24]), climate prediction, and long-term storage of nuclear waste.

The approach requires multiple models developed by independent groups. This approach does not actually provide an estimate of model form uncertainty; it only provides an indication of the similar or dissimilar nature of each model prediction. Because of the cost and time involved, this approach will likely be limited to application in matters of very high priority.

Since in many applications simulating the full physics is very consuming in terms of time and computational resources, a viable approach is to use simplified models for some of the physics components in the system (for example, see [25,26]). For example, in computational fluid dynamics, a popular simplification is the use of the Reynolds-averaged Navier Stokes (RANS) model in place of the more complex large eddy simulation (LES) or even direct numerical simulation

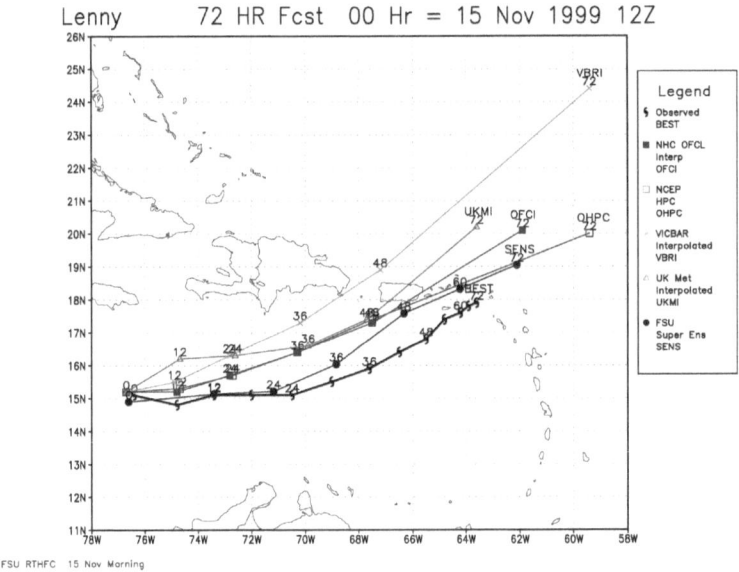

Fig. 6. Superensemble track forecast of Hurricane Lenny with predicted tracks of some member models and associated superensemble track shown [23]

(DNS). These simplified models are often benchmarked or calibrated against the more complex counterparts. Outside the benchmark/calibration regime, the predictions from these two models of different physics fidelity can be compared. As a result, the model form uncertainty in the lower fidelity model could be estimated by comparison of the predictions with the higher fidelity model predictions at a limited number of conditions that are similar to the application conditions of interest. If it is concluded that the lower fidelity model accuracy is judged to be inadequate for the application of interest, then one may (a) increase the modeling fidelity of the lower fidelity model so as to attain the needed accuracy, or (b) characterize the model form uncertainty in some appropriate way so that the predictive uncertainty is recognized by the user of the simulation results.

4   Discussion

4.1   The Role of Model Form Uncertainty

For the case of extrapolations to application domains outside of the validation domain or to a higher level of system model, it is likely that model form uncertainty will dominate the extrapolated uncertainty. Extrapolating model form uncertainty is complex because it is extrapolating the error structure of a model, combined with the uncertainty in the experimental data, in a high dimensional space.

Uncertainty can be reduced, compared with for example, extrapolating a regression fit of the measured system response quantities, by taking advantage of

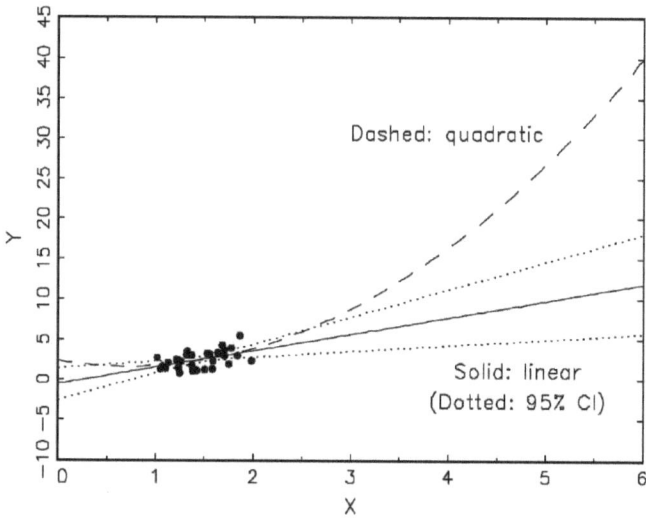

Fig. 7. Model dependence of the extrapolation with equivalent goodness of fit in the validation domain [27]

the physics incorporated into the model [3]. Take for example Fig. 7, where there is no physical basis to the quadratic or linear models used to describe the data, and then extrapolate to large values of x. The fitted models over the range of the data are indistinguishable but in the range of extrapolation, the quadratic falls far outside of the confidence interval (CI) of the linear fit. A physics based model would help constrain the extrapolated uncertainty.

4.2   Use of UQ to Manage Extrapolation Uncertainties

The goal of a simulation is to produce a prediction along with an estimate of the effect of all of the relevant uncertainties on the system response quantities of interest. In addition to estimating the uncertainty of an extrapolation, UQ can serve to reduce that uncertainty through methodologies and mathematical methods for [28]:

- tuning (or calibrating) a simulation model to match with experimental results,
- establishing the integrity of (i.e., validate) a simulation model,
- assessing the region of validity of a simulation model,
- characterizing the output uncertainties of a simulation model,
- identifying the major sources of uncertainties of a model,
- providing information on which additional experiments are needed to improve the understanding of a model.

These methodologies can be used to understand how the model performs over the validation domain and improve it, if necessary. As in Fig. 5, these improvements serve to improve the quality of the extrapolation and reduce the uncertainty in the extrapolation. While there might be significant uncertainty on the magnitude of extrapolated uncertainties, the fact that the uncertainty has been reduced and by what fraction is useful in itself.

4.3   Missing Physics (Unknown-Unknowns)

In distant extrapolations, there is always the potential for missing physics to be present either as new unit mechanisms or as new coupling of mechanisms at higher levels in the validation hierarchy, see Fig. 3. Therefore, we discuss two elements aimed at reducing the unquantified uncertainty arising from missing physics for the case where there is no recognized disagreement between experiment and simulation in the validation domain. (If there is disagreement in the validation domain, as there is in Fig. 5, an unknown-unknown becomes a known-unknown).

The Role of Peer Review to Address Missing Physics. Because extrapolation is more of a physics endeavor than a statistics endeavor, scientific peer review plays a critical role in the extrapolation process. A nearly analogous issue was faced by Theofanous and co-workers in the application of the risk oriented accident analysis methodology (ROAAM) for low probability, high consequence hazards [29]. The basic premise is that once the selected sample of the community of experts in the problem area is convinced that the model reflects to the extent possible all of the relevant physics, the problem may be considered characterized. By this we mean that what is obtained is the best that can be done at the present time with the committed resources. One outcome may be that additional resources need to be committed to the problem or that additional resolution simply cannot be obtained, as there is no known path forward to gain such resolution. This peer review process must be traceable and scrutable [29].

The Importance of Data Assimilation to Mitigate Missing Physics. Unfortunately, as systems become more complex, a point is reached at which whole system models simply cannot be validated, in the sense of comparison with experimental results. The potential for new unit processes or coupling at higher levels of the validation hierarchy must be acknowledged when making an extrapolation [17]. This does not diminish the value of the extrapolated uncertainties but requires additional attention through data assimilation.

Take for example the case of hurricane forecasting (see, for example, [17,30]). These forecasts save millions of dollars by limiting evacuation areas. But in the early stages of the prediction of the track of a storm, models sometimes predict tracks that do not coincide with the actual track that the storm eventually follows. Hurricane forecasters effectively use data assimilation to constantly update their model predictions and uncertainties. As data is assimilated over time, the uncertainties decrease because the period of prediction (i.e., extrapolation) becomes shorter as a hurricane nears the region of interest (i.e., the location where it will make landfall).

The same will be true for prediction of aging of nuclear reactor materials. As materials are irradiated in a real fission environment, an accompanying data assimilation effort must be in place, the model must be constantly updated, missing physics revealed, and as a result, uncertainties in predictions can be expected to decrease.

5   Conclusions

The problem posed by the nuclear industry, referred to as irradiation effects scaling [1], is an ideal example of a high impact uncertainty quantification problem requiring scaling across validation domains and extrapolations to application domains where there is little or no experimental data. The need for predictive material aging models will drive the need for associated uncertainties in the predictions.

The case of extrapolation is more of a physics endeavor than a statistics endeavor. The goal is to produce a prediction with scientifically defensible and acceptable uncertainty. Most extrapolation methods do not deal with missing physics, i.e., they only estimate (extrapolate) the error structure of known-unknowns. Therefore, the process involves using validating experimentation and more detailed physics-based models that capture the essential physics thus enabling the required scaling and extrapolation and reducing uncertainties.

The idea that uncertainty increases when extrapolating outside of the validation domain is clear. However, exactly how the uncertainty increases is not well understood, in addition to being model and situation dependent [17]. This is likely due to the fact that the uncertainty increase is very tightly coupled with the physics-basis of the model. Inaccuracies inherent to models that approximate the relevant physics, i.e., model form uncertainty, will likely dominate the uncertainties. Methods to quantify uncertainties in predictions of models, particularly model form uncertainty, are lacking and are topics requiring further fundamental research.

Acknowledgements. The authors are grateful to (i) Jon Helton of Sandia National Laboratories for his contributions to the document in general and to Sect. 2.2 in particular, (ii) Rob Rechard of Sandia National Laboratories for providing input on the Yucca Mountain Project, (iii) David Higdon of Los Alamos National Laboratory for permission to use Fig. 4 and Fig. 5, (iv) T. N. Krishnamurti of The Florida State University for permission to use Fig. 6, (vi) Gary King of Harvard University for permission to use Fig. 7 (vii) Richard Kochendarfer AREVA Federal Services LLC for a contribution to Sect. 2.2, (viii), Len Lorence of Sandia National Laboratories for providing input on the QASPR project, and (ix) Tony O'Hagan of Sheffield University, William Meeker of Iowa State University, J. Tinsley Oden of The University of Texas at Austin, Habib Najm of Sandia National Laboratories, and Janet Allen of University of Oklahoma for extensive technical input. This work performed under the auspices of the U.S. Department of Energy by Lawrence Livermore National Laboratory under Contract DE-AC52-07NA27344.

References

1. Crabtree, G., Malozemoff, A.: Science for Energy Technology: Strengthening the Link between Basic Research and Industry. U.S. Department of Energy (2010), http://science.energy.gov/~/media/bes/pdf/reports/files/setf_rpt.pdf
2. Roy, C.J., Oberkampf, W.L.: A Comprehensive Framework for Verification, Validation, and Uncertainty Quantification in Scientific Computing. Computer Methods in Applied Mechanics and Engineering 200, 2131–2144 (2011)
3. Oberkampf, W.L., Roy, C.J.: Verification and Validation in Scientific Computing. Cambridge University Press, New York (2010)
4. King, W., Allen, T., Arsenlis, T., Bench, G., Bulatov, V., Fluss, M., Klein, R., McMahon, D., Middleton, C., Morley, M., Turchi, P., Was, G.: Report on the Workshop on Accelerated Nuclear Energy Materials Development. Lawrence Livermore National Laboratory, LLNL-TR-436353 (2010), http://tinyurl.com/ANEMDworkshopreport
5. Fluss, M., King, W.E., Bench, G., Bulatov, V., De Caro, M.S.: Workshop on Science Applications of a Triple Beam Capability for Advanced Nuclear Energy Materials. Lawrence Livermore National Laboratory, LLNL-MI-413125 (2009), http://tinyurl.com/triplebeamworkshop
6. U.S. DOE: Viability Assessment of a Repository at Yucca Mountain. U.S. Department of Energy, Office of Civilian Radioactive Waste Management, DOE/RW-0508 (1998), http://www.osti.gov/bridge/product.biblio.jsp?osti_id=762971
7. U.S. DOE: Yucca Mountain Repository Safety Analysis Report. U.S. Department of Energy, DOE/RW-0573 (2008), http://pbadupws.nrc.gov/docs/ML0907/ML090710096.pdf
8. Sandia National Laboratories: Total System Performance Assessment Model/Analysis for the License Application. U.S. Department of Energy Office of Civilian Radioactive Waste Management, MDL-WIS-PA-000005 Rev. 00, AD 01 (2008)
9. Helton, J.C., Martell, M.-A., Tierney, M.S.: Characterization of Subjective Uncertainty in the 1996 Performance Assessment for the Waste Isolation Pilot Plant. Reliability Engineering and System Safety 69, 191–204 (2000)
10. Rechard, R.P., Tierney, M.S.: Assignment of Probability Distributions for Parameters in the 1996 Performance Assessment for the Waste Isolation Pilot Plant. Part 1: Description of Process. Reliability Engineering and System Safety 88, 1–32 (2005)
11. Rechard, R.P., Tierney, M.S.: Assignment of Probability Distributions for Parameters in the 1996 Performance Assessment for the Waste Isolation Pilot Plant. Part 2. Application of Process. Reliability Engineering and System Safety 88, 33–80 (2005)
12. Myers, S.M., Wampler, W.R.: The Science of QASPR (Qualification Alternatives to the Sandia Pulsed Reactor) in Sandia's Physical, Chemical, and Nano Sciences Center (PCNSC). Sandia National Laboratories, SAND2008-7810P (2008)
13. Paulsen, R., Castro, J., Hoekstra, R., Myers, S., Wampler, W., Romero, V., Rutherford, B., Griffin, P., Giunta, A., King, D.: QASPR Silicon Device Prototype Exercise and Results Summary. Sandia National Laboratories, SAND2010-8043 (2010)
14. Meeker, W.Q., Escobar, L.A.: Pitfalls of accelerated testing. IEEE Trans. Reliab. 47, 114–118 (1998)
15. Nelson, W.: Accelerated Testing: Statistical Models, Test Plans and Data Analyses. Wiley, New York (1990)

16. Oberkampf, W.L., Trucano, T.G.: Verification and Validation in Computational Fluid Dynamics. Prog. Aerosp. Sci. 38, 209–272 (2002)

17. Cacuci, D.G., Ionescu-Bujor, M.: Sensitivity and Uncertainty Analysis, Data Assimilation, and Predictive Best-Estimate Model Calibration. In: Cacuci, D.G. (ed.) Handbook of Nuclear Engineering. Springer, New York (2010)

18. Higdon, D., Klein, R., Anderson, M., Berliner, M., Covey, C., Ghattas, O., Graziani, C., Habib, S., Seager, M., Sefcik, J., Stark, P., Stewart, J.: Uncertainty Quantification and Error Analysis. In: Messina, P., Bishop, A. (eds.) Workshop: Scientific Grand Challenges in National Security: the Role of Computing at the Extreme Scale, pp. 121–142. Department of Energy, Hilton Hotel Washington DC North (2009)

19. Kennedy, M.C., O'Hagan, A.: Bayesian Calibration of Computer Models. Journal of the Royal Statistical Society Series B-Statistical Methodology 63, 425–450 (2001)

20. Eldred, M.S., Burkardt, J.: Comparison of Non-Intrusive Polynomial Chaos and Stochastic Collocation Methods for Uncertainty Quantification. In: 47th AIAA Aerospace Sciences Meeting including The New Horizons Forum and Aerospace Exposition, pp. AIAA-2009-2976. American Institute of Aeronautics and Astronautics (2009)

21. Najm, H.N.: Uncertainty Quantification and Polynomial Chaos Techniques in Computational Fluid Dynamics. Annual Review of Fluid Mechanics 41, 35–52 (2009)

22. Ferson, S., Oberkampf, W.L., Ginzburg, L.: Model Validation and Predictive Capability for the Thermal Challenge Problem. Computer Methods in Applied Mechanics and Engineering 197, 2408–2430 (2008)

23. Vijaya Kumar, T.S.V., Krishnamurti, T.N., Fiorino, M., Nagata, M.: Multimodel Superensemble Forecasting of Tropical Cyclones in the Pacific. Monthly Weather Review 131, 574–583 (2003)

24. Krishnamurti, T.N., Kishtawal, C.M., LaRow, T.E., Bachiochi, D.R., Zhang, Z., Williford, C.E., Gadgil, S., Surendran, S.: Improved Weather and Seasonal Climate Forecasts from Multimodel Superensemble. Science 285, 1548–1550 (1999)

25. Oden, J.T., Prudhomme, S.: Control of Modeling Error in Calibration and Validation Processes for Predictive Stochastic Models. Int. J. Numer. Methods Eng. 87, 262–272 (2011)

26. Qian, Z.G., Seepersad, C.C., Joseph, V.R., Allen, J.K., Wu, C.F.J.: Building Surrogate Models Based on Detailed and Approximate Simulations. Journal of Mechanical Design 128, 668–677 (2006)

27. King, G., Zeng, L.C.: The Dangers of Extreme Counterfactuals. Polit. Anal. 14, 131–159 (2006)

28. Tong, C.: PSUADE User's Manual (Version 1.2.0). Lawrence Livermore National Laboratory, LLNL-SM-407882 (2009)

29. Theofanous, T.G.: On the Proper Formulation of Safety Goals and Assessment of Safety Margins for Rare and High-consequence Hazards. Reliability Engineering & System Safety 54, 243–257 (1996)

30. Kalnay, E.: Atmospheric Modeling, Data Assimilation, and Predictability. Cambridge University Press, New York (2003)

A Note on Uncertainty in Real-Time Analytics

Mladen A. Vouk

Department of Computer Science
North Carolina State University
Raleigh, NC, USA
vouk@csc.ncsa.edu

Abstract. Today real-time analytics of large data sets is invariably computer-assisted and often includes a "human-in-the-loop". Humans differ from each other and all have a very limited innate capacity to process new information in real-time. This introduces statistical and systematic uncertainties into observations, analyses and decisions humans make when they are "in the loop". Humans also have unconscious and conscious biases, and these can introduce (major) systematic errors into human assisted or human driven analytics. This note briefly discusses the issues and the (considerable) implications they can have on real-time analytics that involves humans, including software interfaces, learning, and reaction of humans in emergencies.

Keywords: human mind bandwidth, uncertainty, human bias, real-time analytics, computer-assisted analytics, human-in-the-loop.

1 Introduction

True artificial intelligence is still the "stuff" of science fiction stories and possibly future. To avoid and mitigate mistakes in computer-driven modeling and decision making we often include a human into "the loop" — for example, FAA flight controllers, real-time network and computer security threat identification and mitigation analysts, inclement weather decision making personnel, or scientists involved in very expensive high-end simulations, e.g., [1]. Well designed complex computer analytics and decision making systems, and workflows, include points where humans are inserted into the process to monitor, augment, direct, take over, or stop processes. This makes a lot of sense, and this can also be an issue. People have biases [2] and people make mistakes too, lots of them. An interesting, and often used alternative to "human-in-the-loop" is the "computer-in-the-loop" where analytics is human-driven and the computer only augments human decision making by offering suggestions, second opinions, etc. A recent well publicized example of this approach is the IBM's Watson project and its application in medicine [3].

This note provides an overview of some of the uncertainties that humans can inject into data analysis and interpretations processes. Sect. 2 of this note discusses information processing limitation of human mind in the context of new inputs, and in Sect. 3 it very briefly covers some possible systematic biases. Sect. 4 concludes the note.

A. Dienstfrey and R.F. Boisvert (Eds.): WoCoUQ 2011, IFIP AICT 377, pp. 312–318, 2012.

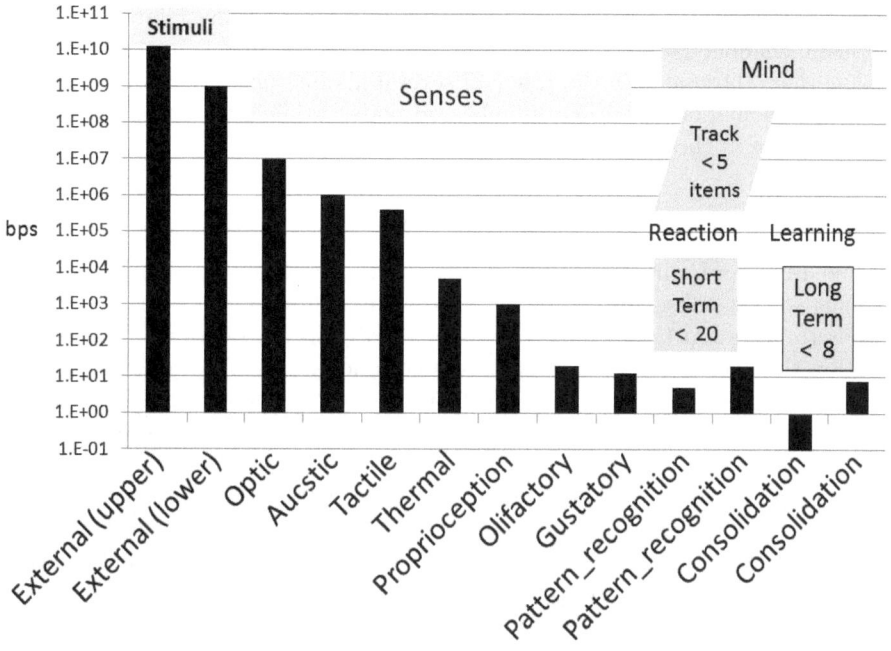

Fig. 1. Input to humans – data reduction is in the giga-fold range (after [4])

2   Uncertainties

From the perspective of external inputs, a human operates in five principal outward facing sensory domains - visual, auditory, tactile, olfactory, and taste, as well as in a number of inward-facing sensory domains [4]. There are a number of models of how sensory inputs end-up being processed, become actionable and possibly become memories, and there is a huge amount of literature regarding the human mind, its capabilities, and its limits. An interesting top-level general book on the topic of memory is by Thompson and Madigan [5].

2.1   Information Flux

Humans are bombarded with a huge amount of stimuli - as much as a terabit per second [4]. Fig. 1 illustrates the information flow rates humans might receive and the principal stages of the related information processing. The scale on the vertical axis is logarithmic, and in units of bits per second (bps). Horizontal axis covers some of the basic senses, or sensors, humans have. Estimated upper and lower bounds are shown for the overall external flux of stimuli, and for short and long term memory information processing. Also shown are approximate maximum rates into different senses.

Humans do not process outside stimuli directly but through a multitude of sensor channels and filters which prepare that information for input into human mind. For example, optic channel may carry as much as 10 megabits per second, acoustic channel as much as a megabit per second, tactile channel is somewhat below that, thermal is in the range of several kilobits per second, as is the proprioception channel (position of limbs and body), while olfactory and gustatory channels are in the 10 to 20 bits per second range.

2.2   Information Processing Capacity

While a human brain has considerable capacity, and has very high internal processing speeds, conscious processing (pattern matching) of information a person needs to act upon (either physically or mentally) is considerable slower. Information that needs to be actively processed lands in short-term memory where it is then processed at the rate of 5 to 20 bits per second. Long-term storing of information (learning), happens at an even slower rate – from zero to perhaps 8 bits per second.

It is obvious that a considerable reduction needs to take place between inputs and the amount of new data a person "works with". This means that a lot of that data is either discarded, or is dynamically codified into patterns that a human acts upon (based on stored patterns) or stores permanently. The question is whether there is a loss of useful information in the process. If the processing is done automatically before presenting data to a human, there could be unintended loss of information that may bias the outcomes.

In addition, humans often have built-in biases brought on by culture, training, education, social environments, intent, and so on, and that can exacerbate the problem.

2.3   Models

As already mentioned, there are a number of models of how sensory inputs end-up being processed, become actionable and possibly become memories. There is a huge amount of literature regarding the human mind, its capabilities, and its limits. For example, one model states that environmental input goes to sensors (vision, auditory, ...), then to short-term temporary working memory, and depending on what happens there, this "chunk" of information may elicit a re-action, may become a permanent memory, may be ignored, etc. [5,6]. Baddeley [7,8,9] developed the theory of working memory that involves the "central executive" — an attention-controlling sub-system, the "visuospatial sketch pad" which manipulates visual images, and the "phonological loop" which stores and rehearses speech-based information.

While very large amount of information comes into sensors, the mind is capable of processing only a limited amount of new information. But, the mind is also very good at on-the-fly abstraction, and mapping of those abstractions onto a huge pool of encoded information in its permanent memory. "Chunks" are associated with the working memory throughput. They were originally proposed by Miller [10,11], and there appears to be a limited number (between 3 and 7) that a human can handle/manage at any one time without making an increasing number of mistakes. For example, humans may have trouble tracking (without error) more than four to five visual or auditory "targets".

Humans appear be able to enumerate quickly up to about five items, beyond that the process is slower and time taken increases linearly [12]. For example, counting 50 stars on a flag may take some time, recognizing a flag takes much less than a second, and retrieving the number of stars that should be on it from permanent memory is also very quick. Humans also appear to handle about four dimensions [13] and up to 5-way interactions reasonably well [14]. All this indicates that it is important to manage complexity of content and interrelations presented to humans [13,15]. Beyond a comfortable limit (and that limit may change with age) errors start happening. Most recent examples of sometimes fatal real-time analytics distractions to humans come from the use of cell phones and other in-car devices by drivers (see e.g., [21,22]).

"Chunks" can contain a very small amount of information – from a single tone, to a very detailed and rich visual, auditory and tactile scene. They can contain raw data, but more often they tend to be abstractions, pattern mappings of complex objects or scenes onto however this impression is stored/encoded in the permanent memory of the subject. Different people very likely extract storable abstractions of the same scene, object, or signal differently. Pattern or "symbol" recognition appears to be the primary way we think about scenes and situations which would otherwise overload our working memory, and the question is how long does it take to map a sensory "chunk" or moment onto a familiar pattern to possibly elicit a conscious reaction.

Chunks also appear to have an expiration time – they are retained in working memory for not more than several minutes and, unless processed through action or permanent memory storage, they are lost. They appear to be constructed from segments that do not exceed about 100 ms each (but could be shorter). Some classical work in that domain was done by Stroud [23,24] who noted that humans appear to have ability to consciously process from about 5 to 20 of such discrete psychological time slices and effect discriminations based on that. In a situation where a reaction to such a slice is binary (e.g., yes or no, or ok, not ok) conscious absorption and reactions rate appears to be in the 5 to 20 bits per second range. However, in order to match a particular pattern (which could consist of more than one "psychological moment"), mind may need to do a large number of comparisons. This number is sometimes known as the Stroud number, and has been used in attempts to describe software development and software fault generation processes, e.g., [25,26]. While 5 or 20 bits may not look like much, to match a 20 bit pattern, for example, may require comparison with as many as 1,000,000 or so internal patterns.

2.4   Limitations

Unfortunately, there appear to be at least three bottlenecks on the input path. For example: i) attentional blink (AB), ii) visual short-term memory capacity (VSTM), and iii) psychological refractory period (PRP) phenomena [11]. Attention blink relates to the following example, *"when subjects attempt to identify two targets in a rapid, serial visual presentation of distractors, they are severely impaired at detecting the second of the two targets when it is presented within 500 ms of the first target [25]. Importantly, the deficit with the second target (T2) is a result of attending to the first target (T1): subjects have no difficulty in reporting T2 when only it is required to be detected"* [11,15]. Part of the problem may lie in the need to move eyes from one spot to another [12,27]. This may also be related to "change blindness" [5,16]. VSTM (or working memory) capacity of about 4 to 7 items has already been mentioned. PRP is related to taking action as a result of a stimulus. This appears to impact our ability to simultaneously take two actions based on a single stimulus (parallelism).

3   Systematic Bias

By nature of their cultural upbringing, educational training, experiences, motivation, trust, religion, politics, perceived and real situational awareness and other mental models, prevailing policies, regulations and laws, etc., humans will often ignore data and information and make conclusions based on other factors, or they might interpret the same information differently. This is often called cognitive bias (see e.g., [2,17]). The list of possible cognitive biases is very long, e.g. [18], and as expected, there are supporters and critics, e.g., [19]. Of course, the issue of systematically biasing information, or its interpretations, to fit a particular purpose has been around since the dawn of time. An iconic phrase in this context is "lies, damned lies, and statistics", e.g. [20], which, as large amounts data become more accessible over internet, is being increasingly used and re-used in many forms. The concept takes an additional dimension when deception is considered in the context of information technology security (see e.g., [28]).

Undeniably, epistemic uncertainty is a very serious issue when doing analytics, and particularly real-time analytics. Ideally, one would like to reduce epistemic uncertainty to aleatory uncertainty.

4   Summary

In general, it seems to be well established that humans have a (very) limited ability to pro-actively process new incoming information and act upon it, and a very high propensity to unconscious or conscious misinterpretation and misrepresentation of new information. Implications on analytics are substantial. Careful studies need to be made to develop appropriate methods for reduction of epistemic uncertainty down to aleatoric, and to allow reduction of new incoming information to a level appropriate for human consumption. This includes training of the analysts to work at the appropriate level of abstraction, developing appropriate computer-based tools and approaches, and for both humans and machines to handle unexpected anomalies in (large) input streams. How to do that appears to be an open question, although ideas abound.

Acknowledgements. This work has been supported in part by the DOE Sci-DAC grant DE-FC02-07-ER25809, by the U.S. Army Research Office (ARO) grant W911NF-08-1-0105 managed by NC State University Secure Open Systems Initiative, by the IBM Shared University Research Program, and by NSF grants CCF-0939081 and CNS-0910767.

References

1. Barreto, R., Klasky, S., Mouallem, P., Podhorszki, N., Vouk, M.: Collaboration Portal for Petascale Simulations. In: Proceedings of the International Conference on Collaboration Technologies and Systems, pp. 384–393 (2009)
2. Gilovich, T., Griffin, D.: Heuristics and Biases: Then and Now. In: Gilovich, T., Griffin, D., Kahneman, D. (eds.) Heuristics and Biases: The Psychology of Intuitive Judgment, pp. 1–4. Cambridge University Press (2002)
3. WellPoint and IBM Announce Agreement to Put Watson to Work in Health Care (2011), http://www-03.ibm.com/press/us/en/pressrelease/35402.wss (accessed October 31, 2011)
4. Leherl, S., Fischer, B.: The Basic Parameters of Human Information Processing: Their Role in the Determination of Intelligence. Person. Individ. Diff. 9(5), 883–896 (1988)
5. Thompson, R.F., Madigan, S.A.: Memory. Princeton University Press (2005)
6. Atkinson, R.C., Shiffrin, R.M.: The Control of Short-term Memory. Scientific American 225, 82–90 (1971)
7. Baddeley, A.D., Hitch, G.J.: Working Memory. In: Bower, G.A. (ed.) Recent Advances in Learning and Motivation, pp. 47–89. Academic Press, New York (1974)
8. Baddeley, A.D.: Working Memory. Science 255(5044), 556–559 (1992)
9. Baddeley, A.D.: Working Memory: Looking Back and Looking Forward. Nature Reviews Neuroscience 4, 829–839 (2003)
10. Miller, G.A.: The Magical Number Seven, Plus or Minus Two: Some Limits on our Capacity for Processing Information. Psychological Review 63, 81–97 (1956)
11. Marois, R., Ivanoff, J.: Capacity Limits of Information Processing in the Brain. Trends in Cognitive Sciences 9(6), 296–305 (2005)
12. Piazza, M., Mechelli, A., Butterworth, B., Price, C.J.: Are Subitizing and Counting Implemented as Separate or Functionally Overlapping Processes? NeuroImage 15, 435–446 (2002)
13. Halford, G.S., Wilson, W.H., Phillips, S.: Processing Capacity Defined by Relational Complexity: Implications for Comparative, Developmental, and Cognitive Psychology. Behavioral and Brain Sciences 21, 803–865 (1998)
14. Halford, G.S., Baker, R., McCredden, J.E., Bain, J.D.: How Many Variables Can Humans Process? Psychological Science 16(1), 70–76 (2005)
15. Lavie, N.: Distracted and Confused? Selective Attention Under Load. Trends in Cognitive Sciences 9(2), 75–82 (2005)
16. Simmons, D.J., Levin, D.T.: Change Blindness. Trends in Cognitive Sciences 1(7), 261–267 (1997)
17. Kahneman, D., Tversky, A.: Subjective Probability: A Judgment of Representativeness. Cognitive Psychology 3(3), 430–454 (1972)
18. List of Cognitive Biases (2011), http://en.wikipedia.org/wiki/List_of_cognitive_biases (accessed October 22, 2011)

19. Gigerenzer, G.: Bounded and Rational. In: Stainton, R.J. (ed.) Contemporary Debates in Cognitive Science, p. 129. Blackwell (2006)
20. Twain, M.: Chapters from My Autobiography. North American Review DXCVIII. Project Gutenberg (1906), http://www.gutenberg.org/files/19987/19987.txt (retrieved May 23, 2007)
21. Strayer, D.L., Drews, F.A.: Cell-Phone–Induced Driver Distraction. Current Directions in Psychological Science 16(3), 128–131 (2007)
22. Statistics and Facts About Distracted Driving (2011),
    http://www.distraction.gov/content/get-the-facts/
    facts-and-statistics.html (accessed December 05, 2011)
23. Stroud, J.M.: The Fine Structure of Psychological Time. In: Quastler, H. (ed.) Information Theory in Psychology, pp. 174–207. The Free Press, Glencoe (1955)
24. Stroud, J.M.: The Fine Structure of Psychological Time. Annals of the New York Academy of Sciences 138, 623–631 (1967)
25. Halstead, M.H.: Elements of Software Science. Operating and Programming Systems Series. Elsevier Science Inc., New York (1977)
26. Coulter, N.S.: Software Science and Cognitive Psychology. IEEE Transactions on Software Engineering 9(2), 166–171 (1983)
27. Rousselet, G.A., et al.: How Parallel is Visual Processing in the Ventral Pathway? Trends in Cognitive Sciences 8, 363–370 (2004)
28. Yuill, J.: Defensive Computer-Security Deception Operations: Processes, Principles and Techniques. PhD Dissertation, North Carolina State University (2006)

Author Index